MATHEMATICAL PROGRAMMING

FOR OPERATIONS RESEARCHERS AND COMPUTER SCIENTISTS

INDUSTRIAL ENGINEERING

A Series of Reference Books and Textbooks

Editor

WILBUR MEIER, JR.
Head, School of Industrial Engineering
Purdue University
West Lafayette, Indiana

Volume 1: Optimization Algorithms for Networks and Graphs, *Edward Minieka*

Volume 2: Operations Research Support Methodology, *edited by Albert G. Holzman*

Volume 3: MOST Work Measurement Systems, *Kjell B. Zandin*

Volume 4: Optimization of Systems Reliability, *Frank A. Tillman, Ching-Lai Hwang, and Way Kuo*

Volume 5: Managing Work-In-Process Inventory, *Kenneth Kivenko*

Volume 6: Mathematical Programming for Operations Researchers and Computer Scientists, *edited by Albert G. Holzman*

Additional Volumes in Preparation

MATHEMATICAL PROGRAMMING
FOR OPERATIONS RESEARCHERS AND COMPUTER SCIENTISTS

edited by

ALBERT G. HOLZMAN

Department of Industrial Engineering,
Engineering Management,
and Operations Research
University of Pittsburgh
Pittsburgh, Pennsylvania

Marcel Dekker, Inc. New York and Basel

Library of Congress Cataloging in Publication Data

Main entry under title:

Mathematical programming for operations researchers and
computer scientists.

(Industrial engineering; v. 6)
Articles which originally appeared in the Encyclopedia of computer science and technology, vols. 5, 8–12.
Includes index.
1. Programming (Mathematics) I. Holzman, Albert George. II. Series.
T57.7.M32 519.7 81-5528
ISBN 0-8247-1499-7 AACR2

These articles originally appeared in the *Encyclopedia of Computer Science and Technology*, Volumes 5, 8, 9–11, and 12, edited by Jack Belzer, Albert G. Holzman, and Allen Kent, © 1976, 1977, 1978, and 1979, respectively, by Marcel Dekker, Inc.

MARCEL DEKKER, INC.

270 Madison Avenue, New York, New York 10016

Current printing (last digit):

10 9 8 7 6 5 4 3 2 1

PRINTED IN THE UNITED STATES OF AMERICA

PREFACE

The stimulus for my assuming the role as editor of this volume is a perceived need for in-depth articles in the area of mathematical programming. The objective is to bridge the gap between introductory operations research textbooks and books which have been devoted exclusively to one particular topic of mathematical programming.

Each of the chapters in this volume is self-contained and is a substantial overview of the topic. It is felt that operations researchers, computer scientists, and engineers will find the level of presentation attractive and meaningful to them.

These articles were prepared originally for the *Encyclopedia of Computer Science and Technology,* and were published as chapters of six separate volumes of this Encyclopedia. It was edited by Jack Belzer, Albert G. Holzman, and Allen Kent. The authors of the articles selected for *Mathematical Programming* have had extensive experience, and have national and international reputations in their area of expertise. Criteria guidelines were submitted to the authors to maintain a consistent level of exposition and rigor; however, it is obviously not feasible to have 12 different authors interpret these criteria in a uniform manner. Their presentations are naturally influenced by their own pedagogical, consulting, and research experiences. It must also be recognized that it is not possible to include all *areas* of mathematical programming in a single volume.

The book starts with a survey of the fundamentals of linear programming in Chapter 1, and terminates with a substantial treatment of nonlinear programming in Chapter 11. In between these two chapters are included: extensions of linear programming to discrete optimization methods, multi-objective functions, quadratic programming, and the more recent thrusts in linear complementarity problems and fixed point computing methods; geometric programming, the elegant solution procedure of relatively recent origin; and classical calculus methods which provide the reader with the fundamentals for solving nonlinear programming problems.

Many people have made significant contributions to Mathematical Programming over the past 35 years. Regrettably, it is not possible to recognize them individually. However, I would be remiss if I did not acknowledge Dr. George B. Dantzig, who must be recognized as the person who pioneered Mathematical Programming in the 1940's and who has continued to advance the state-of-the-art for over 30 years.

Albert G. Holzman

CONTENTS

CONTRIBUTORS

MORDECAI AVRIEL, Faculty of Industrial Engineering and Management, Technion-Israel Institute of Technology, Haifa, Israel

LEON COOPER*, Department of Industrial Engineering and Engineering Management, Southern Methodist University, Dallas, Texas

JAMES S. DYER†, Graduate School of Management, University of California, Los Angeles, California

ALBERT G. HOLZMAN, Department of Industrial Engineering, Engineering Management, and Operations Research, University of Pittsburgh, Pittsburgh, Pennsylvania.

SANG M. LEE, Department of Management, University of Nebraska, Lincoln, Nebraska

WILLIAM F. LUCAS, School of Operation Research and Industrial Engineering, Cornell University, Ithaca, New York

KATTA G. MURTY, Department of Industrial and Operations Engineering, The University of Michigan, Ann Arbor, Michigan

DON. T. PHILLIPS, Department of Industrial Engineering, Texas A & M University, College Station, Texas

*Now deceased

†*Current affiliation:* Department of Management, University of Texas, Austin, Texas

R. SAIGAL*, Technical Staff, Operations Research Center, Bell Telephone Labs, Holmdel, New Jersey

RAKESH K. SARIN, Graduate School of Management, University of California, Los Angeles, California

VINCE SPOSITO, Department of Statistics, Iowa State University, Ames, Iowa

HAMDY A. TAHA, Department of Industrial Engineering, University of Arkansas, Fayetteville, Arkansas

Current affiliation: Department of Industrial Engineering, Northwestern University, Evanston, Illinois

INTRODUCTION

Linear programming and the concomitant Simplex method developed by George Dantzig in 1947 is still one of the most powerful procedures for solving mathematical programming problems. Modifications of the Simplex method have been used to obtain solutions to many nonlinear type problems. In fact, in the book *Nonlinear and Dynamic Programming* by Hadley, the author states in his introduction to Chapter 9: "Every method that we have considered thus far for solving nonlinear programming problems numerically, with the exception of material on classical optimization methods studied in Chapter 3, has been in some sense or other an 'adjacent extreme point' method which employed the simplex algorithm as the fundamental computational tool."

The excitement caused by the recent publication of the research of the Russian mathematician L. G. Khachian on an ellipsoid algorithm for solving linear programming problems in polynomial time has dissipated considerably. While it obviously has theoretical significance, it does not appear to be of practical importance at the present time. George Dantzig conducted an experiment in using both the Simplex method and Khachian's algorithm for solving a linear programming problem. He found that the problem could be solved in about one-half hour using Simplex, but would take about 50 million years if Khachian's algorithm were used.

Linear programming problems having over 10,000 constraints and even more variables are being solved in an economically feasible manner today on a large computer. However, the solution of nonlinear programming problems is much more complex and time consuming. No "standard" method, such as the Simplex, can be applied to the general nonlinear problem. The approach employed is to exploit the structure of a particular nonlinear problem. Frequently, the solution procedure is an hybrid of several methods.

The first Chapter of this book considers the background of the development of linear programming and the fundamental properties of a solution to the LP problem. This leads into the theory of the Simplex and duality theory. The principle of complementary slackness is extended to the fundamental problem of mathematical programming as defined by Dantzig and Cottle. Post optimal analysis in terms of sensitivity analysis and parametric programming is reviewed. Then the relationship between linear programming and the theory of games is shown. The chapter concludes with an introduction to decomposition, bounded variables, and multicriteria decision making.

It is not uncommon in the real world to have problems which require that variables take on discrete values. Chapter 2 addresses the integer programming problem, for which the first finite algorithm was developed by Gomory in 1958. Included under the scope of integer programming applications are pure integer, mixed integer, and binary problems, and a categorization of integer programming applications as direct, coded, and transformed models. The major part of this chapter is devoted to an explanation of branch and bound methods, implicit enumeration methods, and cutting plane methods. The computational problems encountered by the various methods are enumerated, and it is recommended that heuristics be used to obtain good starting solutions.

Interest in game theory dates back to the pioneering work of von Neumann in 1928. In Chapter 3 insights are given on the nature of game theory, types of games, and historical highlights. Since most of the models utilized in the study of games are either in normal form, characteristic function form, or extensive form, these methods are presented first in a summary manner, and then described in considerable detail and clarified by applications and illustrations. The discussion of the normal form goes beyond the two-person zero-sum game to include two-person general-sum games, and n-person noncooperative games. The basic model and theory are developed prior to considering the applications in politics and economics. A discussion of information and existence theorems follows the definitions of commonly used terms used for games in extensive forms.

During the past decade considerable activity in both research and application has been exhibited in the area of multicriteria decision making. Chapters 4 and 5 relate directly to this topic. Goal programming, which is developed in Chapter 4, is based on a modification of the linear programming structure which enables the decision maker to obtain a simultaneous solution to a complex system of competing objectives. Real world problems frequently demand an analysis of multiple and often conflicting organizational objectives and environmental factors. In goal programming the deviations in not achieving the goals are minimized. The author reviews the modified Simplex method of goal programming, and then expands on areas of current research interest.

Chapter 5 extends the horizons of multicriteria decision making beyond that of a linear programming focus. The major part of this chapter is on two areas: decisions under certainty, and decisions under risk. The conditions for existence of an ordinal utility function and additivity are presented prior to a discussion of mathematical programming approaches involving man-machine interaction and generation of efficient solutions. For decisions under risk, a much more difficult problem, the additive utility function and the multiplicative function are introduced.

A special case of the nonlinear programming problem is the quadratic programming problem, where the criterion includes a quadratic function as well as a linear function. The constraints are assumed to be linear, the same as in the LP problem. Chapter 6 addresses this problem. The first part of this chapter relates the Kuhn-Tucker theory for saddle values to quadratic duality. Then the two well-known algorithms of Hildreth and Wolfe for solving quadratic programming problems are discussed in detail. The chapter is concluded with an extensive discussion on solving the regression problem by a quadratic programming formulation.

All linear programming problems, convex quadratic programming problems, and bimatrix games, can be transformed into linear complementarity problems (L.C.P.). Such problems can then be solved by the complementary pivot algorithm. Chapter 7 on Complementarity Problems provides a unified theory for studying all these different problems. Another important aspect is that the arguments used in this algorithm have led to the insight necessary for the development of fixed point computing algorithms. After a discussion on the importance of complementarity problems and the role of computers in solving them, the problem is clearly defined using mathematical notation. Then applications of the complementarity problems are demonstrated for linear programming, quadratic programming, and two-person games. Under the section on algorithms for solving L.C.P. are included a discussion on bases, pivot operations, adjacent almost complementary feasible bases, and the complementary pivot rule. Two numerical examples are presented to clarify the procedure. The theoretical results are then summarized, and a brief introduction is given to the nonlinear complementarity problem which can be transformed into the problem of computing a Kakutani fixed point.

Geometric programming was developed originally by Duffin, Zener, and Peterson in the 1960's to solve certain types of engineering design problems at Westinghouse Electric Corporation. The mathematical formulation of the primal and dual geometric programming problems is given first in Chapter 8. The problem as formulated by Duffin et al. assumed the objective function and the constraints to be of posynomial form (positive polynomial). The use of signum functions to permit the application of geometric programming to polynomials which are not positive is considered in this chapter. The author notes that serious solution problems exist when degrees of difficulty, as defined for the geometric programming problem, are present.

Chapter 9 on Fixed Point Computer Methods is a logical sequel to Chapter 7. At the start two of the most celebrated theorems are presented; they are Brouwer's Theorem and Kakutani's Theorem. The algorithms developed can be made to converge quadratically to a fixed point for a continuously differentiable function whose derivative satisfied a Lipschitz condition. The following two applications of the fixed point computing methods are reviewed: (1) the nonlinear programming problem, and (2) computing economic equilibrium prices. After illustrating the basics of the procedure by considering a fixed point of a continuous mapping, another algorithm is developed such that initiation of the search is not restricted to the boundary of the space. Next the triangulation procedure is discussed concomitant with a very appealing geometric interpretation. The chapter is concluded with a discussion of computational considerations.

Optimization has been an important objective of operations research ever since this approach to decision making was originally postulated. In fact, the idea of maximization and minimization dates back at least 2000 years. While many optimization methods have been used in solving various types of OR problems, classical optimization is restricted to finding maxima and minima of continuous and differentiable functions. Starting with a brief background on the origins of optimization, Chapter 10 on Classical Optimization then presents some of the basic mathematical concepts, including topics such as global and relative maximum, convex and concave functions, Taylor's theorem, and the implicit function theorem. Under the heading of unconstrained optimization, the necessary and sufficient conditions for an optimal point are developed, first considering the functions of a single variable, and then treating functions of several variables. The logical progression to constrained optimization problems follows the section on unconstrained optimization. Examples point up the necessity to have a method in which constraints are taken into account explicitly, rather than implicitly. This leads to the introduction of the Lagrange multiplier method and the Lagrangian function for equality constraints. The last consideration in constrained optimization is an extensive revision of Lagrangian methods to solve problems having inequality constraints. This extension is based on the well-known Kuhn-Tucker conditions. The discussion of classical methods of optimization in this chapter serves as an excellent introduction to Chapter 11.

The final chapter is a definitive and an extraordinarily encompassing survey on Nonlinear Programming. To introduce the topic examples are cited for nonlinear curve fitting and chemical process design. The chapter has two major thrusts: one is on Nonlinear Programming Analysis, and the other on Methods of Nonlinear Programming. Even though solution of optimization problems can be traced to the ancient Greeks and the 17th and 18th century mathematicians, most of the results in this survey article have been derived after 1950.

In the analysis section the necessary and sufficient conditions for optimality are derived, with particular emphasis on Lagrange multipliers and Kuhn-Tucker conditions. This section then considers convexity in nonlinear programming and concludes with a substantial discussion of duality in convex programming.

Under the major heading of methods of solving nonlinear programming problems the author begins with minimization of single-variable functions, then considers nonderivative methods, multidimensional unconstrained optimization using derivatives, penalty methods, and methods of restricted movement.

MATHEMATICAL PROGRAMMING
FOR OPERATIONS RESEARCHERS AND COMPUTER SCIENTISTS

1
LINEAR PROGRAMMING

ALBERT G. HOLZMAN

Department of Industrial Engineering,
Engineering Management,
and Operations Research
University of Pittsburgh
Pittsburgh, Pennsylvania

INTRODUCTION

The birth of linear programming (LP) is usually identified with the development of the Simplex method in 1947 by George B. Dantzig [1]. Even though much of the mathematical theory of linear programming relates directly to the theory of linear inequalities and convex sets [2] which have been formulated over the past century, it was the introduction of the Simplex method concomitant with the advent of the high-speed digital computer that stimulated the interest in practical applications as well as further research in linear programming. Literally hundreds of books and thousands of articles have been published on this topic.

It is interesting to note that at the national TIMS/ORSA meeting in May 1977, Dantzig presented a keynote paper on "Linear Programming—Its Origins and Impact." He stated that his introduction of an "objective function" was the major contribution which precipitated the solution of greatly constrained problems involving inequalities as well as equalities. Dantzig acknowledged the contributions made by the great mathematician John von Neumann, and the outstanding economist Tjalling Koopmans, during 1946-1947 when he was developing the famous Simplex method for solving linear programming problems. Von Neumann was the first one to point out that a game problem can be reduced to a linear programming problem, and that a problem concerning the maximizing of a linear form whose variables are subject to a system of linear inequalities could be replaced by a solution to an extended system of linear inequalities. Koopmans was a joint winner of the 1975 Nobel Prize in economic science. The award was for bridging "the general theory of competitive economics and the normative theory of optimal resources allocation."

In Dantzig's paper at the TIMS-ORSA meeting he alluded to the contributions in economic planning made by the Russian Leonid Kantorovich in the late 1930s.

Kantorovich shared the 1975 Nobel Prize in economic science with Koopmans. The work of Kantorovich was not known in this country until the mid-1950s when Koopmans translated and published his major work.

Dantzig noted that at first he called his optimization procedure "Programming a Linear Structure," but this was soon shortened to "Linear Programming." The term "programming" stemmed from the fact that in the forties programming was synonymous with "planning," which was the activity addressed by Dantzig at that time. Of course, also in the 1940s the term programming had not been applied to the development of computer codes.

Since linear programming is a subset of a more encompassing area called mathematical programming, which is included under the broad field of operations research (OR), background in this area will be provided to put linear programming in its proper perspective. Even though some people trace the origin of OR to the third century B.C. when Archimedes was requested by the King of Syracuse to break the Roman naval siege of his city, an organized form of OR was first introduced in Great Britain at the beginning of World War II. Frequently referred to as a pioneering effort in OR was the work of a team headed by Professor Blackett of the University of Manchester on the detection of ships and submarines by radar equipment in airplanes. The interdisciplinary team was called "Blackett's circus" [3].

Dr. J. B. Conant, who was then Chairman of the National Defense Research Committee of the United States, was primarily responsible for the transfer of OR across the Atlantic in the early forties. Shortly after the United States entered World War II, both the Navy and the Army Air Force started to work in the OR field. The work was extended in 1946 as a result of the Air Force providing 10 million dollars for Project RAND, which later was changed to the RAND Corporation. It was in the early fifties that OR started to take root in industry and academia. Individual courses in this area were offered at first; this was followed by the advent of entire programs in OR, usually offered at the graduate level and frequently housed in industrial engineering departments or schools of business. The Operations Research Society of America (ORSA) was founded in 1952, and a closely allied society, The Institute of Management Sciences, was organized in 1953. The publications of these two societies provide the major source of both theoretical and applied articles in the area of OR and Management Science. These two societies now have joint meetings and joint publications.

COMMON DENOMINATORS OF OPERATIONS RESEARCH

While there have been as many definitions of OR as there are people who attempt to define it, nevertheless the following common denominators are at least implied in most definitions.

1. Problems relative to the attainment of specified objectives or criteria
2. Alternatives; choices exist
3. Optimization; selection of the best alternative for stated criterion
4. Systems perspective; tendency to consider the interrelationship of components in their environment rather than as separate entities

The following classical transportation problem points up the factors enumerated above.

Problem. To determine the least cost shipping assignment of barrels of oil from five supply points to three destination terminals given the following information. (This is an extract from Memorandum on Application of Linear Programming to a Transportation Problem by G. Symonds, Esso Standard Oil Company.)

Supplies (No. 2 Heating Oil)

Origin point	A	10,000 bbl/day
	B	20,000
	C	30,000
	D	80,000
	E	100,000

Requirements

Destination	X	40,000 bbl/day
	Y	80,000
	Z	120,000

Transportation Costs

A to X	57¢/bbl
B to X	60
C to X	55
D to X	54
E to X	62
A to Y	53
B to Y	52
C to Y	50
D to Y	59
E to Y	51
A to Z	58
B to Z	63
C to Z	61
D to Z	56
E to Z	64

The salient features of an OR problem can be easily identified from this problem: (1) the objective is to minimize costs, (2) many possible assignments exist, (3) the best assignment in terms of cost is sought, and (4) it does require recognition of the various potential relationships between origins and destinations.

The optimum solution (in the strict sense) is given in Table 1 which gives the shipping schedule from each origin to each destination in terms of thousands of barrels per day.

It is interesting to note that the best solution does not contain any shipments on the least expensive route (C to Y = 50¢/bbl) but does include an assignment of 20,000 bbl/day on the most costly route (E to Z = 64¢).

In the development of a definition for engineering design by the Engineers Council for Professional Development (ECPD), the following important characteristics are easily identified:

1. Systems approach
2. Decision-making process

TABLE 1
Least Cost Shipping Assignment of Barrels of Oil (1000 bbl/day)

	Origin					
Destination	A	B	C	D	E	Total
X	0	10	30	0	0	40
Y	0	0	0	0	80	80
Z	10	10	0	80	20	120
Total	10	20	30	80	100	240

3. Use of basic sciences, mathematics, and engineering science
4. Optimal conversion of resources
5. Consideration of sociological, economic, and legal aspects

The similarity between the features of engineering design and OR is quite striking.

USE OF OPERATIONS RESEARCH TECHNIQUES

A study was conducted recently to determine "state-of-the-art" utilization of OR methods [4]. In Table 2 the relative use of seven of the more commonly used techniques is shown. It is obvious that linear programming is one of the most frequently used OR methods. The frequency of application (with % shown in parentheses) of each of the seven OR techniques in 11 areas of the production function is given in Table 3. It is of interest to note that linear programming has a relatively high utilization over a broad spectrum of applications, with more than 40% of the respondents indicating that linear programming was used for problems in blending, plant location, and production scheduling.

Mathematical Programming and Linear Programming

Linear programming is categorized under the general heading of mathematical programming in the OR field. The general mathematical programming problem is expressed as maximize the criterion (objective function)

$$Z = f(x_1, \ldots, x_n)$$

TABLE 2
Relative Use of Operations Research Techniques

		Degree of use in %					
Techniques	Number of respondents	Never 1	2	3	4	Very frequently 5	Mean
Regression analysis	74	9.5	2.7	17.6	21.6	48.6	3.97
Linear programming	78	15.4	14.1	21.8	16.7	32.0	3.36
Simulation (in production)	70	11.4	15.7	25.7	24.3	22.9	3.31
Network models	69	39.1	29.0	15.9	10.1	5.8	2.14
Queueing theory	71	36.6	39.4	16.9	5.6	1.4	1.96
Dynamic programming	69	53.6	36.2	7.2	0.0	2.9	1.62
Game theory	67	59.7	25.4	8.9	6.0	0.0	1.61

TABLE 3
Application of Operations Research Techniques in Production

Application areas	Linear programming	Dynamic programming	Network models	Simulation	Queueing theory	Game theory	Regression analysis	Other
Production scheduling	30(41.1)	7(9.6)	6(8.2)	26(35.6)	9(12.3)	0(0.0)	5(6.8)	10(13.7)
Production planning/ control	19(26.0)	3(4.1)	7(9.6)	18(24.7)	4(5.5)	0(0.0)	3(4.1)	3(4.1)
Project planning/ control	10(13.7)	1(1.4)	28(38.4)	9(12.3)	2(2.7)	0(0.0)	0(0.0)	3(4.1)
Inventory analysis/ control	11(15.1)	3(4.1)	3(4.1)	27(37.0)	4(5.5)	1(1.4)	12(16.4)	7(9.6)
Quality control	2(2.7)	0(0.0)	1(4.1)	2(2.7)	0(0.0)	0(0.0)	15(20.5)	9(12.3)
Maintenance and repair	0(0.0)	1(1.4)	3(4.1)	8(11.0)	3(4.1)	1(1.4)	4(5.5)	3(4.1)
Plant layout	13(17.8)	0(0.0)	5(6.8)	19(26.0)	5(6.8)	1(1.4)	2(2.7)	3(4.1)
Equipment acquisition/ replacement	4(5.5)	0(0.0)	1(1.4)	11(15.1)	1(1.4)	0(0.0)	0(0.0)	7(9.6)
Blending	32(43.8)	0(0.0)	1(1.4)	6(8.2)	1(1.4)	0(0.0)	3(4.1)	1(1.4)
Logistics	27(37.0)	1(1.4)	8(11.0)	24(32.9)	3(4.1)	2(2.7)	6(8.2)	2(2.7)
Plant location	32(43.8)	2(2.7)	8(11.0)	23(31.5)	1(1.4)	0(0.0)	5(6.8)	4(5.5)
Other	7(9.6)	1(1.4)	2(2.7)	7(9.6)	1(1.4)	1(1.4)	3(4.1)	4(5.5)

subject to the constraints

$$g_i(x_1, \ldots, x_n) \leq b_i, \qquad \text{for } i = 1, \ldots, m$$

The criterion may be expressed as a minimizing function, and the constraints can be $\geq$ and $=$ as well as $\leq$. The problem has n unknowns $(x_1, \ldots, x_n)$ and m constraints.

The linear programming problem is a special case of the general programming problem where both the objective function and the set of constraints must be linear. Thus in linear programming we

maximize

$$Z = f(x_1, \ldots, x_n) = \sum_{j-1}^{n} c_j x_j$$

subject to the constraints

$$g_i(x_1, \ldots, x_n) = \sum_{j=1}^{n} a_{ij} x_j \leq b_i, \qquad \text{for } i = 1, \ldots, m$$

and also subject to the nonnegativity restrictions

$$x_j \geq 0, \qquad \text{for } j = 1, \ldots, n$$

This problem has n unknowns and m constraints.

Consequently, the LP problem can be characterized by the following essential parts:

Optimize (maximize or minimize) a linear objective function, often called the criterion function, subject to

(A) a set of linear constraints which may be expressed as equalities or inequalities ($\leq$ or $\geq$)

(B) a set of nonnegative restrictions for the unknowns of the problem.

In matrix notation the problem is stated as

maximize[1]

$$Z = cX$$

subject to

$$AX \leq b$$

$$X \geq 0$$

where Z is a scalar, c is a known $(1 \times n)$ vector, X is an $(n \times 1)$ vector of unknowns (variables), A is a known $(m \times n)$ coefficient matrix, and b is a known $(m \times 1)$ constant column.

---

[1]It should be understood throughout the article that "minimize" can be used as well as "maximize."

Expanding the matrix symbolism gives

maximize

$$Z = c_1x_1 + c_2x_2 + \cdots + c_jx_j + \cdots + c_nx_n$$

subject to

$$\begin{array}{llllll}
a_{11}x_1 & + a_{12}x_2 & + \cdots + a_{1j}x_j & + \cdots + a_{1n}x_n & \le b_1 \\
a_{21}x_1 & + a_{22}x_2 & + \cdots + a_{2j}x_j & + \cdots + a_{2n}x_n & \le b_2 \\
\vdots & \vdots & \vdots & \vdots & \vdots \\
a_{i1}x_1 & + a_{i2}x_2 & + \cdots + a_{ij}x_j & + \cdots + a_{in}x_n & \le b_i \\
\vdots & \vdots & \vdots & \vdots & \vdots \\
a_{m1}x_1 & + a_{m2}x_2 & + \cdots + a_{mj}x_j & + \cdots + a_{mn}x_n & \le b_m
\end{array}$$

$$x_1 \ge 0,\ x_2 \ge 0,\ \ldots,\ x_j \ge 0,\ \ldots,\ x_n \ge 0$$

GRAPHICAL EXAMPLE OF LP PROBLEM

As a simple example, suppose that a company produces two products, each of which must be processed on two machines, and that the manager wishes to determine the number of units of each product to produce in order to maximize profit. The profit for one unit of Product 1 is \$4 and the profit for one unit of Product 2 is \$5. It requires 3 hr to make 1 unit of Product 1 on Machine 1, and 6 hr to produce 1 unit of Product 2 on Machine 1. Similarly, it takes 6 hr to process Product 1 on Machine 2, and 5 hr to make Product 2 on Machine 2. Each machine can be operated up to 2100 hr for the period being considered. In LP format this problem is stated as

maximize

$$Z = 4x_1 + 5x_2$$

subject to

$$3x_1 + 6x_2 \le 2100$$

$$6x_1 + 5x_2 \le 2100$$

$$x_1 \ge 0$$

$$x_2 \ge 0$$

where x_1 = number of units of Product 1 to be made
x_2 = number of units of Product 2 to be made

(It is assumed that fractional parts are acceptable. If not, an integer programming problem must be formulated.)

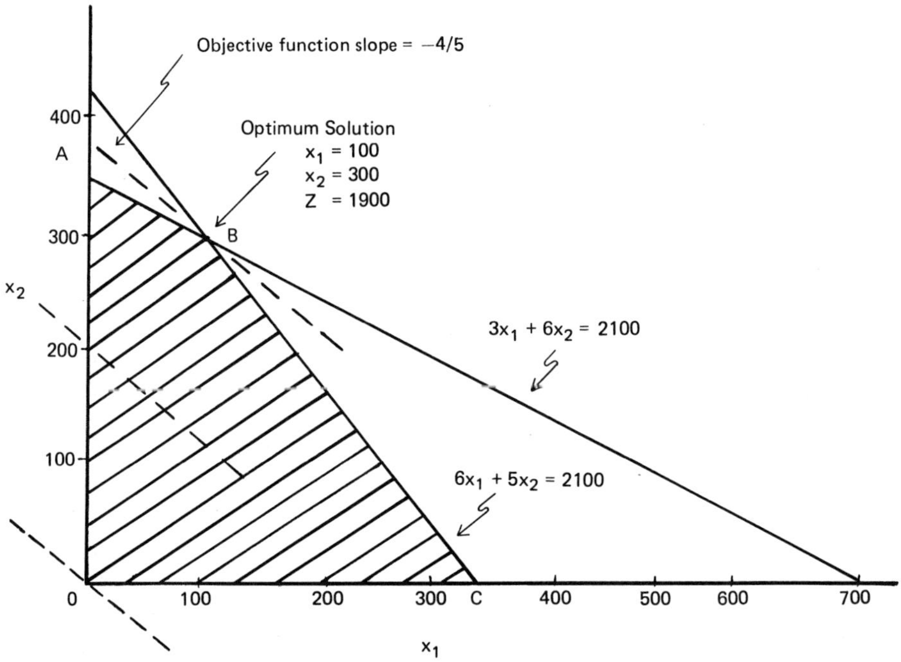

FIG. 1. Simple LP problem solved by graphical means.

This simple problem could be solved quite easily by graphical means, as depicted in Fig. 1. The feasible solution area for this problem is denoted by the convex polyhedron OABC, with the optimum solution located at vertex B. This is the point where the objective function attains its maximum value. Problems of interest for LP solution are of higher dimensional space and cannot be solved graphically; consequently, analytical methods must be employed. However, it is desirable to point out salient aspects of the LP problem from the geometrical presentation.

1. Usually an infinite number of feasible solutions exist, as is true for the problem depicted in Fig. 1. A feasible solution can occur anywhere on the boundary or the interior of the feasible solution set.
2. The optimum solution is located at one or more of the vertices of the feasible solution area; thus only the vertex points must be checked to determine the optimum solution. The vertex points can be calculated by solving systems of linear equations.
3. The feasible solution set is always a convex set, i.e., given any two points in the set, all points on the line joining these points must also lie in the feasible solution area.
4. It is possible to have more than one optimal solution to the LP problem.

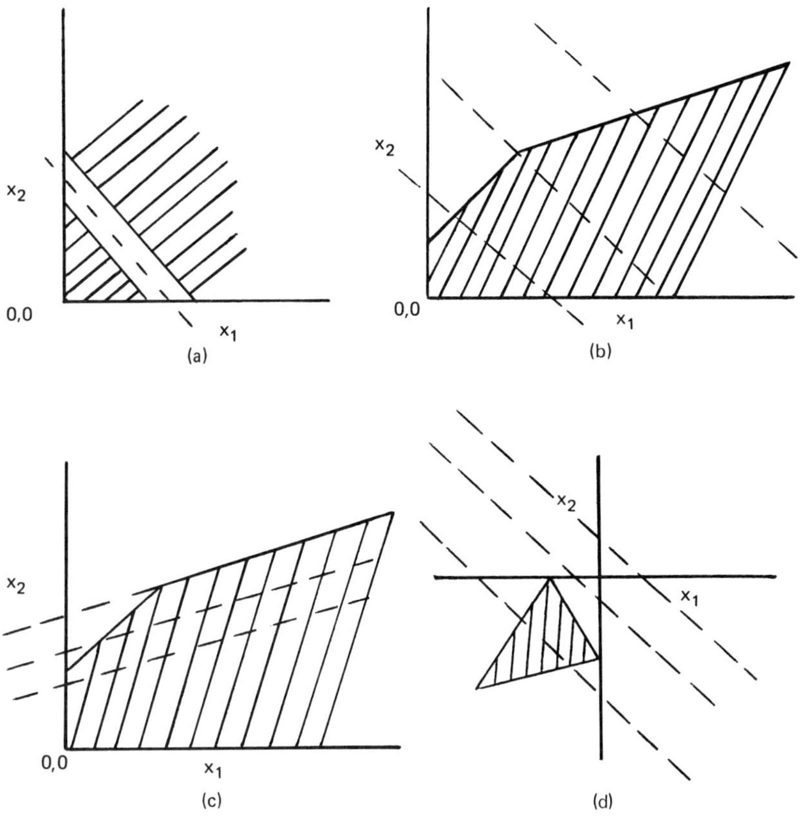

FIG. 2. Possible outcomes of an LP problem: (a) inconsistent; (b) unbounded; (c) arbitrarily large variables; and (d) no feasible solution.

If in the example problem the objective function were changed to $Z = 6x_1 + 5x_2$, then the slope would be exactly the same as that for the constraint $6x_1 + 5x_2 = 2100$, and the most extreme location of the objective function would be superimposed on this constraint. The result would be that both vertices B and C in Fig. 1 would represent optimal solutions, and likewise, any point on the line joining these two vertices would also be an optimal solution.

Other possible outcomes that may occur are: no solution, where the constraints are inconsistent, as shown in Fig. 2(a); an unbounded solution, where the value of the objective function can be made arbitrarily large, as depicted in Fig. 2(b); arbitrarily large variables, all of which give the maximum value, as pictured in Fig. 2(c); and no feasible solution because no points satisfying the constraints also satisfy the nonnegativity restrictions, as shown in Fig. 2(d). (The broken lines are the slopes of the objective function.) It is comforting to point out that the Simplex method has the capability of identifying these exceptional problems.

FUNDAMENTAL PROPERTIES OF A SOLUTION TO THE LP PROBLEM

It is important to distinguish among the following various types of solutions to a linear programming problem [5] .

A feasible solution is a vector $X = (x_1, x_2, \ldots, x_n)$ which satisfies both the constraints and the nonnegativity restrictions.

A basic solution is obtained by setting n - m variables equal to zero and solving the constraint set for the remaining m variables. These m variables are referred to as the variables in the basis.

A basic feasible solution is a basic solution which also satisfies the nonnegativity restrictions. (For a detailed discussion refer to the article on Basic Feasible Solutions in Vol. 3 [6]).

A nondegenerate basic feasible solution is a basic feasible solution which has exactly m positive values in the solution vector.

A degenerate basic feasible solution is a basic feasible solution which has one or more variables in the basis at the zero level.

In the Simplex method of LP an acceptable solution is either a nondegenerate basic feasible solution or a degenerate basic feasible solution.

As stated previously, the solution of a LP problem involves solving systems of linear equations. Therefore, it is essential that all inequalities be converted to equalities. This is accomplished by adding a slack variable to an inequality of the $\leq$ type, and subtracting a surplus variable to an inequality of the $\geq$ type. For example, suppose that the constraint set had the following two inequalities:

$$8x_1 + 2x_2 \leq 85$$

$$2x_1 + 2x_2 \geq 8$$

A slack variable x_3 would be added to the first constraint, and a surplus variable x_4 would be subtracted from the second constraint to give the equalities

$$8x_1 + 2x_2 + x_3 = 85$$

$$2x_1 + 3x_2 - x_4 = 8$$

The coefficients in the objective function for slack and surplus variables are usually assigned a value of zero.

To solve a system of m equations and m unknowns it is necessary that the $m \times m$ matrix have a rank m, i.e., the determinant of the matrix must not vanish. In solving the LP problem the Gaussian elimination procedure is usually employed. This requires that the vectors in the basis form an identity matrix. In the above example only the vector corresponding to x_3 is a unit vector. We would then add an artificial variable, say x_5, to the second constraint; now the vectors corresponding to variables x_3 and x_5 form an identity matrix. Similarly, it is usually necessary to add an artificial variable to a constraint which originally was an equality. Thus, suppose that the original problem were given as follows:

maximize

$$Z = 5x_1 + 2x_2$$

subject to

$$8x_1 + 2x_2 \le 85$$
$$2x_1 + 3x_2 \ge 8$$
$$x_1 + x_2 = 10$$
$$x_1 \ge 0,\ x_2 \ge 0$$

The augmented problem would be expressed as

maximize

$$Z = 5x_1 + 2x_2 + 0x_3 + 0x_4 - Mx_5 - Mx_6$$

subject to

$$\begin{aligned} 8x_1 + 2x_2 + x_3 \qquad\qquad\qquad\qquad &= 85 \\ 2x_1 + 3x_2 \qquad - x_4 + x_5 \qquad &= 8 \\ x_1 + x_2 \qquad\qquad\qquad + x_6 &= 10 \end{aligned}$$

with the nonnegativity restriction holding for all variables $x_1, \ldots, x_6$. In the augmented problem, x_3 is a slack variable, x_4 a surplus variable, x_5 an artificial variable, and x_6 an artificial variable. It should be noted that a large penalty coefficient (-M) is associated with each artificial variable in the objective function. Consequently, the variables in the basis for the first solution would be x_3, x_5, and x_6. The large penalty is added in the objective function for each artificial variable so that all artificial variables will be forced out of the basis at a positive level if a basic feasible optimum solution exists for the problem. For a minimizing problem a (+M) would be assigned as the coefficient in the objective function for an artificial variable.

The Simplex method is an iterative procedure whereby at each successive iteration the value of the objective function is at least as large (maximizing) as the value of the objective function for the previous iteration. The procedure is to move from one extreme point (vertex) of the convex set of constraints to an adjacent extreme point. An algorithm is utilized in determining the variable to be removed from the basis at each iteration to ensure that the nonnegativity restrictions are always satisfied. Thus it is never necessary to formally express the nonnegativity restrictions as constraints.

Conceivably, it would be possible to obtain the optimal solution by a "brute force" procedure. A basic solution could be obtained from each vertex by solving a system of equations. Repeat this procedure for all extreme points and select for the optimal solution the feasible solution which gives the largest value for the objective function. However, the number of systems of equations required to be solved by this method would be the combination of the n variables taken m at a time $\left(C_m^{\ n} = \frac{n!}{n!(n - m)!}\right)$. This would be a huge computational problem for most realistic size problems, and is not a satisfactory method, despite the calculation power of present computers.

ON THE THEORY OF THE SIMPLEX METHOD

The fact that the set of all feasible solutions to the LP problem is a convex set assures us that the optimum is located at an extreme point of the convex set. It can be proved easily that the feasible solution set is convex: it must be shown that every convex linear combination of any two feasible solutions is also feasible. Suppose we select two feasible solutions X_1 and X_2. Then

$$AX_1 \leq b, \quad \text{for } X_1 \geq 0$$

$$AX_2 \leq b, \quad \text{for } X_2 \geq 0$$

Let X_3 be given by the convex linear combination of X_1 and X_2, that is,

$$X_3 + \alpha X_1 + (1 - \alpha)X_2$$

where $0 \leq \alpha \leq 1$. Therefore,

$$AX_3 = A[\alpha X_1 + (1 - \alpha)X_2]$$

$$= \alpha AX_1 + (1 - \alpha)AX_2$$

Since $AX_1 \leq b$ and $AX_2 \leq b$, then

$$AX_3 \leq \alpha b + (1 - \alpha)b = b$$

Now that it is established that the feasible solution set is convex for the LP problem, it can now be proved that the optimum solution (maximum or minimum) will be located at an extreme point of the convex set. We assume that the convex set has a finite number of extreme points. The objective function is denoted by f(X), the extreme points by $\bar{X}_1, \bar{X}_2, \ldots, \bar{X}_q$, and the maximum feasible solution by X_0.

Assume that X_0 is not an extreme point; then X_0 can be expressed as a convex combination of q extreme points, that is,

$$X_0 = \sum_{i=1}^{q} \alpha_i \bar{X}_i$$

with $\alpha_i \geq 0$, and

$$\sum_{i=1}^{q} \alpha_i = 1$$

We can now state the value of the objective function for X_0 as

$$f(X_0) = f\left(\sum_{i=1}^{q} \alpha_i \bar{X}_i \right) = \alpha_1 f(\bar{X}_1) + \alpha_2 f(\bar{X}_2) + \cdots + \alpha_q f(\bar{X}_q) = m$$

where m is the maximum value of f(X).

Suppose we let $f(\bar{X}_m) = \max_i f(\bar{X}_i)$, then

$$f(X_0) \leq \sum_{i=1}^{q} \alpha_i f(\bar{X}_i) = f(\bar{X}_m)$$

since we assumed at the start that X_0 was the maximizing solution, which means that

$$f(X_0) \geq f(X)$$

for all X in the convex polyhedron.

However, we have just proven that $f(X_0) \leq f(\bar{X}_m)$, where $f(\bar{X}_m)$ was the largest value of the objective function among the extreme points; therefore $f(X_0)$ must be equal to $f(\bar{X}_m)$.

Other significant properties that follow are:

1. A set of m linearly independent vectors from the given set of n vectors is associated with each extreme point.
2. X can be an extreme point of the convex polyhedron if and only if the positive x's are coefficients of linearly independent vectors $\bar{a}_j$ in the expression

$$\sum_{j=1}^{n} x_j \bar{a}_j = b$$

In the previous section it was shown how to convert inequalities to equalities by the addition of slack variables for the $\leq$ type, and by the subtraction of surplus variables for the $\geq$ type. It was also shown how to ensure that the vectors in the basis of the initial solution form an identity matrix, which is a set of linearly independent vectors. This usually requires the addition of an artificial variable for each greater than or equal to, or equal to type constraint. For the augmented problem, then, we are assured that the rank of the matrix is equal to the number of constraints, denoted by m.

A significant aspect of the Simplex method is that we can determine if the original constrained problem has a feasible solution, or if any of the constraints are redundant. The artificial variables are added so that we will always have a basic feasible solution to the augmented problem. By the Simplex procedure we are able to determine if a basic feasible solution exists for the original problem [7].

Therefore, the constraints for any linear programming problem, augmented as appropriate, can be written as

$$AX = b$$

which is a set of m simultaneous equations in n unknowns.

$$\underset{A}{\begin{bmatrix} \bar{a}_1 & \bar{a}_2 & \cdots & \bar{a}_j & \cdots & \bar{a}_n \\ a_{11} & a_{12} & \cdots & a_{1j} & \cdots & a_{1n} \\ a_{21} & a_{22} & \cdots & a_{2j} & \cdots & a_{2n} \\ \vdots & \vdots & \vdots & \vdots & \vdots & \vdots \\ a_{i1} & a_{i2} & \cdots & a_{ij} & \cdots & a_{in} \\ \vdots & \vdots & \vdots & \vdots & \vdots & \vdots \\ a_{m1} & a_{m2} & \cdots & a_{mj} & \cdots & a_{mn} \end{bmatrix}} \underset{X}{\begin{bmatrix} \\ x_1 \\ x_2 \\ \vdots \\ x_j \\ \vdots \\ x_n \end{bmatrix}} = \underset{b}{\begin{bmatrix} \\ b_1 \\ b_2 \\ \vdots \\ b_i \\ \vdots \\ b_m \end{bmatrix}}$$

In summation symbolism form we can write this system of equations as

$$\sum_{j=1}^{n} a_{ij} x_j = b_i, \quad i = 1, \ldots, m$$

Another way of expressing this system of equations is by scalar multiplication and vector addition of each of the vectors, denoted by $[\bar{a}_1, \bar{a}_2, \ldots, \bar{a}_j, \ldots, \bar{a}_n]$, where

$$\bar{a}_j = \begin{bmatrix} a_{1j} \\ a_{2j} \\ \vdots \\ a_{ij} \\ \vdots \\ a_{mj} \end{bmatrix}$$

Thus

$$x_1\bar{a}_1 + x_2\bar{a}_2 + \cdots + x_j\bar{a}_j + \cdots + x_n\bar{a}_n = \sum_{j=1}^{n} x_j\bar{a}_j = b$$

From the n vectors, a set of m linearly independent vectors is identified as the basis, $B = (b_1 b_2 \ldots b_i \ldots b_m)$. Therefore, any other vector $\bar{a}_j$ can be expressed as a linear combination of these basic vectors, giving

$$\bar{a}_j = y_{1j}b_1 + y_{2j}b_2 + \cdots + y_{ij}b_i + \cdots + y_{mj}b_m = B\bar{y}_j$$

or

$$\bar{y}_j = B^{-1}\bar{a}_j$$

Likewise, the solution vector, expressed as the column vector $X_B = [x_{B1} x_{B2} \ldots x_{Bi} \ldots x_{Bm}]$, is equal to $B^{-1}b$.

The coefficients of the objective function that correspond to the variables in the basis are defined as the row vector $c_B = (c_{B1}, \ldots, c_{Bi}, \ldots, c_{Bm})$.

For each j-th column in A we can calculate a value for the j-th variable defined as

$$Z_j = \sum_{i=1}^{m} c_{Bi} y_{ij} = c_B \bar{y}_j$$

To clarify the above symbolic notation, consider the following problem, which was depicted in Fig. 1.

Maximize

$$4x_1 + 5x_2$$

subject to

$$3x_1 + 6x_2 \leq 2100$$

$$6x_1 + 5x_2 \leq 2100$$

$$x_1 \geq 0, \; x_2 \geq 0$$

Converting the constraints to equalities we have by adding slack variables x_3 and x_4,

maximize

$$4x_1 + 5x_2 + 0x_3 + 0x_4$$

subject to

$$3x_1 + 6x_2 + x_3 = 2100$$

$$6x_1 + 5x_2 + x_4 = 2100$$

with the nonnegativity restrictions holding.

These constraints can also be written as

$$x_1\bar{a}_1 + x_2\bar{a}_2 + x_3\bar{a}_3 + x_4\bar{a}_4 = b$$

or

$$x_1 \begin{bmatrix} 3 \\ 6 \end{bmatrix} + x_2 \begin{bmatrix} 6 \\ 5 \end{bmatrix} + x_3 \begin{bmatrix} 1 \\ 0 \end{bmatrix} + x_4 \begin{bmatrix} 0 \\ 1 \end{bmatrix} = \begin{bmatrix} 2100 \\ 2100 \end{bmatrix}$$

Since the vectors $\bar{a}_3$ and $\bar{a}_4$ form the identity matrix, we will identify these as the basis vectors

$$B = (b_1, b_2) = \begin{bmatrix} 1 & 0 \\ 0 & 1 \end{bmatrix}$$

Therefore

$$X_B = B^{-1}b = Ib = \begin{bmatrix} 2100 \\ 2100 \end{bmatrix}$$

The solution for this set of basis vectors is $x_3 = 2100$, $x_4 = 2100$. The variables not in the basis have a value of zero.

Suppose that in the next iteration the vector $\bar{a}_2$ replaces the vector $\bar{a}_3$ in the basis. (At each iteration of the Simplex method, one vector is removed from the basis and another vector comes into the basis.) Now the vectors in the basis are $b_1 = \bar{a}_2$ and $b_2 = \bar{a}_4$, and the new solution vector is

$$X_B = \begin{bmatrix} 6 & 0 \\ 5 & 1 \end{bmatrix}^{-1} \begin{bmatrix} 2100 \\ 2100 \end{bmatrix} = \begin{bmatrix} 1/6 & 0 \\ -5/6 & 1 \end{bmatrix} \begin{bmatrix} 2100 \\ 2100 \end{bmatrix} = \begin{bmatrix} 350 \\ 350 \end{bmatrix}$$

The solution now is $x_2 = 350$, $x_4 = 350$.

Also,

$$\bar{y}_1 = B^{-1}\bar{a}_1 = \begin{bmatrix} 1/6 & 0 \\ -5/6 & 1 \end{bmatrix} \begin{bmatrix} 3 \\ 6 \end{bmatrix} = \begin{bmatrix} 1/2 \\ 7/2 \end{bmatrix}$$

and $\bar{a}_1$ can be expressed as a linear combination of the vectors in the basis as follows

$$\bar{a}_1 = \frac{1}{2} \begin{bmatrix} 6 \\ 5 \end{bmatrix} + \frac{7}{2} \begin{bmatrix} 0 \\ 1 \end{bmatrix} = \begin{bmatrix} 3 \\ 6 \end{bmatrix}$$

We can calculate the value for Z_1 as

$$Z_1 = c_B\bar{y}_1 = [5 \quad 0] \begin{bmatrix} 1/2 \\ 7/2 \end{bmatrix} = \frac{5}{2}$$

The value of the objective function for a basic feasible solution $X_B = B^{-1}b$ is determined as

$$Z_0 = c_B X_B$$

For the above problem, with the basic variables being x_2 and x_4, then

$$Z_0 = [5 \quad 0] \begin{bmatrix} 350 \\ 350 \end{bmatrix} = 1750$$

If for a particular vector $\bar{a}_j$, not in the basis, the corresponding $c_j > Z_j$, or $(Z_j - c_j) < 0$ when maximizing, and at least one element of $\bar{y}_j$ is greater than zero, then this j-th vector can enter the basis, replacing one of the vectors in the present basis, to obtain a new basic feasible solution which results in a new value for the criterion function. This value will be larger than the existing value, assuming the solution is not degenerate. Regardless, the new value for the objective function will not be inferior to the present value. When minimizing, the criterion used for bringing a vector in the basis is $(Z_j - c_j) > 0$, in contrast to $(Z_j - c_j) < 0$ for the maximizing problem.

The Simplex method enables us to determine very easily if the problem is unbounded: if for a basic feasible solution there exists a nonbasic column vector $\bar{a}_j$ such that, when maximizing, $(Z_j - c_j) < 0$, with $y_{ij} \leq 0$ for $i = 1, 2, \ldots, m$, then the solution to the given problem is unbounded, the result being that the value of the objective function can be made arbitrarily large. An example is shown in Fig. 2(b). For the minimizing problem, the only difference is that $(Z_j - c_j) > 0$.

An optimal basic feasible solution is obtained at a particular iteration of the Simplex method if $(Z_j - c_j) \geq 0$ for every vector $\bar{a}_j$, when maximizing, or if $(Z_j - c_j) \leq 0$ for every vector $\bar{a}_j$, when minimizing.

It was stated previously that it is possible to have more than one optimal solution to the LP problem. If, for an optimal basic feasible solution, $(Z_j - c_j) = 0$ for a vector $\bar{a}_j$ not in the basis and at least one y_{ij}, for $i = 1, 2, \ldots, m$, is positive, then this vector can be brought into the basis, replacing an existing basis vector. This new solution will give the same value for the objective function. Also, any convex combination of these alternate optimal basic feasible solutions will be a solution for which the value of the objective function is optimum. However, these solutions will not be <u>basic</u> feasible solutions since they do not occur at an extreme point.

An infinite number of optimal solutions is possible without having two or more basic feasible optimal solutions. If for some $\bar{a}_j$ not in the basis, we have $(Z_j - c_j) = 0$, and $y_{ij} \leq 0$ for $i = 1, 2, \ldots, m$, then an infinite number of optimal solutions exists. This type of problem is pictured in Fig. 2(c).

When one or more artificial vectors are in the basis of the optimal solution, at a positive level, then no solution to the problem exists. This is due to either inconsistent constraints or violation of the nonnegativity restrictions. These two cases are shown in Figs. 2(a) and 2(d), respectively.

The presence of one or more artificial vectors in the basis of the optimal solution at a zero level indicates degeneracy or redundancy in the original constraints. If one or more artificial variables are in the basis of the optimal solution at the zero level, a feasible solution exists and the original constraints are consistent. Suppose that for some $\bar{a}_j$ not in the basis, $y_{ij} \neq 0$, where i corresponds to the i-th artificial vector in the basis at the zero level. The new vector $\bar{a}_j$ can enter the basis at a zero level, replacing the artificial vector. This is the degenerate case.

We repeat this process, removing all artificial vectors, unless we eventually arrive at the state where $y_{ij} = 0$ for all $\bar{a}_j$ not in the basis, with i corresponding to the i-th artificial vector in basis. Now it is not possible to replace the i-th artificial vector with $\bar{a}_j$, and the i-th constraint of the original system is redundant.

Sometimes a two-phase method is employed when problems have artificial variables. Simply stated, the purpose of the first phase is to drive all artificial variables to zero by considering only the artificial variables in the objective function. If a basic feasible solution is obtained in Phase 1, then the original objective function is used to obtain the optimum solution to the original problem. If the artificial variables cannot be driven to zero in Phase 1, then no solution exists to the original problem.

COMPUTATIONAL PROCEDURE OF THE SIMPLEX METHOD

Since a detailed description of the computational aspects of the Simplex method is given by Bazaraa in Vol. 7 of this Encyclopedia, the reader interested in the algorithms employed to solve a linear programming problem should consult this article [8]. In addition to the regular Simplex method, the author discusses the Revised Simplex method and Product Form of the Inverse.

For a person with very little mathematical background, programmed instruction material on matrix algebra and linear programming by Holzman, Schaefer, and Glaser should provide a good working knowledge of these areas [9].

DUALITY THEORY AND ITS APPLICATIONS

The theory of duality is one of the most important concepts in linear programming from both the theoretical and applied viewpoints. If we call the original problem the primal problem, then associated with this problem is another linear programming problem called the dual problem. Suppose that the <u>primal</u> linear programming problem is given in the form

maximize

$$cX$$

subject to

$$AX \leq b$$

$$X \geq 0$$

Then the <u>dual</u> problem is defined as

minimize

$$wb$$

subject to

$$wA \geq c$$

$$w \geq 0$$

This is referred to as the symmetric dual problem.

The new set of variables defined by the row vector $w = (w_1, w_2, \ldots, w_m)$ is called the dual vector.

It is desirable to note the following relationship between the two problems [10]:

1. The criterion has been changed from "maximize" in the primal to "minimize" in the dual. If the primal problem were formulated as a minimizing problem, then the dual problem would be a maximizing one.
2. The coefficients of the objective function of the primal problem are the right-hand side constants of the dual, and the right-hand side constants of the primal are the objective function coefficients of the dual.
3. The constraint inequalities have been reversed. If the original constraints were of the $\geq$ type, then the constraints of the dual would be of the $\leq$ type.
4. The number of constraints in the dual problem is equal to the number of primal variables, since each column in the primal corresponds to a constraint in the dual.
5. Since each constraint in the primal corresponds to a column in the dual, there is one dual variable for every primal constraint.
6. The primal problem can be obtained by taking the dual of the dual.

The dual theorem states that if either the primal or the dual has a finite optimum solution, then a finite optimum solution exists for the other problem and the values of the objective function for the primal and the dual problems are equal. If either the primal or the dual has an unbounded solution, then the other problem has no feasible solution.

We can easily prove that maximum cX = minimum wb. Given the primal and dual constraints

$$AX \leq b$$

$$wA \geq c$$

premultiply both sides of the primal constraints by the vector w and postmultiply both sides of the dual constraints by the vector X to obtain

$$wAX \leq wb$$

$$wAX \geq cX$$

Therefore

$$cX \leq wAX \leq wb$$

$$cX \leq wb$$

However, since the objective of the primal problem is to obtain the largest value for cX, and the criterion of the dual problem is to seek the smallest value of wb, the optimum value for both problems occurs when $cX = wb$.

It can be shown that the optimal values for the dual variables can be obtained from the expression $w^0 = c_B B^{-1}$, where c_B is the row vector of objective function coefficients corresponding to the variables in the basis for the primal problem, and B is the $m \times m$ matrix of vectors in the basis for the optimal solution of the primal problem. Thus the optimal solution for the dual problem is available from the primal solution. In fact, the values for w_i are contained in the $(Z_j - c_j)$ row of the optimal solution to the primal problem. The optimal value for w_i would be equal to the $(Z_j - c_j)$ value for the slack variable associated with the i-th constraint. This is true because

$$Z_j = c_B y_j = c_B B^{-1} \bar{a}_j$$

Since $\bar{a}_j$ is a unit vector e_i for the slack variable, and since the coefficient c_j for a slack variable is zero, then

$$(Z_j - c_j) = (c_B B^{-1} e_i - 0) = w_i$$

The dual variable is also referred to as a shadow price or imputed value in the language of the economist, i.e., w_i is the value for an additional unit of resource (constraint) i in the primal problem. If the i-th resource is not utilized completely in the primal problem, then the corresponding dual variable has a value of zero in the optimum solution.

Let us consider the example problem which was presented earlier.

Maximize

$$4x_1 + 5x_2$$

subject to

$$3x_1 + 6x_2 \leq 2100, \quad \text{Machine 1 resource}$$

$$6x_1 + 5x_2 \leq 2100, \quad \text{Machine 2 resource}$$

$$x_1 \geq 0$$

$$x_2 \geq 0$$

The decision variables, denoted by x_1 and x_2, are the number of units of Product 1 and Product 2 to be produced in order to maximize the objective function, which gives the profit for each product. Both products require the resources of two machines; the coefficients in the constraints represent the hours to produce a unit of a particular product on a specific machine. A total of 2100 hr is available on each machine.

The dual problem is formulated as

minimize

$$2100w_1 + 2100w_2$$

subject to

$$3w_1 + 6w_2 \geq 4$$

$$6w_1 + 5w_2 \geq 5$$

$$w_1 \geq 0$$

$$w_2 \geq 0$$

The dual variables w_1 and w_2 are associated with Machine 1 resource and Machine 2 resource, respectively.

By applying the Simplex method the optimal solution determined for the primal is

$$x_1 = 100$$

$$x_2 = 300$$

and for the dual is

$$w_1 = \frac{10}{21}$$

$$w_2 = \frac{3}{7}$$

The value for the objective function of the primal problem is $(4)(100) + (5)(300) = \underline{1900}$; for the dual problem it is $(2100)\frac{10}{21} + (2100)\frac{3}{7} = \underline{1900}$. Thus max cX = min wb at the optimum for both problems.

The imputed values (dual variables) are interpreted as follows:

1. If an additional hour of Machine 1 time could be obtained, the profit for the primal problem would increase by the amount $w_1 = 10/21$.
2. Similarly, if an additional hour of Machine 2 time could be obtained, the profit for the primal problem would increase by the amount $w_2 = 3/7$.

The fact that w_1 and w_2 are positive indicates that there is no idle time (slack) on the two machines. If the optimum solution for the primal problem did have a slack variable for the i-th constraint in the basis at a positive level, then the corresponding dual variable w_i would equal zero. This condition can be stated more formally by the important principle of "complementary slackness," which will be discussed in the following section.

The unsymmetric dual problem is defined as follows:

<u>Primal</u>

$$\begin{aligned} &\text{maximize} && cX \\ &\text{subject to} && AX = b \\ &&& X \geq 0 \end{aligned}$$

<u>Dual</u>

$$\begin{aligned} &\text{minimize} && wb \\ &\text{subject to} && wA \geq c \end{aligned}$$

The difference between the unsymmetric and symmetric problems is that the constraints of the primal problem are equalities rather than inequalities and that the nonnegativity restriction does not hold for the dual variables. Since an equality can be expressed as two inequalities, an unsymmetric problem can be formulated as a symmetric problem. It is also possible to express a primal problem in mixed form ($\leq$, $\geq$, $=$) as a dual problem [11].

Usually, when speaking of the Simplex method, we refer to an iterative method which always maintains feasibility ($X \geq 0$), and works toward satisfying the optimality criterion ($Z_j - c_j \geq 0$, when maximizing). This is called the primal Simplex method in contrast to the dual Simplex method by which the optimality criterion is satisfied at each iteration, but the feasibility conditions are not met until the optimum solution is obtained. Even though the dual Simplex method is related to the dual problem, it should not be confused with the formulation of the dual problem. The dual problem is usually solved by the primal Simplex method.

COMPLEMENTARY SLACKNESS

Let us consider the symmetric dual problems

Primal

maximize cX

subject to $AX \leq b$

$X \geq 0$

Dual

minimize wb

subject to $wA \geq c$

$w \geq 0$

The complementary slackness theorem states that if X_0 and w_0 are feasible solutions, then X_0 and w_0 are optimal for the primal and dual problems, respectively, if and only if

$$(w_0 A - c)X_0 + (b - AX_0)w_0 = 0$$

Let us represent the slack variables for the primal problem by the vector ν_0 and the surplus variables in the dual problem by the vector μ_0 for the solutions X_0 and w_0. Then

$$AX_0 + \nu_0 = b, \quad \text{for } X_0 \geq 0,\ \nu_0 \geq 0$$

$$w_0 A - \mu_0 = c \quad \text{for } w_0 \geq 0,\ \mu_0 \geq 0$$

Premultiply both sides of the first equation by w_0 and postmultiply the second equation by X_0 to give

$$w_0 AX_0 + w_0 \nu_0 = w_0 b$$

$$w_0 AX_0 - \mu_0 X_0 = cX_0$$

Subtracting we obtain

$$w_0 \nu_0 + \mu_0 X_0 = w_0 b - cX_0$$

If we assume that X_0 and w_0 are optimal, then $cX_0 = w_0 b$ and we have

$$w_0 \nu_0 + \mu_0 X_0 = 0$$

where

$$\nu_0 = b - AX_0 \quad \text{and} \quad \mu_0 = w_0 A - c$$

The solution vectors X_0 and w_0 are optimum since now

$$w_0 b - cX_0 = 0$$

$$w_0 b = cX_0$$

Another way of stating the complementary slackness principle is that

$$w_{i_0} \nu_{i_0} = 0, \quad \text{for } i = 1, 2, \ldots, m$$

$$\mu_{j_0} x_{j_0} = 0 \quad \text{for } j = 1, 2, \ldots, n$$

The principle of complementary slackness has been used to develop primal-dual algorithms to solve problems having artificial variables [7].

THE FUNDAMENTAL PROBLEM

The principle of complementary slackness is directly related to what is defined as the Fundamental Problem by Cottle and Dantzig [12]. This has resulted in the identification of a class of problems referred to as Linear Complementarity Problems [13]. Lemke has contributed significantly to the development of an algorithm for solving this problem [14]. Even though the experience used in the solution of problems formulated in this manner is limited, there is an indication that the complementary pivot algorithm developed for solving this problem may be an improvement over the Simplex method in solving linear programming problems [15, 16]. The significance of the Fundamental Problem formulation is that it can be used to solve quadratic programming and bimatrix games problems as well as the linear programming problem. The Fundamental Problem is defined as follows.

Given a $p \times p$ matrix M and a p vector q, find vectors w and z satisfying

$$w = q + Mz$$

$$wz = 0$$

$$w \geq 0, \quad z \geq 0$$

It is important to note that the problem has no objective function.

We will consider the expression of the linear programming and quadratic programming problems in this form.

The symmetric primal-dual problem is defined as

<u>Primal</u>

$$\begin{aligned} &\text{maximize} && cX \\ &\text{subject to} && AX \leq b \\ &&& X \geq 0 \end{aligned}$$

<u>Dual</u>[2]

$$\begin{aligned} &\text{minimize} && \lambda b \\ &\text{subject to} && \lambda A \geq c \\ &&& \lambda \geq 0 \end{aligned}$$

By the dual theorem the maximum of the primal equals the minimum of the dual when the primal and dual systems are feasible.

The inequality constraints are converted to equations by the use of slack and artificial vectors. Thus

---

[2]The vector for the dual variables will be denoted by λ rather than w which was used in our earlier discussion of the dual problem. The reason is that the vector w has a different interpretation for the Fundamental Problem.

$AX + \nu = b, \quad X \geq 0, \; \nu \geq 0$

$\lambda A - \mu = c, \quad \lambda \geq 0, \; \mu \geq 0$

or

$\nu = b - AX$

$\mu = -c + \lambda A$

To transform this problem to the Fundamental Problem, we define

$$w = \begin{bmatrix} \nu \\ \mu \end{bmatrix}, \quad q = \begin{bmatrix} b \\ -c \end{bmatrix}, \quad M = \begin{bmatrix} 0 & -A \\ A^T & 0 \end{bmatrix}, \quad z = \begin{bmatrix} \lambda \\ X \end{bmatrix}$$

Therefore, for the LP problem

$$\begin{matrix} w & = & q & + & M & z \end{matrix}$$

$$\begin{bmatrix} \nu \\ \mu \end{bmatrix} = \begin{bmatrix} b \\ -c \end{bmatrix} + \begin{bmatrix} 0 & -A \\ A^T & 0 \end{bmatrix} \begin{bmatrix} \lambda \\ X \end{bmatrix}$$

$$wz = [\nu \, \mu] \begin{bmatrix} \lambda \\ X \end{bmatrix} = \nu\lambda + \mu X = 0$$

$X \geq 0, \quad \nu \geq 0, \quad \lambda \geq 0, \quad \mu \geq 0$

Now let us consider the quadratic programming problem

maximize

$cX + \frac{1}{2}X'DX$

subject to

$AX \leq b$

$X \geq 0$

Expressed as a Lagrangian function we have

$L(X, \lambda) = cX + \frac{1}{2}X'DX + \lambda(b - AX)$

Applying the Kuhn-Tucker necessary conditions

$$\frac{\partial L}{\partial X} = c + DX - \lambda A \leq 0, \quad \text{and if } < \text{ holds, } X = 0$$

$$\frac{\partial L}{\partial \lambda} = b - AX \geq 0, \quad \text{and if } > \text{ holds, } \lambda = 0$$

After introduction of slack and surplus variables we have

$c + DX - \lambda A + \mu = 0$

$b - AX \qquad - \nu = 0$

$\mu X + \nu\lambda = 0$

This can be easily placed in the format of the Fundamental Problem,

$$w = q + M z$$

$$\begin{bmatrix} \nu \\ \mu \end{bmatrix} = \begin{bmatrix} b \\ -c \end{bmatrix} + \begin{bmatrix} 0 & -A \\ A^T & -D \end{bmatrix} \begin{bmatrix} \lambda \\ X \end{bmatrix}$$

$$wz = [\nu \mu] \begin{bmatrix} \lambda \\ X \end{bmatrix} = \nu\lambda + \mu X = 0$$

$$X \geq 0, \quad \nu \geq 0, \quad \lambda \geq 0, \quad \mu \geq 0$$

It is interesting to point out that complementary pivoting methods used to solve the Fundamental Problem can be generalized so as to be able to determine approximate Brouwer and Kakutani fixed points. The result is that fixed point computing methods have been applied to solving problems in the areas of economic equilibrium, nonlinear programming, nonlinear boundary values, and complex polynomials [17].

SENSITIVITY ANALYSIS AND PARAMETRIC PROGRAMMING (POSTOPTIMAL ANALYSIS)

In most real-life problems the input data for the model are not known with certainty, as is assumed in the case of the linear programming model. The coefficients of the objective function, the coefficient matrix of the constraints, and the right-hand side vector are all considered to be deterministic in solving the LP problem. Variation in these values may affect the optimal solution.

Sensitivity analysis, which starts with the optimal solution for a particular problem, can be used to consider changes in the coefficients of the objective function, the right-hand side, and the constraint matrix, and the addition of a new activity or a new constraint.

As an example, it may be desirable to determine the range of the values of the objective function c_1 for which the current vectors remain in the optimal solution of the primal problem that was presented in the section on Duality Theory and Its Applications. The optimum tableau for this problem is presented as Table 4.

It is obvious that this solution is optimal since we are maximizing, and all $(Z_j - c_j) \geq 0$.

A change in the value of c_1 will not change the values for $(Z_1 - c_1) = 0$ and $(Z_2 - c_2) = 0$, since they correspond to the variables in the basis. However, the values for $(Z_3 - c_3)$ and $(Z_4 - c_4)$ do change if c_1 changes, since

TABLE 4
Optimum Table

Vectors in Basis	c_B	X_B	$\bar{y}_3$	$\bar{y}_4$	$\bar{y}_1$	$\bar{y}_2$
$\bar{a}_2$	$c_2 = 5$	300	2/7	-1/7	0	1
$\bar{a}_1$	$c_1 = 4$	100	-5/21	2/7	1	0
$(Z_j - c_j)$			10/21	3/7	0	0

$$Z_3 = c_B\bar{y}_3 = [5,\ c_1]\begin{bmatrix} 2/7 \\ -5/21 \end{bmatrix} = 10/7 - 5/21c_1$$

and

$$Z_4 = c_B\bar{y}_4 = [5,\ c_1]\begin{bmatrix} -1/7 \\ 2/7 \end{bmatrix} = -5/7 + 2/7c_1$$

The maximum change permitted before a new vector enters the basis can be calculated by setting $Z_3 - c_3 = 0$ and $Z_4 - c_4 = 0$. Thus, since $c_3 = 0$ and $c_4 = 0$,

$$Z_3 - c_3 = \frac{10}{7} - \frac{5}{21}c_1 - 0 = \frac{10}{7} - \frac{5}{21}c_1 = 0$$

and

$$Z_4 - c_4 = -\frac{5}{7} + \frac{2}{7}c_1 - 0 = -\frac{5}{7} + \frac{2}{7}c_1$$

From these two expressions we find that for the vectors $\bar{a}_2$ and $\bar{a}_1$ to remain in the optimum solution, the range for c_1 is

$$\frac{5}{2} \le c_1 \le 6$$

If c_1 is less than 5/2, then the vector $\bar{a}_4$ will enter the basis, and if c_1 is greater than 6, the vector $\bar{a}_3$ will enter the basis.

For this same problem let us now consider the determination of the range for an element in the right-hand side, the b vector. Assume that we would like to know how much b_2 (2100 hr of Machine 2) can vary and still have the same vectors, $\bar{a}_2$ and $\bar{a}_1$, in the basis of the optimum solution. Since the elements in the $(Z_j - c_j)$ now will not be affected by a change in the b vector, we must be concerned only with the requirement that the nonnegativity restriction holds for all variables in the basis, i.e., $X_B \ge 0$.

At any stage of the Simplex method, $X_B = B^{-1}b$, and thus we can determine the range in b_2 values for the example problem as follows:

$$X_B = B^{-1}b \ge 0$$

$$= \begin{bmatrix} \frac{2}{7} & -\frac{1}{7} \\ -\frac{5}{21} & \frac{2}{7} \end{bmatrix} \begin{bmatrix} 2100 \\ b_2 \end{bmatrix} = \begin{bmatrix} 600 - \frac{1}{7}b_2 \\ -500 + \frac{2}{7}b_2 \end{bmatrix} \ge 0$$

This gives the range in Machine 2 hours as

$$1750 \le b_2 \le 4200$$

Thus vectors $\bar{a}_2$ and $\bar{a}_1$ will be in the basis of the optimal solution as long as the available hours on Machine 2 are no less than 1750 and no more than 4200. Of course, it is true that the value of the objective function will change as the number of available hours varies.

Sensitivity analysis is also used to evaluate the following changes in the coefficient matrix A:

1. Addition of a new variable
2. Changes in the resource requirements for a particular product (variable)
3. Addition of a new constraint

When we add a new variable, the existing solution remains optimal if for this variable $(Z_j - c_j) \geq 0$, when maximizing. If $(Z_j - c_j) < 0$, then this new variable must enter the basis, and consequently, we have a different set of vectors in the basis.

Changes in the resource requirements for a nonbasic variable can be handled by recalculating the $(Z_j - c_j)$ value for the j-th variable which has had changes made in the resource requirements, and by applying the same criteria noted above for adding a new variable. When changes are made in the coefficients of a basic variable, it is usually desirable to solve the problem over again after making the changes in the initial tableau.

If, for a new constraint added, the values for the existing optimal solution satisfy this constraint, then the present optimal solution remains optimal. However, if the new constraint is not satisfied by the existing optimal solution, then we add the new constraint to the current optimal tableau and apply the dual Simplex procedure to obtain the new optimal solution.

Parametric programming is an extension of sensitivity analysis. In our discussion of changes in the cost coefficients and right-hand side elements, we considered only one element at a time. In parametric programming we permit more than one element to be changed at a time, with the changes expressed as a function of a single parameter [7]. This is also a postoptimal analysis procedure.

Let us first consider changes in the cost coefficients. We replace the vector c by the vector

$$c^* = c + \phi V$$

where ϕ is a nonnegative scalar parameter and V is an arbitrarily specified vector which indicates the simultaneous changes that are being considered. For example, if we have four cost coefficients and wish to obtain the effect of increasing c_2 at twice the rate of change of c_1, then

$$V = [v_1 v_2 v_3 v_4] = [1\ 2\ 0\ 0]$$

We wish to determine the largest value of ϕ for which the given solution remains optimal. This critical value is obtained by solving

$$\phi_c = \min_j - \frac{(Z_j - c_j)}{(V_B \bar{y}_j - v_j)}, \quad \text{for } (V_B \bar{y}_j - v_j) < 0$$

where the vector V_B represents the elements of the arbitrarily specified vector corresponding to the variables in the basis of the optimum solution. The range in the c_j values can be determined from the expression

$$c^* = c + \phi_c V$$

In a similar manner we can address changes in the right-hand side, often referred to as the requirements vector. We now define a new vector

$$b^* = b + \phi R$$

where $\phi \geq 0$, and R is an arbitrarily specified vector. The procedure is to determine the largest ϕ for which the optimal basis gives a feasible solution, i.e., no negative values in the requirements vector. Thus

$$X_B^* = B^{-1}b^* = B^{-1}b + \phi B^{-1}R$$

$$= X_B + \phi Xr, \quad \text{with } Xr = B^{-1}R$$

For each i-th row which has $x_{ir} < 0$, we set $x_{i0}^* = x_{i0} + \phi x_{ir} = 0$. Therefore, the critical value for the parameter is

$$\phi_c = \min_i - \frac{x_{i0}}{x_{ir}}, \quad \text{for } x_{ir} < 0$$

The range in the b vector is obtained by solving the equation

$$b^* = b + \phi_c R$$

It is possible to determine the optimal solutions for the whole range of ϕ by making the necessary changes in the vectors in the basis of the optimal solution.

THEORY OF GAMES AND LINEAR PROGRAMMING

The famous two-person, zero-sum game as conceived by von Neumann [18] can be expressed as a linear programming problem and solved by the Simplex method. Game theory is concerned with the class of problems in which there is conflict or competition between two or more parties, and where each party has some control over the outcome of the competing interests [19]. The classical interpretation of competing parties would be two players opposing each other in a game; however, game theory applies to many other situations where conflict or competition are present, such as business competitors, opponents in a war, marketing campaigns, rivals for space, political campaigns, or clashes with nature.

Life is replete with conflict and competition. Game theory is a mathematical theory that has its roots in certain problems abstracted from real-life situations, where the decision-makers' preferences are not in consonance with each other.

The underlying theory dates back to 1928 when the great mathematician John von Neumann published the first article on the subject. Many books and articles have been published on game theory since then, but the fundamental work of von Neumann is still the basis for most treatments of the subject.

The basic concepts of game theory relate to two-person zero-sum games: these games involve two adversaries, as the name suggests, who are diametrically opposed to each other so that what one competitor wins, the other loses. Because we are concerned here with game theory as related to linear programming, we will not consider the areas involving n-person games, infinite games, and cooperative games [20].

As an introduction to game theory, consider the game of matching pennies. In the game are two players, designated as P_1 and P_2. Each player chooses either heads or tails. After the choice is made by each player without the knowledge of the other player, their choices are made known. If the pennies match, Player P_1 wins, and P_2 pays P_1 one penny. If the pennies do not match, P_2 wins, and P_1 pays

		PLAYER P_2	
		Heads	Tails
PLAYER P_1	Heads	+1	-1
	Tails	-1	+1

FIG. 3. Payoff matrix.

P_2 one penny. A payoff matrix is set up in game theory to show the results of the strategies of the competing players. The payoff matrix for the penny matching game is shown in Fig. 3. This matrix shows a +1 payment when the pennies match. Thus, if both players choose heads or tails, P_2 pays P_1 one penny. The matrix also shows a -1 payment when the pennies do not match, which indicates a payment from P_1 to P_2. The plus and minus signs indicate the direction of the payment, where the plus payment shows payment to P_1 from P_2, and the minus payment indicates a payment to P_2 from P_1.

The objectives of the game theoretic approach are:

1. To choose the optimum strategy for one or both players from among various alternative strategies
2. To determine the value of the game

By optimum strategy is meant the course of action that will give a player the best expected value (mathematical expectation) of the payoff. The value of the game is the expected value of the payoff.

These objectives will be interpreted in terms of the penny matching game. Refer to Fig. 4. Player P_1 should choose heads one-half of the time and tails one-half of the time. It is called a mixed strategy because more than one strategy is played as the game is repeated. As can be seen in Fig. 4, the optimum strategy of P_2 is also to play heads one-half of the time and tails one-half of the time.

The second objective in solving a problem in game theory is to determine the value of the game. The value of a game can be determined for various different strategies, but usually it means the value of the game when each player is using the optimum strategy. The value of the penny matching game is zero; that is, if each player plays his optimum strategy, the expected value of the payoff is zero.

		P_2	
		.5 H	.5 T
P_1	.5 H	+1	-1
	.5 T	-1	+1

FIG. 4. Mixed strategies with payoff matrix.

Player 2

$$Y = [y_1 \quad y_2 \quad \cdots \quad y_n]$$

Player 1

$$X = \begin{bmatrix} x_1 \\ x_2 \\ \vdots \\ x_m \end{bmatrix} \quad \begin{bmatrix} a_{11} & a_{12} & \cdots & a_{1n} \\ a_{21} & a_{22} & \cdots & a_{2n} \\ \vdots & \vdots & & \vdots \\ a_{m1} & a_{m2} & \cdots & a_{mn} \end{bmatrix} = A$$

	$[y_1 = .5$	$y_2 = .5]$
$x_1 = .5$	+1	-1
$x_2 = .5$	-1	+1

FIG. 5. Matrix symbolism (top) and penny matching game (bottom).

The computation of the optimum strategy and the value of the game will be considered later.

A two-person zero-sum game is frequently expressed in general symbolic form as a matrix, shown in Fig. 5. The payoff function is represented by any $m \times n$ matrix A, where m denotes the number of strategies for Player P_1, and n denotes the number of strategies for Player P_2. The elements of matrix A, such as a_{11} and a_{12}, are any real numbers which indicate the payoff corresponding to the strategy selected by each player. For example, element a_{12} represents the payoff when Player P_1 plays Strategy 1, and Player P_2 plays Strategy 2. A plus sign attached to a number in the payoff matrix indicates the payment is from P_2 to P_1, whereas a minus sign denotes the payment is from P_1 to P_2. In game theory, the strategies may be represented by row or column vectors. The strategy for Player P_1 is usually denoted by a column vector, as is shown by vector X in Fig. 5. When the strategy for Player P_1 is shown as a column vector, the strategy for Player P_2 is shown as a row vector denoted by Y in the figure. The probabilities associated with each of the elements of matrices X and Y are the proportions of time that each strategy is used by the respective players. Thus, referring to the penny matching game shown in Fig. 5, Player P_1 would play Strategy 1 one-half of the time and Strategy 2 also one-half of the time. The probabilities for each player are, in statistical language, probability density functions. This means that the probability value for each strategy is equal to or greater than zero, and the summation of the probabilities for each player must equal 1.

Using this mathematical notation, the expected value of the game is

$$E(X, Y) = X'AY' = \sum_{i=1}^{m} \sum_{j=1}^{n} x_i a_{ij} y_j$$

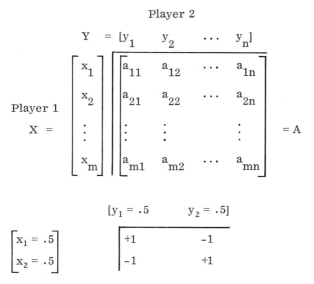

If it turns out that one strategy is better than any of the other strategies, regardless of which strategy the opponent plays, the player has what is called a pure strategy. This strategy should be played for every game. However, often there is no single strategy that should be played to the exclusion of the other strategies. The best plan of action then is to alternate strategies, with a certain relative frequency (proportion) assigned to each strategy. This is called a mixed strategy. We cannot play a mixed strategy in a single game; we can play only one strategy in a game. The term "mixed strategy" refers to the play of successive games.

We will now relate the two-person, zero-sum game (also called "matrix game") to the linear programming problem. The Fundamental Theorem of Matrix Games can be stated as follows:

For every matrix game $\max_X \min_Y E(X, Y)$ and $\min_Y \max_X E(X, Y)$ exist, and are equal to the value of the game.

This means that every two-person, zero-sum game has a solution $[X_0, Y_0; v]$, with the saddle point condition satisfied, i.e.,

$$E(X, Y_0) \le E(X_0 Y_0) = v \le E(X_0, Y)$$

The equivalence of the two-person, zero-sum game problem to linear programming is given by the following primal and dual problems.

<u>Player 1</u>

$$\begin{aligned}
&\text{maximize} \quad v = \text{minimize } -v \\
&\text{subject to:} \quad a_{11}x_1 + a_{21}x_2 + \cdots + a_{m1}x_m \ge v \\
&a_{12}x_1 + a_{22}x_2 + \cdots + a_{m2}x_m \ge v \\
&\vdots \\
&a_{1n}x_1 + a_{2n}x_2 + \cdots + a_{mn}x_m \ge v \\
&x_1 + x_2 + \cdots + x_m = 1 \\
&x_i \ge 0, \quad i = 1, \ldots, m
\end{aligned}$$

<u>Player 2</u>

$$\begin{aligned}
&\text{minimize} \quad v = \text{maximize } -v \\
&\text{subject to:} \quad a_{11}y_1 + a_{12}y_2 + \cdots + a_{1n}y_n \le v \\
&a_{21}y_1 + a_{22}y_2 + \cdots + a_{2n}y_n \le v \\
&\vdots \\
&a_{m1}y_1 + a_{m2}y_2 + \cdots + a_{mn}y_n \le v \\
&y_1 + y_2 + \cdots + y_n = 1 \\
&y_j \ge 0, \quad j = 1, \ldots, n
\end{aligned}$$

The probability density function conditions are satisfied for Player 1 by the constraint $\sum_{i=1}^{m} x_i = 1$, and by the nonnegativity restriction $x_i \geq 0$, $i = 1, \ldots, m$. In a similar manner for Player 2, $\sum_{j=1}^{n} y_j = 1$, and $y_j \geq 0$, $j = 1, \ldots, n$. It is desirable to note that by the dual theorem the maximum value of the objective function for Player 1 is equal to the minimum value of the objective function for Player 2. The value of the game v is an unrestricted variable in both problems; however, this can be handled without difficulty in the application of the Simplex method.

The optimum solution to the LP problem for Player 1 will give the value of the game and the proportion of times that each i-th strategy (x_i) should be played. In a similar manner, Player 2 will obtain the value of the game and the proportion of times that each j-th strategy (y_j) should be played. If for a particular strategy, x_i or y_j is equal to zero in the optimum solution, then that is a poor strategy and should never be played.

DECOMPOSITION AND DECENTRALIZED PLANNING

Despite the tremendous speed and capacity of high-speed computers, it is not unusual to have a real life LP problem which either exceeds the machine capacity or else cannot be solved efficiently by the regular Simplex procedure. Dantzig and Wolfe developed a decomposition procedure to solve large-scale problems having a special structure [21]. In the process of their structuring this procedure that takes place inside a computer, they recognized that it could be translated to the decentralized decision-making function which has a similar structure.

This is referred to as the "block diagonal" structure, in which the decision variable vector X is decomposed into mutually exclusive subsets, $X_1, X_2, \ldots, X_q$. This particular structure for an LP problem is

maximize

$$c_1X_1 + c_2X_2 + \cdots + c_qX_q$$

subject to

$$A_1X_1 + A_2X_2 + \cdots + A_qX_q \leq b \tag{1}$$

$$B_1X_1 \leq b_1 \tag{2}$$

$$B_2X_2 \leq b_2 \tag{3}$$

$$\vdots$$

$$B_qX_q \leq b_q \qquad \text{(r)}$$

$$X_1 \geq 0, \quad X_2 \geq 0, \quad \ldots, \quad X_q \geq 0$$

It should be pointed out that the vector c has been decomposed into the subsets $c_1, c_2, \ldots, c_q$, and matrix A decomposed into $A_1, A_2, \ldots, A_q, B_1, B_2, \ldots, B_q$, and thus correspond to the decomposition for the decision variable vector X.

The set of constraints represented by (1) is referred to as the master or executive problem, and the q sets of constraints denoted by (2), (3), ..., (r) are called the subproblems or the division problems. In terms of decentralized decision-making, (1) is the set of corporate constraints spanning all divisions, whereas (2) is the set of constraints for division 1 only, (3) is the set of constraints for division 2 only, and (r) is the set of constraints for division q only. The matrices B_1, B_2, ..., B_q are the coefficient matrices for each of the divisions.

A macrodescription of the procedure is now presented [22].

1. Each division receives the set of prices corresponding to its decision variables, and is asked to solve its own subproblem. For example, at the start division 1 would receive the set of prices denoted by the vector c_1, and would solve for the optimum solution considering only the constraints $B_1X_1 \leq b_1$.
2. After each division obtains the optimum solution for the prices received, the optimum solutions are submitted to the executive group.
3. These values are plugged into the executive problem to determine the impact of all the division demands on the corporate resources, thus determining what benefits or burdens the individual solutions have for all divisions.
4. The price vector is revised by the addition of a bonus or penalty to account for the benefits or burdens that have accrued.
5. These revised prices are sent down to the divisions, and division optimum solutions using the new prices are calculated and again sent back to the corporate group.
6. The process is repeated, with the executive group determining a new set of revised prices, and giving them to the divisions.
7. Since there is only a finite number of basic feasible solutions for each division, and since each cycle of the procedure is designed to increase the overall objective function, it will arrive at the optimal solution for the overall problem in a finite number of cycles.

The revised prices are in effect provisional dual prices for the executive problem. In the solution procedure for the executive program, the problem is to determine weighted averages for each of the optimum solutions submitted by the various divisions. Thus the variables in the executive problem are weights which must satisfy the conditions for a convex linear combination of vectors.

For a two division problem, where each division has already submitted three proposed optimum solutions, the executive program would be formulated as follows.

Maximize

$$c_1X_1^a + c_2X_2^a$$

subject to

$$A_1X_1^a + A_2X_2^a \leq b$$

$$\mu_1 + \mu_2 + \mu_3 = 1$$

$$\nu_1 + \nu_2 + \nu_3 = 1$$

$$\mu_1 \geq 0, \quad \mu_2 \geq 0, \quad \mu_3 \geq 0$$

$$\nu_1 \geq 0, \quad \nu_2 \geq 0, \quad \nu_3 \geq 0$$

where for division 1

$$\begin{array}{llll} & \text{(1st opt.)} & \text{(2nd opt.)} & \text{(3rd opt.)} \\ X_1^a = & \mu_1 X_{11} + & \mu_2 X_{12} + & \mu_3 X_{13} \end{array}$$

and for division 2

$$\begin{array}{llll} & \text{(1st opt.)} & \text{(2nd opt.)} & \text{(3rd opt.)} \\ X_2^a = & \nu_1 X_{21} + & \nu_2 X_{22} + & \nu_3 X_{23} \end{array}$$

Provisional dual prices, $\pi_1, \pi_2, \ldots, \pi_m$, are calculated for each of the m corporate resources, and these are used to revise the profit coefficients for the divisions' objective functions. For example, the revised price for the first price coefficient of division 1, denoted by $c_{11}{}^1$, is equal to

$$c_{11} - (\pi_1 a_{11} + \pi_2 a_{21} + \cdots + \pi_m a_{m1})$$

The provisional dual prices can be determined to correspond to any basic feasible solution, and unlike optimal values for the dual variables, can be positive, zero, or negative.

The dual prices associated with the weight constraints for each division are used to determine whether or not the optimal solution has been obtained for the entire problem consisting of the corporate and division constraints. Denoting these dual prices by $\bar{\pi}_1$ and $\bar{\pi}_2$, for division 1 and division 2 respectively, the optimal solution is obtained when the total revised profit contribution equals $\bar{\pi}_1$ for division 1 and $\bar{\pi}_2$ for division 2.

For a more extensive treatment of decomposition and partitioning methods, the reader is referred to the article in this Encyclopedia by Himmelblau [23] and to the book by Lasdon [24].

BOUNDED VARIABLES

In many linear programming problems there are restrictions on the values of the variables, e.g., $x_j \geq$ lower bound, $x_j \leq$ upper bound, or $\sum_j x_j \leq$ upper bound. Since these are constraints in the LP problem formulation, which in some cases could substantially increase the size of the problem, it is desirable to seek methods to solve such problems without requiring explicit representation of these constraints.

Lower bound inequalities can be handled very easily without increasing the size of the model. For a lower bound L_j on the j-th variable, then

$$x_j \geq L_j$$

Introducing a nonnegative slack (surplus) variable s_j we have

$$x_j - s_j = L_j$$

or

$$x_j = L_j + s_j$$

We then substitute $L_j + s_j$ in the LP model, and the number of variables and constraints remains the same.

For the upper bound constraint the procedure is more difficult. Considering the simple upper bound (SUB), denoted by U_j for the j-th variable,

$$x_j \leq U_j$$

Since the slack variable on an upper bound constraint must have the same upper bound as the variable x_j, then the slack will be denoted as the complement of the variable by $\bar{x}_j$.

Therefore,

$$x_j + \bar{x}_j = U_j$$

$$x_j = U_j - \bar{x}_j$$

$$\bar{x}_j = \bar{U}_j - x_j$$

Three types of iterations must be considered when using simple upper bounds on the variables.

1. A variable in the basis is decreased to zero as the incoming nonbasic variable is increased. The regular Simplex procedure is employed for the iteration.
2. A basic variable is increased to its upper bound as the incoming nonbasic variable is increased. The basic variable is replaced by its complement, and the usual Simplex method is used for the iteration.
3. The incoming nonbasic variable reaches its upper bound as it is increased. This variable is replaced by its complement and this completes the iteration.

In order to know which of the three types of iterations is to be used at a given stage, we must determine the minimum of the following three parameters. (It should be noted that the vector entering the basis at this stage has already been determined by the usual method.)

1. $\min\limits_i \dfrac{x_{Bi}}{y_{ik}} = \theta_1, \quad \text{for } y_{ik} > 0$

The basis variable to reach zero first is thus identified.

2. $\min\limits_i \dfrac{(x_{Bi} - U_{Bi})}{y_{ik}} = \theta_2, \quad \text{for } y_{ik} < 0$

This parameter equals infinity if there is no upper bound or if no $y_{ik} < 0$.

3. U_k = upper bound on incoming variable = θ_3

The method of upper bounds has been generalized by Dantzig and Van Slyke [25] to accommodate a problem in which the first m constraints are arbitrary, with additional restrictions that limit positive, linear sums of the variables. This is called generalized upper bounding (GUB) and has been found to have applications in areas such as resource allocation, transportation, and production-distribution.

INTEGER PROGRAMMING

Integer programming, a special case of the linear programming problem, requires that the decision variables be nonnegative integers. It is defined as

maximize

cX

subject to

$AX \leq b$

$x_j = 0, 1, 2, \ldots \quad j = 1, \ldots, n$

It is amazing that the introduction of the integer constraint causes such difficult computational problems. Unfortunately, the success in solving large-scale integer programming problems cannot come close to that achieved by the Simplex method for solving LP problems with continuous variables. The first finite algorithm, called the cutting plane method, was developed by Ralph Gomory in 1958.

There are various types of integer programming problems. The problem defined above is an all-integer case, also referred to as the pure integer programming problem, in which all variables are required to be integers. In other problems a subset of the variables is required to be an integer, with the remaining variables permitted to be continuous. This is called a mixed integer problem. It is possible that the variable is required to take on only one of two integer values, 0 or 1. This is referred to as binary or zero-one programming.

This Encyclopedia has substantial articles which describe the important methods and applications of integer programming. The topics and authors are Bivalent Programming by Implicit Enumeration, E. Balas [26], Branch and Bound Techniques, I. B. Turksen [27], Cutting Plane Methods, L. A. Wolsey [28], and Integer Programming, H. A. Taha [29].

NETWORK PROBLEMS

The classical transportation problem was a prelude to a broad spectrum of network types of problems that could be structured in linear programming form. An example of the transportation problem is given above in the section entitled Common Denominators of Operations Research. This was the problem of determining the least cost shipping route for shipping barrels of oil from five supply points to three destination terminals. This problem can be formulated as a typical linear programming problem for solution by the Simplex method. The available capacity at each origin represents a constraint, and the demand at each destination is also a constraint.

However, because of the special structure of this problem, methods that are more efficient than the Simplex method have been formulated. The values of the dual variables can be obtained by simple addition, and as a result, it is easy to ascertain whether or not a basic feasible solution is optimal. The "stepping-stone" method of Charnes and Cooper was one of the first computational algorithms developed to solve the transportation problem [30].

The pioneering work of Ford and Fulkerson set the stage for the application of linear programming to a large class of problems involving flows in networks [31].

Graph theory provides the necessary underpinning for network-type problems. It is important to note that many methods other than linear programming have been developed to solve the broad array of network problems. The book by Minieka discusses a number of algorithms for solving network problems [32].

MULTIPLE-CRITERIA DECISION MAKING

The linear programming problem as defined by Dantzig has a single objective or criterion function, e.g., maximize profit, minimize cost, maximize production, minimize machine idle time. However, many practical problems today cannot be satisfactorily characterized by a single criterion, and in order to portray the real-life conditions, demand that multiple, conflicting objectives be considered.

This is not a new problem addressed to linear programming. In fact, in the early 1960s Charnes and Cooper focused on this problem, which they called goal programming, when they considered the analysis of contradictions in nonsolvable problems [30]. They stated that management sometimes sets goals, even though they may be unattainable within the limits of available resources. These goals may be utilized to provide incentives, to judge performance, or to ensure that immediately attainable objectives do not blot out long-run considerations.

The goals are expressed as constraints which are to be achieved "as close as possible." This means that the slack variable, which is added to each goal expressed as a constraint, is a free variable, and can thus take on a positive, negative, or zero value. Thus the slack variable represents the deviation from the stated goal. The objective then is to minimize the value of the slack variable associated with each goal constraint. Goal programming requires that an ordinal ranking (preemptive priority) of the objectives be made by the decision maker. If there are multiple goals for a priority level, then differential weights must be assigned to each of the goals. A modified Simplex method of linear programming attempts to satisfy the highest priority goal first, then the next highest priority goal, etc.

Suppose that we have a problem with two goals which we would like to achieve "as close as possible." The problem would be stated as

minimize

$$s_1^- + s_1^+ + s_2^- + s_2^+$$

subject to

$$\sum_{j=1}^{n} a_{1j}x_j + s_1^- - s_1^+ = b_1, \qquad \text{Goal 1}$$

$$\sum_{j=1}^{n} a_{2j}x_j + s_2^- - s_2^+ = b_2, \qquad \text{Goal 2}$$

$$x_j \geq 0, \quad s_1^- \geq 0, \quad s_1^+ \geq 0, \quad s_2^- \geq 0, \quad s_2^+ \geq 0$$

where $(s_1^- - s_1^+)$ represents the deviation from Goal 1

$(s_2^- - s_2^+)$ represents the deviation from Goal 2

If Goal 1 has the higher priority, then a modified Simplex method would attempt to satisfy this goal completely before considering Goal 2.

A book on goal programming related to accounting for control was written by Ijiri in 1965 [33], but interest was somewhat dormant until the early 1970s when activity in this area became widespread and intensive. Lee did much to stimulate interest through the popular book which he published in 1972 [34]. A more recent book (1976) on goal programming was authored by Ignizio [35].

A basic and often cited paper on multiple objective functions was written by Roy [36]. Given a set of feasible alternatives or decisions concomitant with an objective or measure for each decision, he addresses the question on how a "best" feasible decision can be made, and what methods can be used or experimented with to reach some decision. About the same time Benayoun and Tergny developed what they called the Progressive Orientation Procedure for solving linear programming problems with multiobjective functions [37]. It is a sequential procedure which does not demand the decision maker to explicitly establish weights for the objective functions. The decision maker is integrated into a man-model system by making partial choices at intermediate stages of the process.

Belenson and Kapur showed that an efficient solution to the multicriteria problem can be obtained by applying a linear programming approach for the solution of two person, zero-sum games [38].

Geoffrion formulated the vector maximum problem as the constrained maximization of an implicit preference function [39]. It is an interactive procedure with the decision maker, based on an intimate partnership between the normative theory of rational choice and mathematical programming. It is assumed that the preference function is concave, increasing, and differentiable, with the constraint functions being concave and differentiable. This approach has been applied to the operation of an academic department [40].

Zionts and Wallenius developed an interactive method which chooses an arbitrary set of positive multipliers or weights at the start, and then generates a composite objective function or utility function using these multipliers [41]. It is assumed that each of the objectives to be minimized is a concave function of the decision variables, and the constraints are convex functions. The decision-maker's utility function is considered to be a linear function of the objective function. An extensive test of this method was made on a real-life corporate planning model.

A different approach to the multiple objective problem is one which extends the range of applicability of decision analysis. It is not a mathematical programming procedure; instead it is a prescriptive approach which brings together uncertainty analysis and preference (or value or utility) analysis.

The basic ideas developed by Raiffa were extended by Keeney. The best source book on this procedure and its applications was written by Keeney and Raiffa and published in 1976 [42].

COMPUTER CODES

One of the earliest linear programming problems solved by a digital computer occurred in 1952 on SEAC, the National Bureau of Standards computer.

Up through the 1960s it was felt to be a reasonable task to identify computer codes for solving linear programming problems. However, at the present time computer codes using the Simplex method are available for practically all general-purpose computers. Many companies and universities have developed their own Simplex and related-type codes for handling LP problems. Independent software development companies and service bureaus have also been prolific sources for LP codes.

The range of sophistication and size of problem that can be accommodated varies greatly among the codes. The available computer configuration dictates to a large extent the ultimate capability of a code that can be developed for implementation on a particular system. Ingenious methods have also been developed to exploit the computer's capability in solving linear programming problems.

Some of the computer procedures used are very complex, and their structure has been the result of the integration of an intimate knowledge of the theory of linear programming with expertise in computer programming and computer architecture. It is possible to solve LP problems having thousands of constraints and thousands of variables. Even though the development of large models may be quite straightforward, the data collection problems are usually staggering. Obtaining reliable data for the LP model is usually a very time-consuming and difficult task.

One of the pioneers in the development of computing techniques for linear programming is Orchard-Hays. To improve the efficiency and accuracy of the Simplex method, he has developed computer methods for dealing with ill-conditioned matrices, structured data formats for computer programs, and simplified input and output transformations. His book on advanced linear-programming computing techniques gives a complete discussion on algorithms for the dual Simplex method, postoptimal ranging, generalized upper bounding, and parametric programming [43].

Other excellent books relevant to the development of codes for linear programming have been written by Beale [44] and Driebeek [45]. The Special Interest Group for Mathematical Programming (SIGMAP) of the Association for Computing Machinery periodically publishes updated lists of computer algorithms for mathematical programming problems.

REFERENCES

1. G. B. Dantzig, Linear Programming and Extensions, Princeton University Press, Princeton, New Jersey, 1963.
2. H. W. Kuhn and A. W. Tucker, Linear Inequalities and Related Systems, Princeton University Press, Princeton, New Jersey, 1956.
3. J. F. McCloskey and F. N. Trefethen, Operations Research for Management, Johns Hopkins Press, Baltimore, Maryland, 1954.
4. W. N. Ledbetter and J. R. Cox, Are OR techniques being used?, Ind. Eng. 9, 19-21 (February 1977).
5. S. I. Gass, Linear Programming, 4th ed., McGraw-Hill, New York, 1975.
6. G. E. Bennington, Basic feasible solutions, in Encyclopedia of Computer Science and Technology, Vol. 3 (J. Belzer, A. G. Holzman, and A. Kent, eds.), Dekker, New York, 1976.
7. G. Hadley, Linear Programming, Addison-Wesley, Reading, Massachusetts, 1962.
8. M. S. Bazaraa, Dantzig simplex method, in Encyclopedia of Computer Science and Technology, Vol. 7 (J. Belzer, A. G. Holzman, and A. Kent, eds.), Dekker, New York, 1977.

9. A. G. Holzman, H. H. Schaefer, and R. Glaser, Mathematical Bases for Management Decision Making, Encyclopaedia Britannica Educational Corp., Chicago, 1962.
10. D. T. Phillips, A. Ravindran, and J. J. Solberg, Operations Research, Wiley, New York, 1976.
11. M. S. Bazaraa and J. J. Jarvis, Linear Programming and Network Flows, Wiley, New York, 1977.
12. R. W. Cottle and G. B. Dantzig, Complementary pivot theory of mathematical programming, in Mathematics of the Decision Sciences, Part 1, American Mathematical Society, Providence, Rhode Island, 1968.
13. K. Murty, Complementarity problems, in Encyclopedia of Computer Science and Technology, Vol. 5 (J. Belzer, A. G. Holzman, and A. Kent, eds.), Dekker, New York, 1976.
14. C. E. Lemke, Bimatrix equilibrium points and mathematical programming, Manage. Sci. 11(7), 681-689 (1965).
15. A. Ravindran, A comparison of the primal simplex and complementary pivot methods for linear programming, Nav. Res. Logistics Q. 20, 95-100 (March 1973).
16. A. Ravindran, A computer routine for quadratic and linear programming problems, Commun. ACM 15, 818-820 (September 1972).
17. R. Saigal, Fixed point computing methods, in Encyclopedia of Computer Science and Technology, Vol. 8 (J. Belzer, A. G. Holzman, and A. Kent, eds.), Dekker, New York, 1977.
18. J. von Neumann and O. Morgenstern, Theory of Games and Economic Behavior, Princeton University Press, Princeton, New Jersey, 1947.
19. A. G. Holzman, Game theory, in Encyclopedia of Library and Information Science, Vol. 9 (A. Kent, H. Lancour, and J. E. Daily, eds.), Dekker, New York, 1973.
20. R. D. Luce and H. Raiffa, Games and Decisions, Wiley, New York, 1957.
21. G. B. Dantzig and P. Wolfe, Decomposition principle for linear programs, Oper. Res. 8(1), 101-111 (1960).
22. W. J. Baumol and T. Fabian, Decomposition, pricing for decentralization, and external economics, Notes for course presented by Mathematica, Berkeley, California, April 24-26, 1963.
23. D. M. Himmelblau, Decomposition methods, in Encyclopedia of Computer Science and Technology, Vol. 7 (J. Belzer, A. G. Holzman, and A. Kent, eds.), Dekker, New York, 1977.
24. L. S. Lasdon, Optimization Theory for Large Systems, Macmillan, New York, 1970.
25. G. B. Dantzig and R. M. Van Slyke, Generalized upper bounding techniques, J. Comput. Syst. Sci. 1, 213-226 (1967).
26. E. Balas, Bivalent programming by implicit enumeration, in Encyclopedia of Computer Science and Technology, Vol. 2 (J. Belzer, A. G. Holzman, and A. Kent, eds.), Dekker, New York, 1975.
27. I. B. Turksen, Branch and bound technique, in Encyclopedia of Computer Science and Technology, Vol. 4 (J. Belzer, A. G. Holzman, and A. Kent, eds.), Dekker, New York, 1976.
28. L. A. Wolsey, Cutting plane methods, in Encyclopedia of Computer Science and Technology, Vol. 7 (J. Belzer, A. G. Holzman, and A. Kent, eds.), Dekker, New York, 1977.

29. H. A. Taha, Integer programming, in Encyclopedia of Computer Science and Technology, Vol. 9 (J. Belzer, A. G. Holzman, and A. Kent, eds.), Dekker, New York, 1978.
30. A. Charnes and W. W. Cooper, Management Models and Industrial Applications of Linear Programming, Vol. 1, Wiley, New York, 1961.
31. L. R. Ford and D. R. Fulkerson, Flows in Networks, Princeton University Press, Princeton, New Jersey, 1962.
32. E. Minieka, Optimization Algorithms for Networks and Graphs, Dekker, New York, 1977.
33. Y. Ijiri, Management Goals and Accounting for Control, Rand-McNally, Chicago, 1965.
34. S. M. Lee, Goal Programming for Decision Analysis, Auerbach, Philadelphia, 1972.
35. J. P. Ignizio, Goal Programming and Extensions, Heath, Lexington, Massachusetts, 1976.
36. B. Roy, Problems and methods with multiple objective functions, Math. Programming 1, 239-266 (1971).
37. R. Benayoun and J. Tergny, Mathematical programming with multiple objective functions: A solution by P.O.P. (Progressive Orientation Procedure), Rev. Metra 9, 279-299 (1970).
38. S. M. Belenson and K. C. Kapur, An algorithm for solving multi-criterion linear programming problems with examples, Oper. Res. Q. 24(1), 65-77 (1973).
39. A. M. Geoffrion, Vector Maximal Decomposition Programming, Working Paper Number 164, Western Management Science Institute, University of California, Los Angeles, 1970.
40. A. M. Geoffrion, J. Dyer, and A. Feinberg, An interactive approach for multicriterion optimization with an application to the operation of an academic department, Manage. Sci. 19(4), 357-368 (1972).
41. S. Zionts and J. Wallenius, An interactive programming method for solving the multiple criteria problem, Manage. Sci. 22(6), 652-663 (1976).
42. R. L. Keeney and H. Raiffa, Decisions with Multiple Objectives: Preferences and Value Tradeoffs, Wiley, New York, 1976.
43. W. Orchard-Hays, Advanced Linear-Programming Computing Techniques, McGraw-Hill, New York, 1968.
44. E. M. L. Beale, Mathematical Programming in Practice, Wiley, New York, 1968.
45. N. J. Driebeek, Applied Linear Programming, Addison-Wesley, Reading, Massachusetts, 1969.

2
INTEGER PROGRAMMING

HAMDY A. TAHA

Department of Industrial Engineering
University of Arkansas
Fayetteville, Arkansas

HISTORY OF INTEGER PROGRAMMING[1]

Integer programming can be considered a by-product of the impressive developments in the area of linear programming. Prior to the formal introduction of linear programming in the late 1940s, developments in discrete optimization problems were restricted to purely theoretical investigations including primarily the solution of systems of linear or nonlinear equations in discrete variables and the integer optimization of a linear (or nonlinear) function subject to a single linear (or nonlinear) equality constraint. The coloring problem is perhaps the most renowned integer problem that was tackled by mathematicians prior to the formal introduction of integer programming in 1958. It seeks the determination of the minimum number of colors needed to color a (planar) map such that regions (e.g., states) with a common boundary (other than a point) must have different colors.

Unfortunately, earlier research results in discrete mathematics and optimization were rather limited. In fact, it can be stated safely that, with very few and perhaps minor exceptions, the contributions of these results to present-day developments in integer programming are practically negligible.

The real interest in the solution of the general integer optimization problem was stimulated by the tremendous success of the simplex method for solving linear

---

[1]The Notes section at the end of this article includes selective references representing only the fundamental works in integer programming. The article is written under the assumption that the reader has adequate knowledge of the theory of linear programming.

programs. This interest was further enhanced by the rapid growth in the variety and number of integer programming applications. Immediate attempts to solve the integer problem by using linear programming-based methods produced limited results, mainly because convergence of the new techniques was not guaranteed. It was only in 1958 that Ralph Gomory developed the first finite algorithm called the cutting-plane method.

Gomory's success gave hope that the integer problem could be solved. This success stimulated other theoretical developments for developing other solution methods. Noted among these is the branch-and-bound method introduced in 1960 by A. Land and A. Doig, and the implicit enumeration method introduced in 1965 by Egon Balas. Unfortunately, in spite of the tremendous advances in the theory of integer programming, the computational efficacy of the developed techniques has been rather disappointing. Although the computation power of the digital computer has been vastly improved, it is still almost impossible to predict with a degree of certainty whether or not the optimum solution to an integer problem, regardless of the size, is attainable.

The fact of the matter is that although finite convergence is guaranteed in theory, the accumulation of machine round-off error has such a pronounced effect that it may distort the original data of the problem. Even in the cases where round-off error is controllable, the number of solution iterations may be so large (though finite) that an answer cannot be secured using a reasonable amount of computer time.

In spite of the erratic behavior of the developed algorithms in solving the general integer problem, success has been achieved in devising special purpose algorithms for solving integer problems with special structures, such as the knapsack problem, the set covering problem, the fixed-charge problem, and the traveling salesman problem.

SCOPE OF INTEGER PROGRAMMING APPLICATION

Most integer programming developments, with the exception of problems with quadratic objective function and problems in binary (zero or one) variables, have been confined to the integer linear programs. A mathematical definition of the linear problem is the following:

Maximize (or minimize)

$$z = \sum_{j=1}^{n} c_j x_j$$

subject to

$$\sum_{j=1}^{n} a_{ij} x_j = b_i, \quad i = 1, 2, \ldots, m$$

$$x_j \geq 0, \quad j = 1, 2, \ldots, n$$

$$x_j \text{ integer}, \quad j \in I \subseteq \{1, 2, \ldots, n\}$$

When $I = \{1, 2, \ldots, n\}$, this indicates that all the variables x_j are restricted to integer values. In this case the problem is referred to as a pure integer program. On the other hand, if $I \subseteq \{1, 2, \ldots, n\}$, only some of the variables are restricted to integer values, and the problem is known as a mixed integer program.

In both the pure and mixed integer problems, the deletion of the integer restriction reduces the problem to an ordinary linear program in which all the variables are continuous. This observation, as shown later, serves as the key idea for solving the integer problem.

A special case of the general integer problem is the binary problem in which the integer problems are restricted to the values zero and one. Special interest has developed in the binary problem because every general integer problem can be converted into a zero-one program as follows. Suppose x_j is a nonnegative integer variable whose maximum integer value is K_j, and assume y_{jk} are binary variables for all k. Then, for r_j defined such that $2^{r_j+1} \geq K_j + 1$, the substitution

$$x_j = y_{j0} + 2y_{j1} + 2^2 y_{j2} + \cdots + 2^{r_j} y_{jr_j}$$

defines all the feasible values of x_j (namely $0 \leq x_j \leq K_j$) in terms of the binary variables y_{jk}. The idea behind this substitution lies in the possibility of developing efficient solution methods for the binary problem. This automatically would lead to solving the general integer problem efficiently. Unfortunately, the tremendous increase in the number of binary variables (as a result of using this substitution) normally counteracts the computational efficiency of the zero-one algorithm.

An interesting categorization of integer programming applications classifies integer models as direct, coded, and transformed. Direct models are those representing problems in which the variables are naturally integers. Examples of such models exist in situations where the variables represent number of planes flying a given route or number of workers operating a certain job.

Coded models arise in situations where the variables are qualitative in nature. The "yes" or "no" decisions provide a typical example of qualitative variables. In order to express the model mathematically, these (qualitative) variables are coded numerically so that when the decision is "yes" the variable is assigned the value "one"; otherwise, a "no" decision is represented by a "zero" value. In this manner the "yes-no" variable is now coded numerically as a binary (zero-one) variable, thus allowing the use of analytic functions in the model representation.

Transformed models arise in situations where integer variables are introduced primarily to make the model more analytically amenable. A typical example occurs in situations where the constraints of the problem specify that only one of two given constraints can hold. For example, consider scheduling problems where several jobs are processed on the same facility. Suppose a_j and a_k are the processing times for jobs j and k, and assume x_j and x_k are their starting times. Since it is not known in advance whether or not x_j precedes x_k, the precedence restriction is represented by one of the two (mutually exclusive) constraints

$$x_j + a_j \leq x_k \quad \underline{\text{or}} \quad x_k + a_k \leq x_j$$

The constraints must be expressed as simultaneous restrictions before the problem is solved. To achieve this, let $y = (0, 1)$ be a binary variable and assume M is a very large positive value. The requirement that only one of the two constraints can be active is then expressed by the simultaneous constraints

$$x_j + a_j \leq x_k + My$$

$$x_k + a_k \leq x_j + M(1 - y)$$

If $y = 0$, the first constraint is active, and the second is inactive because its right-hand side is very large and hence nonbinding. Similarly, when $y = 1$, the first constraint will be inactive while the second becomes active.

It is important to notice that prior to the introduction of the binary variable, the problem has nothing to do with integer programming. Moreover, the binary variable is actually extraneous to the problem in the sense that the specific value of the binary variable in the final solution provides redundant information.

Using the above categorization, direct and coded models represent situations where the original problem is modeled as an integer program, while the transformed models represent cases where integer variables are used primarily to convert ill-conditioned mathematical programs to ones that are analytically amenable. The formulation of the direct integer models is straightforward and does not differ from ordinary continuous programs except for the fact that the variables are integers. However, the applications of the coded and transformed models offer interesting formulations which will be reviewed here briefly.

A classic application of coded models is the capital budgeting problem in which n projects are competing for limited resources (such as equipment, manpower, and money), and the objective is to schedule the projects that yield the largest profit while satisfying the specified limitations. If x_j is defined as a binary variable representing the j-th project so that $x_j = 1(0)$ if the j-th project is sheduled (not scheduled). The model then becomes

maximize

$$z = \sum_{j=1}^{n} c_j x_j$$

subject to

$$\sum_{j=1}^{n} a_{ij} x_j \leq b_i; \qquad i = 1, 2, \ldots, m$$

$$x_j \in \{0, 1\}; \qquad j = 1, 2, \ldots, n$$

where c_j is the profit contribution of the project j and a_{ij} is its utilization of the i-th resource for which the maximum availability is b_i.

The knapsack problem is a special case of the capital budgeting problem. The formulation is concerned with the determination of the most "valuable" mix from among n items to be packed in a knapsack providing that the total allotted volume of selected items does not exceed the capacity of the knapsack. The problem thus has one constraint of the type

$$\sum_{j=1}^{n} a_j x_j \leq b$$

where $x_j = 1$ or 0 depending on whether or not item j is selected. The constant

a_{ij} (> 0) represents the volume requirement of item j of the total knapsack volume b. The knapsack problem is also known by the suggestive names cargo-loading problem and fly-away kit problem stemming from similar applications of loading a vessel or deciding on the most valuable items to be packed for use on a plane. Aside from its immediate practical applications, the knapsack problem has received considerable attention in the literature because of its potential advantageous use in solving the generalized integer problem.

Another important integer formulation is the covering problem. An illustrative application of the problem occurs in delivering orders from a warehouse to m different destinations. Each destination receives its order in one delivery. A carrier may combine at most k orders for simultaneous delivery. Determination of a feasible combination depends on the route assigned to each carrier. Since each destination may be served by more than one route, the objective is to determine the total cost of delivery.

The problem is modeled mathematically as follows. Define an activity as a single combination of one, two, ..., or k orders, and assume n is the total number of possible activities. Suppose c_j is the cost of the j-th activity and x_j equals one or zero, depending on whether or not the activity is chosen. The covering problem formulation then becomes

minimize

$$z = \sum_{j=1}^{n} c_j x_j$$

subject to

$$\sum_{j=1}^{n} e_j x_j = e_0$$

$$x_j \in \{0, 1\}$$

where e_j is a column vector with m entries, with the i-th entry being one if activity j delivers order i and zero otherwise. All the elements of the column vector e_0 are 1's. The constraints thus guarantee that each destination receives one order only.

(The above model generalizes a problem in graph theory from which the name "covering problem" was acquired. The problem occurs in a graph with N nodes and E edges, with each edge joining certain pairs of nodes. The objective is to find a "cover" with the minimum number of edges, where a cover is defined as a subset of edges such that each of the N nodes is incident to some edge of the subset.)

In the above examples, integer programming is used directly in formulating the model. There are other interesting applications where integer programming is used to render a more analytically amenable formulation of an ill-conditioned problem. A typical example occurs in using integer programming to approximate a single-variable nonlinear function by a piecewise linear function. Figure 1 illustrates a nonlinear function f(x) and its piecewise linear approximation. Using the symbols defined in the figure, the piecewise linear approximation of f(x) is given by

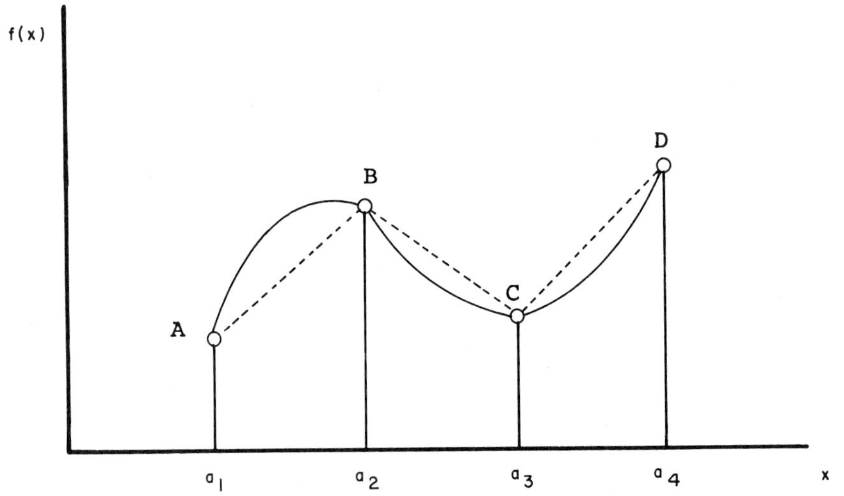

FIGURE 1.

$$f(x) \cong f(a_1)t_1 + f(a_2)t_2 + f(a_3)t_3 + f(a_4)t_4$$

where

$$0 \le t_1 \le y_1$$
$$0 \le t_2 \le y_1 + y_2$$
$$0 \le t_3 \le y_2 + y_3$$
$$0 \le t_4 \le y_3$$
$$y_1 + y_2 + y_3 = 1$$
$$t_1 + t_2 + t_3 + t_4 = 1$$
$$y_j \in \{0, 1\}, \quad j = 1, 2, 3$$

The new variables of the problem are now t_i and y_i, $i = 1, 2, 3, 4$. The value of x can be obtained from

$$x = a_1t_1 + a_2t_2 + a_3t_3 + a_4t_4$$

The approximation can be made as accurate as desired by making the breaking points a_1, a_2, ..., closer.

The idea of the approximation is that the value of the function must now follow the broken line ABCD. The system of constraints automatically provides for that. For example, suppose the value of x lies between a_2 and a_3 in Fig. 1. In this case, only t_2 and t_3 can be positive simultaneously, while t_1 and t_4 must be zero. It is evident from the given constraints that this condition is satisfied by making $y_2 = 1$. This automatically sets $y_1 = y_3 = 0$ and hence $t_1 = t_4 = 0$ as desired.

The above transformation can be employed with separable programming problems in which all the objective and constraint functions are (or can be) expressed as the sum of single-variable functions. The transformed problem has the advantage of being converted into a mixed integer linear program instead of a nonlinear program, which may be easier to manipulate computationally. Although separable programs can be solved by using a version of the simplex method of linear programming, the resulting solution generally guarantees a local optimum (of the approximate problem). The integer programming approximation, on the other hand, guarantees a global optimum.

Another application where integer programming is used to transform the problem into a more manageable form is the fixed-charge problem. The problem occurs in production facilities where a fixed cost is incurred for setting up the machines in preparation for starting production. The incurred cost is independent of the amount (number of units) produced by the facility. A typical production cost function is shown in Fig. 2, where the cost of producing items after the initial fixed charge is assumed directly proportional (i.e., linear) with the number of units produced.

Mathematically, the cost function is expressed as

$$c(x) = \begin{cases} 0, & \text{if } x = 0 \\ K + ax, & \text{if } x > 0 \end{cases}$$

where x (≥ 0) is the amount produced.

In a typical application, a production facility handles a number of units with the cost function $c_j(x_j)$ representing the j-th product. Thus the problem entails the minimization of $\sum_{j=1}^{n} c_j(x_j)$, where n is the number of products, subject to an appropriate set of constraints. The difficulty occurs in obtaining the solution since $c_j(x_j)$ has a discontinuity of $x_j = 0$. Integer programming can alleviate this complication as follows. Let y_j be a binary variable, then the minimization $\sum_{j=1}^{n} c_j(x_j)$ is equivalent to

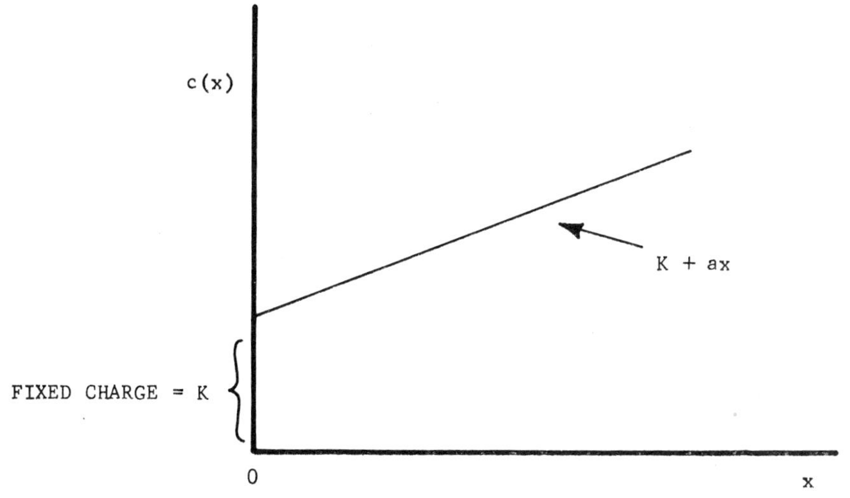

FIGURE 2.

minimize

$$z = \sum_{j=1}^{n} c_j x_j + \sum_{j=1}^{n} K_j y_j$$

subject to

$$0 \leq x_j \leq My_j; \quad j = 1, 2, \ldots, n$$

where M is a very large value. The new formulation shows that if $x_j > 0$, then $x_j \leq My_j$ shows that y must equal one. This automatically adds K_j to the objective function. If $x_j = 0$, then $y_j = 0$ or 1, but since the objective function is minimized and $K_j > 0$, y_j must equal zero in this case.

METHODS OF INTEGER PROGRAMMING

The most obvious (and, perhaps, naive) method for solving integer programs is to exhaustively enumerate all possible feasible combinations for the variables. The inadequacy of such a procedure can be quickly demonstrated by considering a simple integer program with one constraint and ten variables given that the feasible range for each variable lies between zero and ten. This means that there are 10^{10} possible integer combinations to examine. Using very conservative estimates, it can be shown that the IBM 370/55 computer would require 30 hr of computer time to solve this very modest-size problem. The difficulty is magnified at an exponential rate with an increase in the number of variables.

One of the techniques for solving integer programs actually draws on the idea of enumeration but with the additional use of clever devices that aim at examining explicitly only a "fewer" number of integer combinations while automatically accounting for the remaining combinations implicitly. The end effect of both the implicit and explicit enumerations is to account for all possible combinations. These techniques have acquired the suggestive name "implicit enumeration." A less suggestive, but widely circulated, name is "branch-and-bound." Both names actually apply to enumeration-type techniques although, for apparently purely historical reasons, the "implicit enumeration" method is used by many to refer to the solution of binary problems, while the "branch-and-bound" method is used in conjunction with the general integer problem.

The other distinct technique for solving integer programs is the <u>cutting-plane</u> method. The technique calls initially for dropping the integer constraints and using the continuous optimum as a starting point for the succeeding calculations. Specially designed constraints are imposed on the continuous space, one at a time, which effectively cut (hence the name <u>cutting-plane</u>) into the solution space without ever deleting any feasible integer points. After each cut is added, the problem is resolved, and the new continuous optimum is checked for integrality. Because of the special properties of the successive cuts, the application of a finite number of such cuts should eventually produce a continuous optimum that satisfies the integrality condition.

A common characteristic among both cutting plane and branch-and-bound (or enumeration) techniques is that each technique relies primarily on converting the solution space of the integer problem to an equivalent continuous one by dropping all the integer conditions. Both techniques then attempt to locate the optimum integer solution by applying secondary constraints whose primary purpose is to eliminate parts of the continuous regions of the solution space containing no integer feasible points. These are the same continuous regions that were added initially by dropping the integer conditions.

The logic of the two techniques is thus based on a built-in redundancy that first imposes continuity on the integer solution space and then proceeds to identify the optimum integer by again deleting (some of) the added continuous regions. Although this redundancy might account for the reason the available integer methods are computationally inefficient, the need for imposing continuity is actually dictated by the lack of adequate theory in discrete mathematics that can support the development of efficient methods for solving integer programs. The result is that researchers had to rely on the more advanced theory of continuous mathematics in their attempt to solve the integer problem. Indeed, in both cutting-plane and branch-and-bound methods, the procedure for solving the integer problem reduces to solving a series of continuous programs. Even in the case of zero-one (binary) problems where the use of continuous programs may not be explicit, the method of solution can generally be traced back to a continuous version.

Branch-and-Bound Methods

As mentioned earlier, all integer algorithms are based on converting the integer solution space to a continuous one by initially dropping the integer conditions. The branching and bounding operations in the branch-and-bound methods are designed to locate efficiently the optimum integer solution in the continuous space. Specifically, applying branching imposes additional restriction on the continuous problem whose ultimate effect is to delete the continuous regions that do not include integer values. This normally results in creating smaller continuous problems whose combined solution spaces still include all the feasible integer points of the original problems. The name "branching" thus implies that the original continuous problem is "branched" into smaller continuous problems from which the optimum integer can be identified more readily.

Bounding is a mechanism designed to reduce the amount of computations by developing a "bound" on the optimum objective value of the integer problem. Any of the created subproblems whose value of the objective function falls outside this bound may be automatically discarded as nonpromising. It is evident that tightening the bound leads to more efficient computations. Indeed, the efficiency of the branch-and-bound method relies principally on the tightness of the bound.

The specifics of the branch-and-bound procedure can be illustrated by the following numerical example.

Maximize

$$z = 2x_1 + 3x_2$$

subject to

$$5x_1 + 7x_2 \leq 35$$

$$4x_1 + 9x_2 \leq 36$$

$$x_1, x_2 \geq 0 \text{ and integer}$$

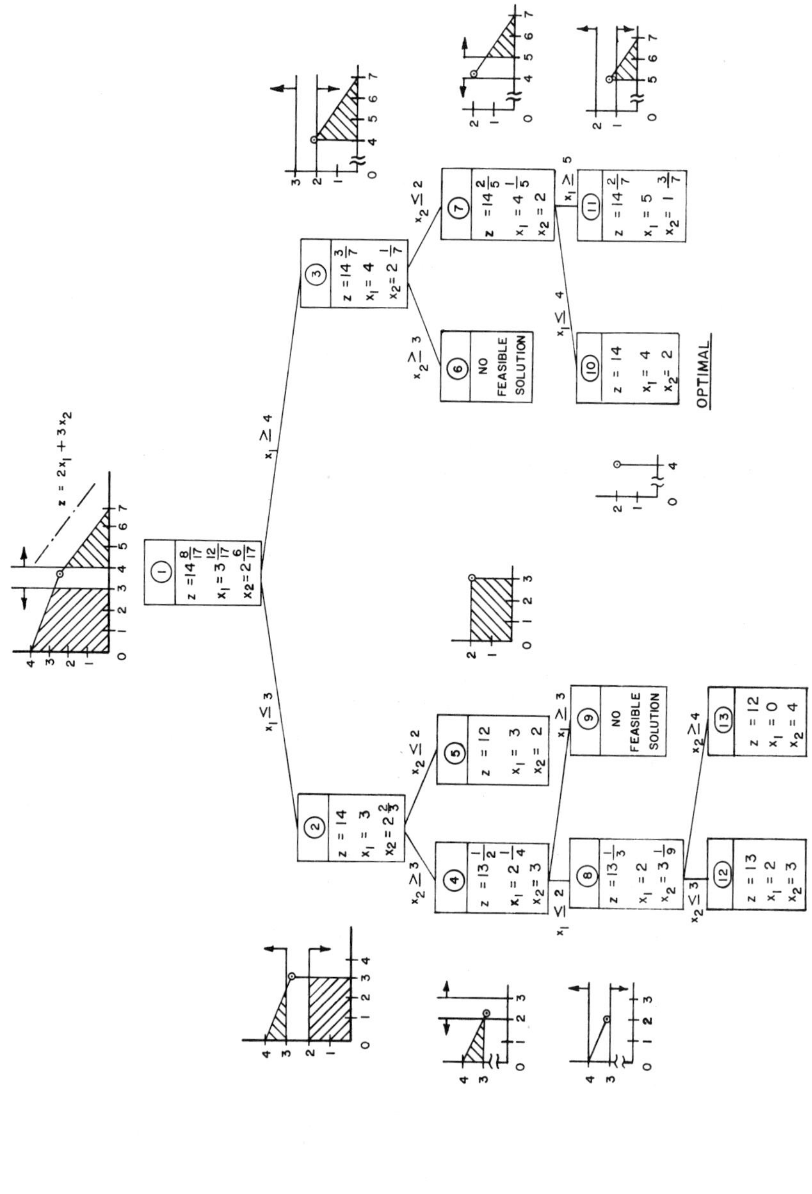
z = 2x1 + 3x2
1
z = 14 8/17
x1 = 3 12/17
x2 = 2 6/17
x1 ≤ 3
x1 ≥ 4
2
z = 14
x1 = 3
x2 = 2 2/3
3
z = 14 3/7
x1 = 4
x2 = 2 1/7
x2 ≥ 3
x2 ≤ 2
4
z = 13 1/2
x1 = 2 1/4
x2 = 3
5
z = 12
x1 = 3
x2 = 2
6
NO
FEASIBLE
SOLUTION
7
z = 14 2/5
x1 = 4 1/5
x2 = 2
x1 ≤ 2
x1 ≥ 3
8
z = 13 1/3
x1 = 2
x2 = 3 1/9
9
NO
FEASIBLE
SOLUTION
x1 ≤ 4
x1 ≥ 5
10
z = 14
x1 = 4
x2 = 2
OPTIMAL
11
z = 14 2/7
x1 = 5
x2 = 1 3/7
x2 ≤ 3
x2 ≥ 4
12
z = 13
x1 = 2
x2 = 3
13
z = 12
x1 = 0
x2 = 4

FIGURE 3.

Figure 3 summarizes the solution steps. The graphical solution is shown next to each of the nodes (subproblems) created by the branching process. The circled dot (⊙) on each graphical solution represents the optimum (continuous) point.

The solution procedure starts by dropping the integer restrictions on x_1 and x_2. This results in Problem ① whose optimum solution is $x_1 = 3\frac{12}{17}$, $x_2 = 2\frac{6}{17}$ with $z = 14\frac{8}{17}$. Since x_1 is required to be an integer, and since its continuous optimum value is $x_1 = 3\frac{12}{17}$, it is evident that the range $3 < x_1 < 4$ can be deleted from the continuous solution space without deleting any feasible integer values. Stated differently, the two mutually-exclusive constraints $x_1 \leq 3$ and $x_2 \geq 4$ can be applied to Problem ① to effect the deletion of the region $3 < x_1 < 4$ from the continuous solution space. This results in the two problems numbered ② and ③. Notice that the collection of feasible integer solutions of Problems ② and ③ are exactly those contained in the original solution space of Problem ①.

The general rule for branching may be summarized as follows. Let x_i be an integer variable whose fractional value at the current note is x_i^*. If $[x_i^*]$ is defined as the largest integer not exceeding x_i^*, then the region $[x_i^*] < x_i < [x_i^*] + 1$ contains no feasible integer values, and two new problems (branches) can be created by imposing the restrictions $x_i \leq [x_i^*]$ and $x_i \geq [x_i^*] + 1$ on the current problem.

One may ask at this point why x_1, rather than x_2, was chosen to effect branching. The answer is that it is optional to choose any of the integer variables to effect branching. Indeed, Problem ① could have been "branched" (that is, partitioned) by imposing the restrictions $x_2 \leq 2$ and $x_2 \geq 3$. Of course, this may lead to a completely different sequence of created subproblems, and hence better or worse computational efficiency depending on the number and the properties of generated subproblems. But there is no way to predict which choice will be best, and indeed this difficulty is one of the unresolved complications in the branch-and-bound procedures. This means that the procedure does not have the capability to foresee whether a chosen branching variable will be a good choice from the standpoint of future computations.

The idea of branching applied at Problem (or Node) ① must now be repeated at Nodes ② and ③ by selecting a branching variable. For example, at Node ②, branching must be effected on x_2 since x_1 already has an integer value. The same point applies at ③. In general, the branching variable is selected as the one having the largest fraction at the current node. This rule-of-thumb assumes that such a variable would result in the largest change in the value of the objective function (as compared with that of the parent node), a condition which may be favorable for effecting efficient computations.

The importance of obtaining a good bound at the early stages of computations on the efficiency of the procedure cannot be overemphasized. For example, the different nodes in Fig. 3 can actually be generated in any order (the given numeric order represents only one possibility). Suppose these problems were generated and examined (for satisfying the integer restriction) in the order ① → ③ → ⑦ → ⑩. (This actually means that the problems of Nodes ②, ⑥, and ⑪ are not immediately examined as they are generated, but are stored in a list for later scanning.) At Node ⑩ an integer feasible solution is obtained with the value of the objective function equal to 14. This value now sets a lower bound on the optimum objective value for the integer solution, meaning that unless a node promises to produce a better value of the objective function than the lower bound, it should be automatically discarded. Using this result in Fig. 3, it follows that Nodes ②, ⑥, and ⑪ are automatically discarded. Since there are no more nodes left unexamined, the branch-and-bound procedure terminates with the information that Node ⑩ yields the optimum.

Suppose, on the other hand, the solution examines the nodes in the order (1)→(2)→(5) while storing Nodes (3) and (4) for later examination. At Node (5), a lower bound, $z = 12$, is obtained. However, because it is a "loose" one, it will be necessary to examine every node shown in Fig. 3 before the optimum is again encountered at Node (10). Thus the looseness of the bound, $z = 12$, has resulted in solving 11 problems as compared to only 4 when the bound $z = 14$ was encountered at an early stage of the computations.

The example given above points to three difficulties which are prime sources of computational inefficiency:

1. Proper selection of the next node to be examined
2. Proper selection of the branching variable
3. Complete solution of a continuous problem at each node

The first and second difficulties concern the total number of subproblems that must be tested before optimality is verified, while the third difficulty deals with the amount of "local" computations at each node.

The rules that have been developed for the selection of the next node to be examined and the branching variable are mainly heuristics. The effectiveness of these heuristics is mainly dependent on securing a tight bound at an early stage of the computations in order to develop rules for "predicting" the potential advantage of selecting specific node and branching variable in locating a better solution quickly.

The complete solution of the continuous problem at each node actually entails costly computations, particularly if the associated value of the objective function is the only information needed to decide whether to discard or to retain a node. The problem is alleviated by developing techniques for quickly <u>estimating</u> the optimum value of the objective function at a node by using efficient computation. These estimates are <u>never worse</u> than the true optimum value. Thus, if the estimated optimum objective value at a node is worse than the best available bound, the node is discarded as nonpromising. In general, if a node is not discarded by this test, it may be necessary to solve the continuous problem completely before a decision is made as to whether or not the node is discarded.

The method for estimating the optimum objective value has been developed for integer <u>linear</u> programs only. The idea is illustrated in Fig. 4. The solid piecewise linear curve shows the exact variation of the optimum objective value of z of a maximization linear program in the range $[x_k^*] \leq x_k \leq [x_k^*] + 1$, where x_k is the branching variable at the current node. Point A represents the optimum value of the objective function at the current node. Points B and C represent changes in the basic solution of the linear program.

If the two linear programs created by $x_k \leq [x_k^*]$ and $x_k \geq [x_k^*] + 1$ are solved exactly, the associated optimum values of the objective function equal z_d (Point E) and z_u (Point D), respectively. However, estimates of these values will be $\bar{z}_d$ (Point E) and $\bar{z}_u$ (Point F), respectively. Essentially, these two points are determined by extending the two line segments AB and AE originating at the continuous optimum point A associated with the current node. Points E and F can be determined directly from the linear programming tableau by using simple computations. The difference between z^* and $\bar{z}_u$ is usually known as the "up penalty," P_u. Similarly the "down penalty," P_d, defines the difference between z^* and $\bar{z}_d$. This is the reason the technique is known as the "method of penalties."

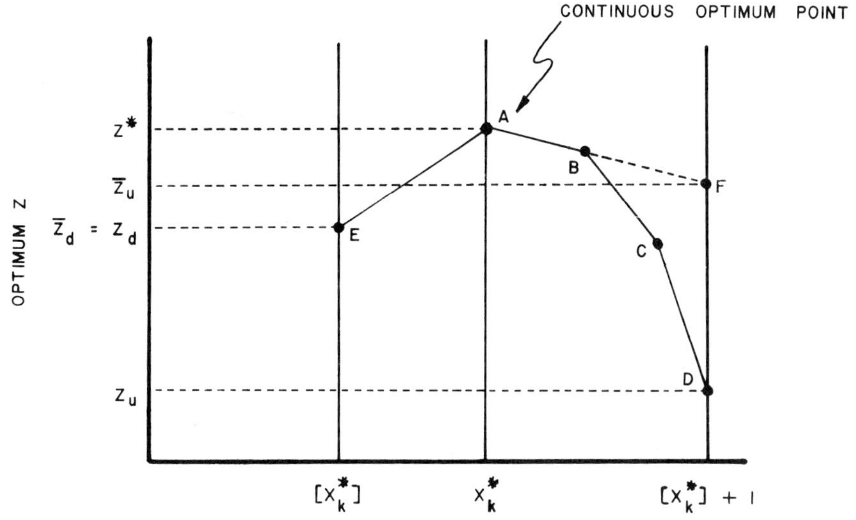

FIGURE 4.

In Fig. 4 the exact optimum value z_d for the branch $x_k \leq [x_k^*]$ happen to coincide with the estimated value $\bar{z}_d$, while for $x_k \geq [x_k^*] + 1$, z_u and $\bar{z}_u$ are different. This shows that the "goodness" of the estimates $\bar{z}_u$ and $\bar{z}_d$ depends on the manner optimum z varies with the value of x_k. In general, the only way to predict the closeness of the estimate and the exact values is by solving the linear programs exactly for $x_k \leq [x_k^*]$ and $x_k \geq [x_k^*] + 1$.

Although the estimates given in Fig. 4 are easy to determine numerically, computational experience has shown the obtained estimates are usually too loose to be effective. In actual practice, these estimates are supplemented by heuristics that greatly improve the sharpness of the developed tests. (See the section entitled Computations in Integer Programming below.)

It is important to realize that the branch-and-bound algorithm applies equally to the mixed integer problems. This is achieved simply by never selecting the continuous variables for branching. The algorithm terminates when the best solution satisfying the integer restrictions is found.

Another important point is that the branch-and-bound procedure outlined above applies also to nonlinear integer problems. The only difficulty is that global optimality for nonlinear problems is guaranteed only in special cases. The procedure fails when the global optimum cannot be secured at every node.

Implicit Enumeration Methods

Implicit enumeration deals with the solution of integer problems in which all integer variables are binary. Although the method became known some 5 years after branch-and-bound was introduced into the literature, implicit enumeration may actually be regarded as a special case of the more general branch-and-bound technique. Perhaps the basic difference originates from the fact that each integer variable can assume only the values zero and one. This, in addition to simplifying the branching process, allows the development of (almost) elementary rules for discarding (or fathoming) nodes. Also, the use of binary variables allows one to represent the entire branch-and-bound tree by a single vector with its number of elements equal to the number of variables (cf. the general branch-and-bound tree in Fig. 3).

The interesting outcome of restricting the integer variables to binary values is that the branching process fixes the branching variable at either zero or one (compare with the general procedure in Fig. 3 where branching specifies a range of values). Thus, as the level of branching increases, the number of unfixed variables becomes progressively smaller, thus effectively reducing the size of the problem. The consequence of this is the development of simple (almost intuitive) rules (called fathoming rules) which can be used to discard nodes. Such rules discard nodes on the basis of not having the potential to produce a feasible solution whose objective value is better than the current best bound.

To illustrate the fathoming rules, consider the problem

minimize

$$z = \sum_{j=1}^{n} c_j x_j; \quad c_j > 0 \text{ for all } j$$

subject to

$$\sum_{j=1}^{n} a_{ij} x_j + S_i = b_i; \qquad i = 1, 2, \ldots, m$$

$$x_j \in \{0, 1\}; \qquad j = 1, 2, \ldots, n$$

$$S_i > 0; \qquad i = 1, 2, \ldots, m$$

The variables S_i represent the slack variables. The requirement $c_j > 0$ can be satisfied by using the substitution $x_j = 1 - x_j'$ if $c_j < 0$, where $x_j' \in \{0, 1\}$. This requirement shows that to obtain min z, it is advantageous to keep all x_j at zero level unless the feasibility of the constraints requires elevating a variable to level one.

Suppose z_t is the value of z after fixing the values of t variables by the branching process [all nonfixed (or free) variables are assumed to be temporarily at zero level since all $c_j > 0$]. Suppose further that $\bar{z}$ is a known upper bound on the optimum (minimum) value of z.

Fathoming Rule 1. Suppose x_j is a free variable. Then x_j cannot be used as a branching variable if

$$z_t + c_j > \bar{z}$$

since it cannot lead to an improved solution. If all the free variables at a node satisfy this rule, the node must be discarded as nonpromising.

Fathoming Rule 2. Suppose S_i^t is the value of the i-th slack variable after t binary variables are fixed (again assume all free variables are tentatively at level zero). Suppose further that $S_i^t < 0$ and that R_t defines the set of free variables at the current node, then the node must be discarded if

$$\sum_{j \in R_t} \min\{0, a_{ij}\} > S_i^t$$

This follows since this node can never lead to a solution that compensates for the negativity (infeasibility) of the slack variable S_i; that is, it can never lead to a feasible solution.

It is evident that the fathoming rules are actually heuristic in nature. The power (or efficiency) of the implicit enumeration technique lies primarily in the fathoming power of the developed rules.

The main weakness of the above fathoming rules stems from considering the variables and constraints one at a time. Other heuristic rules have been developed which account for more than one variable and more than one constraint simultaneously. The most well-known of these heuristics is the surrogate constraint which combines the effect of several constraints for the purpose of extracting more selective information about the problem. To illustrate the idea, consider the two constraints

$$2x_1 - x_2 \leq -1$$

$$-x_1 + 2x_2 \leq -1$$

$$x_1, x_2 \in \{0, 1\}$$

By considering each constraint separately, one cannot conclude that the two constraints do not have a feasible solution. However, if the two constraints are added to one another, the resulting surrogate constraint

$$x_1 + x_2 \leq -2$$

shows decidedly that the problem cannot have a feasible solution.

In general, a surrogate constraint is formed as a nonnegative linear combination of the original constraints. Thus, if $\mu_j \geq 0$, a surrogate constraint that combines all m constraints of the problem is defined as

$$\sum_{i=1}^{m} \mu_i \left(\sum_{j=1}^{n} a_{ij} x_j \right) \leq \sum_{i=1}^{m} \mu_i b_i$$

It is evident that every feasible solution of the original problem must also satisfy the surrogate constraint. Moreover, if the surrogate constraint has no feasible solution, then neither does the original problem. This information can be used to fathom nodes in the implicit enumeration search.

Several efforts have been directed toward developing the strongest surrogate constraints by attempting to determine intelligent values for the weights μ_j. The underlying definition of strength in all these developments is that the chosen weights yield the most restrictive surrogate constraints.

The binary linear algorithm can be extended to cover nonlinear (polynomial) binary problems defined as

minimize

$$z = \sum_{j=1}^{n} c_j \prod_{k \in K_j} y_k; \quad K_j \subseteq \{1, 2, \ldots, p\}$$

subject to

$$\sum_{j=1}^{n} a_{ij} \prod_{k \in K_j} y_k \leq b_i; \quad i = 1, 2, \ldots, m$$

$$y_k \in \{0, 1\}; \quad k \in \{1, 2, \ldots, p\}$$

The substitution

$$x_j = \prod_{k \in K_j} y_k$$

shows that x_j is a binary variable. By using this substitution, the polynomial problem may be written as a master linear binary problem with polynomial secondary constraints. The binary linear algorithm can be modified to search through all the promising feasible solutions of the master problem and determine the optimum as the best feasible solution of the master problem which also satisfies the secondary polynomial constraints.

It is also possible to linearize the secondary constraints so that the original polynomial problem can be replaced by a completely linear problem. This is done by replacing the polynomial term

$$x_j = \prod_{k \in K_j} y_k$$

by

$$\sum_{k=1}^{|K_j|} y_k - (|K_j| - 1) \leq x_j \leq \frac{1}{|K_j|} \sum_{k=1}^{|K_j|} y_k$$

where $|K_j|$ = number of elements in the set K_j.

The computational efficiency of the two proposed methods depends primarily on the resulting number of "linear" variables in the linearized version of the problem. Thus, if the number of polynomial terms (and hence the number of x_j variables) is less than the number of polynomial (y_k) variables, the first method tends to be more efficient.

The algorithms for the pure zero-one problems do not apply directly to the mixed zero-one problem. Several algorithms exist for the mixed problems. Their basic idea is to separate between the continuous and binary variables by effectively partitioning the solution space of the original problem.

Cutting-Plane Methods

The "oldest" of the integer programming methods is the cutting-plane method. It primarily applies to linear integer problems although there have been limited developments in the quadratic case. As in the branch-and-bound method, the technique starts by first dropping the integer conditions and solving the resulting continuous linear program. By using the properties of the optimum continuous tableau, a constraint can be constructed which "slices off" part of the solution space such that no feasible integer solution is ever deleted. The continuous linear program is then solved subject to the new solution space. If the new optimum is integer, the process ends, otherwise a new cutting plane is constructed from the new tableau and reapplied to slice off another part of the new continuous space. The end effect of generating and applying these special constraints is that the optimum extreme point solution of the modified solution space should satisfy the integrality condition of the discrete variables.

Figure 5 shows pictorially how the cuts may operate on the continuous solution space. Point A gives the optimum solution before any cuts are applied. Point B is the optimum after Cut I is effected and Point C is the optimum after Cut II (and I) is augmented. Since at C both x_1 and x_2 are integers, the process ends.

In spite of the intriguing concept of the cut, one cannot really tell in advance how many cuts should be generated before the integer conditions are realized. Perhaps the only statement that can be made is that it takes at least one cut to solve an integer linear problem. Unfortunately, this statement cannot be refined any further to reflect the effect of the size of the problem (number of variables and constraints) on the speed of acquiring the solution.

To really fathom the complexity of the situation from the computational standpoint, one must understand how cuts are generated algebraically. Suppose that the continuous optimum tableau of the linear program is given as follows:

maximize

$$a = \bar{c}_0 - \sum_{j=1}^{n} \bar{c}_j w_j$$

subject to

$$x_i + \sum_{j=1}^{n} \alpha_{ij} w_j = \beta_i; \quad i = 1, 2, \ldots, m$$

$$x_i, w_j \geq 0 \text{ for all i and j}$$

The variables x_i, $i = 1, 2, \ldots, m$ are the basic variables while w_j, $j = 1, 2, \ldots, n$ are the nonbasic variables. Thus the continuous optimum is $z = \bar{c}_0$ and $x_i = \beta_i$, $i = 1, 2, \ldots, m$.

Suppose x_k is an integer variable but its continuous value β_k is not integer. The equation for x_k in the tableau is given as

$$x_k + \sum_{j=1}^{n} \alpha_{kj} w_j = \beta_k$$

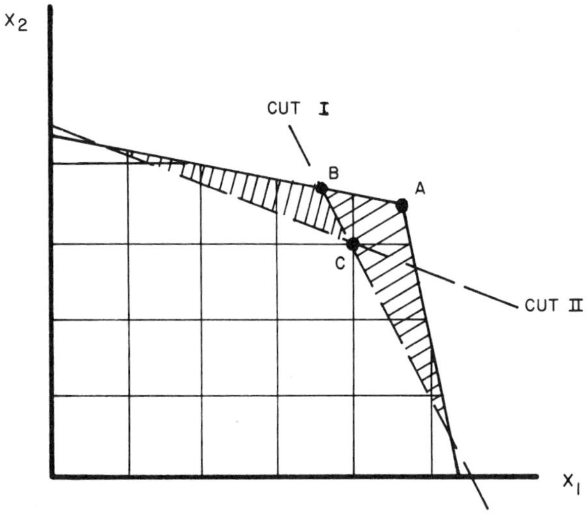

FIGURE 5.

Let f = d - [d] be defined such that [d] is the largest integer included in d. This means $0 \leq f < 1$. Thus the above equation can be written as

$$x_k - [\beta_k] + \sum_{j=1}^{n} [\alpha_{kj}]w_j = f_k - \sum_{j=1}^{n} f_{kj}w_j$$

In order for x_k and w_j for all j to be integer, the left-hand side and, hence, the right-hand side of the above equation must be integer. Thus, under integer conditions, the inequality

$$f_k - \sum_{j=1}^{n} f_{kj}w_j \leq 0 \quad \text{(generated cut)}$$

must be satisfied. This gives a necessary condition for x_k and all w_j to be integer and provides the necessary constraint which may now be applied to the given optimum tableau to modify the solution space as desired.

The difficulties with the generated cuts are as follows:

1. The effectiveness of the generated cut is measured in terms of how deep it cuts into the solution space. Since such a cut may be constructed from any Row i (provided β_i is not integer) of the tableau, different rows may yield different cuts, and there is no definite way of predicting which cut will be the strongest without exhaustively generating all possible cuts. Also, a weaker cut applied now may prove advantageous later since it may lead to stronger cuts (cf. a similar difficulty in the selection of branching variable in branch-and-bound methods in the section of that title). Although rules-of-thumb have been established for selecting the cut advantageously, none of them is uniformly superior.
2. The coefficients of the generated cut, as defined above, are necessarily fractional. This leads to the tremendous problem of machine round-off error which

could completely impair the capability of the technique to reach the optimum integer in a finite number of iterations. Although other types of cuts have been developed which restrict computations to integer coefficients, the developed cuts are necessarily weaker and hence less efficient.

3. Experience has shown that even a simple rearrangement of the constraints prior to the solution of the continuous linear program could convert a computationally manageable problem to a formidable one. This mysterious and erratic behavior of the cutting methods cannot be explained rationally.
4. Although theoretical finite convergence is proven for the cutting methods, experience has shown that only the first few cuts are effective in moving the optimum extreme point toward the optimum integer solution. The technique then simply loses potency, perhaps due to the accumulated machine round-off error.

Although several attempts were made to produce cuts that alleviate some of these problems, none of them can be claimed uniformly efficient. What is worse is that they cannot even be classified in terms of the types of integer problem they can solve more efficiently.

Although branch-and-bound techniques apply equally to pure and mixed integer problems, the cutting methods require special cuts for each of the two cases. The mixed cut is again developed so that no feasible points are deleted when the cut is applied to the continuous space.

Earlier cutting plane methods were mainly of the dual type. This means that no integer solution will be available for the problem if computations are stopped prematurely. To alleviate this difficulty, primal cutting methods were developed which guarantee that every cut will produce an integer feasible solution.

Another method for solving integer problems, called the asymptotic or group-theoretic algorithm, is inspired by the cutting-plane method. The idea of the method is based on the observation that the convex hull of all integer feasible points necessarily has integer extreme points. (See Fig. 6 for an illustration of a convex hull.)

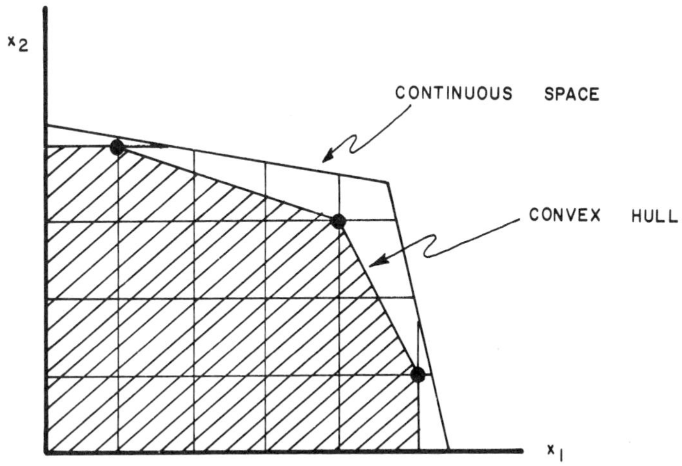

FIGURE 6.

Thus, if the cuts defining the convex hull can be determined, the problem can be solved as a regular linear program.

The difficulty with this idea is that the determination of the convex hull is not a simple matter. The problem is overcome partially by determining only some of the cuts bordering the convex hull. In some situations, these imposed cuts will produce an extreme point solution which directly solves the original integer problem. However, in general, this will not be the case, and it will be necessary to produce additional computations before the optimum integer solution is realized.

Unfortunately, the new method does not alleviate the computational difficulty associated with the cutting plane method. Only when the "partial" convex hull yields the optimum integer solution directly does the method appear efficient. However, there is no way to predict a priori when a certain problem can be solved in this manner.

Specialized Algorithms

Specialized algorithms have been developed to take advantage of the special structures of certain well-known problems with the objective of improving the efficiency of computation. These problems include the knapsack problem, the fixed-charge problem, the covering problem, and the traveling salesman problem. (see the section entitled Scope of Integer Programming Application for brief descriptions of these problems.)

Most of the developed methods are of the branch-and-bound type. However, some methods are, in the apparent sense, based on cutting methods but actually can be traced back to the spirit of the branch-and-bound methods. Apparently, however, the developed methods produce better results than when the generalized algorithms outlined above are applied to these problems.

COMPUTATIONS IN INTEGER PROGRAMMING

The unreliability of almost all integer programming algorithms in solving integer problems has forced some users to be satisfied with solving the problem as a continuous program and then round the resulting solution to the closest integer. This "method" cannot be refuted completely, particularly in the absence of a reliable exact procedure. However, it has its limitations. By using the categorization of integer problems given in the section entitled Scope of Integer Programming Application, one can see that only direct integer models may be susceptible to the idea of rounding since its use as an approximation may be logically acceptable. But in the cases of the coded and transformed models, there is no logical foundation that allows the use of approximation, particularly since the variables initially are quantifications of some nonnumerical codes. This observation underscores the need for the development of efficient exact methods for solving integer programs.

The cutting-plane methods are not generally dependable methods for solving the integer problem. Experience has shown that only the first few cuts are effective in eliminating portions of the continuous space and the continuous solution toward integrality. Subsequent cuts are usually ineffective due to machine round-off error.

The branch-and-bound methods, on the other hand, have been consistently more effective in solving the integer problem. Indeed, most of the available commercial codes (such as IBM's MPSX, Control Data's UMPIRE, CDC's OPHIELE) are all based on some version of the branch-and-bound method.

In spite of the fact that cutting methods cannot, in general, solve the integer problem, they possess important properties which can be used advantageously in conjunction with the branch-and-bound methods. For example, prior to branching at a node, one can first apply two or three cuts to the problem associated with the node, thus improving its bound relative to the true optimum integer solution. This usually assists in reducing the number of nodes and hence the computational effort of the problem.

The branch-and-bound methods, although consistently more reliable than the cutting methods, have their shortcomings. A major difficulty occurs with the computer storage requirement since it is difficult to predict in advance how many nodes will be generated. Another difficulty is that, depending on the difficulty of the problem, the sole use of the penalties (see the section entitled Branch-and-Bound Methods) to estimate the bound at each node is usually not effective in determining the next node to be considered or in discarding nonpromising nodes. In practice, the penalties approach is supplemented by heuristics which not only improve the effectiveness of estimating bounds using the penalties, but also improve the capability of the algorithm to "look ahead" and choose the sequence of nodes that are most likely to produce the optimal solution quickly.

Although the determination of a good bound at the early stages of computations is of paramount importance in improving the efficiency of the branch-and-bound method, the procedure outlined in the section entitled Branch-and-Bound Methods shows that such a bound is determined only as a by-product of the calculations. It is thus beneficial to locate a good bound by independent means. One such method is to search directly in the continuous space by using a heuristic method. Figure 7 outlines the moves of the direct search as applied to the graphical example shown. The overall idea is to start from the origin and move as far as possible, without violating feasibility, in the positive direction of the variable that has the largest coefficient in the objective function. When that point is reached, another promising variable is selected, and a move in its positive direction is made. When a point is reached where no further movements can be made (such as Point C in Fig. 7), the procedure allows one (or more) of the variables to decrease in value in hope that such a perturbation will lead to a future, more advantageous move.

The actual rules of the heuristics are more complex than outlined above. Also, some heuristics attempt to start the search from the continuous optimum rather than from the origin.

Although the indicated heuristics do not produce a verifiable optimal solution, they normally produce good solutions. Such solutions can then be used as an initial bound in the branch-and-bound procedure.

Another important factor distinguishing computations in integer programming from those in other techniques (such as linear programming) is the difficulty of securing the optimum integer without manual intervention during the course of computations. Specifically, the solution steps must be monitored to ensure that progress is being made toward identifying the optimum solution. If not, the user should have the option to invoke a different solution strategy, depending on the information he receives during the course of computations.

Basically, this means that the user should be provided with a number of solution strategies which are designed to alleviate certain computational problems when they arise. For example, if the penalty approach is found to produce poor estimates

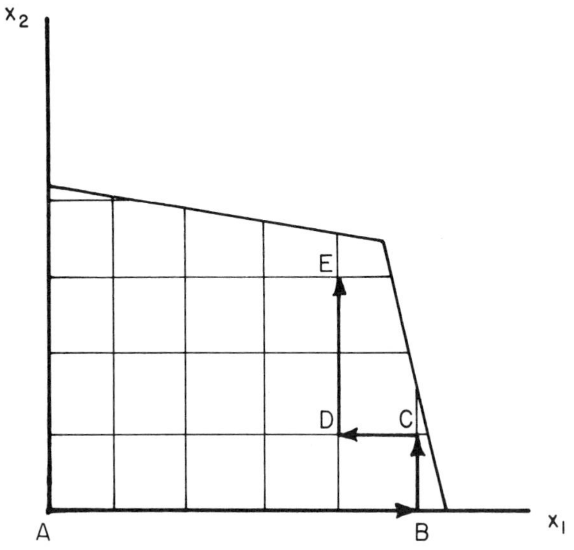

FIGURE 7.

of the bounds at a node, perhaps one can apply one or two cuts to the current problem, thus producing a different continuous optimum and hence a different set of penalties.

It appears that, for the foreseeable future, the solution of integer problems using the available techniques will rely on trial and error and the user's manual intervention to affect the course of computations. In this respect, accumulated experience in solving integer problems becomes an essential component in the successful application of integer programming algorithms.

Because of the expected computational difficulty in solving integer problems, certain points must be taken into account during the formulation stage of the integer model. Perhaps the most important of these is to limit the number of integer variables in the model. Since there is a variety of ways for formulating a mathematical model, one must always seek the formulation that involves the least number of integer variables. In fact, it may be possible to approximate some integer variables by continuous ones provided their numerical solution is expected to assume relatively large values.

Another important point is to restrict the feasible range of the integer variables as much as is practicable. This is necessarily equivalent to reducing the computations associated with the search for the optimum.

CONCLUDING REMARKS

In spite of the impressive advances in the computational power of the digital computer and the fact that finite algorithms have been developed for the integer problem, there is no way of predicting whether a given integer problem is solvable numerically. It seems that future work in this area will, and in fact should, concentrate on developing efficient heuristics. At present, further theoretical advances along the lines developed previously for integer programming appear futile and are not expected to produce a breakthrough.

Also, it is conceivable that a new generation of computers will be developed which will allow simultaneous processing and comparison of several problems in the branch-and-bound tree. Presently, nodes are considered sequentially, and this does not allow one to take advantage of important information (such as tighter bounds) that may be obtained from other nodes should simultaneous (or parallel) processing of problems become possible.

NOTES

History of Integer Programming

Recent books on integer programming include Hu [47], Greenberg [41], Garfinkel and Nemhauser [25], Salkin [62], and Taha [69]. The last three books include extensive bibliographies. The basic reference for this article is Taha [69]. A survey article of the developments in discrete mathematics prior to 1958 is found in Saaty [61].

The simplex method of linear programming was developed by Dantzig [16]. Dantzig [15] was the first to propose the use of secondary constraints to force a linear programming solution toward an integer solution. However, convergence was not guaranteed.

Gomory [37] developed the first converging algorithm for the integer linear problem. Almost simultaneously, Land and Doig [49] developed a different (converging) algorithm for the same problem. Balas [2] developed the implicit enumeration technique dealing with problems in binary variables only.

Scope of Integer Programming Applications

The categorization of integer models as direct, coded, and transformed is due to Taha [69].

Application of "either-or" constraints in the scheduling problem is due to Manne [53]. Other applications in scheduling theory may be found in Bowman [10], Wagner [74], Conway et al. [13], Pritsker et al. [59], and Taha [66].

The capital budgeting problem was first proposed by Weingartner [76]. Hillier [44] and Peterson and Launghhunn [58] treat the probabilistic version of the problem.

Gilmore and Gomory [29, 30] use the knapsack model to solve the cutting-stock problem. The use of the knapsack model in connection with the solution of the general integer problem was proposed by Gomory [40] and later in a completely different context by Elmaghraby and Wig [20].

Other applications of the covering problem include political districting [24], information retrieval [17], line balancing [63], switching theory [46, 60], and capital investment [73].

Piece-wise linear approximation is used in solving separable programming problems [12]. Other approximations are given by Markowitz and Manne [55] and Healey [43].

A closely related problem to the fixed-charge problem is the plant location problem. An excellent exposition of the plant location model is given by Elshafei [21].

Methods of Integer Programming

General survey articles on integer programming methods were written by Balinski [6], Balinski and Spielberg [7], and Geoffrion and Marsten [28]. The relationship between some binary algorithms and continuous linear programs was first recognized by Taha [67]. A related result was introduced by Cabot and Hurter [11].

Branch-and-Bound Methods. The basic article for branch-and-bound methods is due to Land and Doig [49]. The procedure presented here is based on an important modification of the Land-Doig method by Dakin [14].

The general branch-and-bound principle was formalized by Balas [4] and later by Mitten [56].

The use of penalties to estimate the optimum value of the objective function in a linear program was first conceived by Driebeek [18] in connection with the solution of mixed zero-one programs. The idea was extended to the general branch-and-bound algorithm by Beale and Small [8]. Improvements in the effectiveness of using penalties were made by Tomlin [70].

Implicit Enumeration Methods. The first implicit enumeration algorithm for the zero-one problem is due to Balas [2]. Although the algorithm is now recognized as a special case of the branch-and-bound algorithm, it was originally presented as a different type of solution procedure. Glover [32] was the first to introduce important modifications in the implicit enumeration method including the "one-vector" representation of the entire branch-and-bound tree. Geoffrion [26] showed how Balas' algorithm can be superimposed on Glover's enumeration scheme to produce a more simplified algorithm. The relationship between Balas' algorithm and linear programming is given by Taha [67].

Glover's original enumeration scheme is based on the last-in first-out (LIFO) rule in the scanning of stored nodes. Tuan [72] allowed more flexibility in selecting the next node to be examined.

The concept of surrogate constraints was first introduced by Glover [32]. Balas [3] and Geoffrion [27] gave different versions of the surrogate constraint, but were mainly inspired by Glover's initial work. Geoffrion's paper provides computational experience. Glover [34] gives a comparison of the three methods.

Hammer and Rudeanu [42] were the first to propose a method for solving zero-one polynomial programs. The method is not computationally effective. Linearization of the polynomial zero-one terms was proposed, apparently independently, by Fortet [23], Balas [1], and Watters [75]. Glover and Woolsey [36] show how linearization may be effected efficiently with a smaller number of additional variables. Taha [68] proposed a method based on Balas' linear zero-one algorithm and inspired by ideas in the paper by Hammer and Rudeanu [42]. Taha's paper includes computational experience.

Other methods for tackling the polynomial problem include that of Lawler and Bell [51], which is extremely inefficient. The special case of quadratic objective function and linear constraints (which normally arises in capital budgeting problems) was introduced by Laughhunn [50].

Algorithms for the mixed zero-one problem were first proposed by Driebeek [18]. Benders' decomposition approach [9] is the basis for another partitioning algorithm which decomposes the original problem into separate binary and continuous problems. Benders' algorithm was modified by Lemke and Spielberg [52].

Cutting-Plane Methods. The use of cuts to solve integer linear programs was first proposed by Dantzig [15]. However, his method would not support a converging algorithm. The first converging algorithm (called the fractional algorithm) was introduced by Gomory [37] for solving pure integer programs. To overcome the problem of rounding which plagued the fractional algorithm, Gomory [38] introduced his all-integer integer algorithm. The mixed integer algorithm was also introduced by Gomory [39].

The cutting plane method for quadratic integer programs was developed by Kunzi and Oettli [48].

Because all of Gomory's algorithms are of the dual type, primal methods which generate feasible integer solutions during the course of computation were introduced by Young [79] and simultaneously by Glover [33].

Other types of cutting-plane algorithms include the bound-escalation cut by Glover [32] and intersection cuts by Young [80] and Balas [5]. Glover [35] generalized Young's and Balas' cuts by developing the convexity cut which applies to other mathematical programming problems as well.

The asymptotic algorithm was developed by Gomory [40]. Hu [47] developed a network algorithm for solving the asymptotic problem. Shapiro [64] proposed a branch-and-bound method for solving the same problem. An extension of the method to the mixed problem was proposed by Wolsey [77, 78].

Specialized Algorithms. A summary of bibliography for specialized algorithms is not possible in this limited space. See Taha [69], pp. 263-341.

Computations in Integer Programming

An important paper providing computational experience for the cutting methods is due to Trauth and Woolsey [71].

Several papers discuss the efficiency of the branch-and-bound and implicit enumeration techniques. The more important papers include Petersen [57], Shaw [65], and Tomlin [70]. An excellent paper discussing the effectiveness of using penalties as well as heuristics in branch-and-bound methods is due to Forrest et al. [22].

Many heuristics have been developed for the general integer linear problem. Two such methods are due to Echols and Cooper [19] and Hillier [45].

REFERENCES

1. Balas, E., Extension de l'algorithme additif en nombre entiers et a la programmation nonlineare, C. R. Acad. Sci. 258, 5136-5139 (1964).
2. Balas, E., An additive algorithm for solving linear programs with zero-one variables, Oper. Res. 13, 517-546 (1965).
3. Balas, E., Discrete programming by the filter method, Oper. Res. 15, 915-957 (1967).
4. Balas, E., A note on the branch-and-bound principle, Oper. Res. 16, 442-445 (1968); errata, Oper. Res. 16, 886 (1968).
5. Balas, E., Intersection cuts—A new type of cutting planes for integer programming, Oper. Res. 19, 19-39 (1971).
6. Balinski, M. L., Integer programming: Methods, uses, computations, Manage. Sci. 12, 253-313 (1965).

7. Balinski, M. L., and K. Spielberg, Methods for integer programming: Algebraic, combinatorial and enumerative, in Progress in Operations Research—Volume III (J. Aronofsky, ed.), Wiley, New York, 1969, pp. 195-292.
8. Beale, E. M. L., and R. E. Small, Mixed integer programming by a branch-and bound technique, in Proceedings IFIP Congress, New York, May 1965.
9. Benders, J. F., Partitioning procedures for solving mixed-variables programming problems, Numer. Math. 4, 239-252 (1962).
10. Bowman, E. H., The schedule-sequencing problem, Oper. Res. 7, 621-624 (1959).
11. Cabot, A V., and A. P. Hurter, An approach to 0-1 integer programming, Oper. Res. 16, 1206-1211 (1968).
12. Charnes, C., and C. E. Lemke, Separable programming, Nav. Res. Logistics Q. 1, 23-28 (1954).
13. Conway, R. W., W. L. Maxwell, and L. W. Miller, Theory of Scheduling, Addison-Wesley, Reading, Massachusetts, 1967.
14. Dakin, R. J., A tree-search algorithm for mixed integer programming problems, Comput. J. 9, 250-255 (1965).
15. Dantzig, G. B., Notes on solving linear programs in integers, Nav. Res. Logistics Q. 6, 75-76 (1959).
16. Dantzig, G. B., Linear Programming and Extensions, Princeton University Press, Princeton, New Jersey, 1963.
17. Day, R. H., An optimal extracting from a multiple file data storage system: An application of integer programming, Oper. Res. 13, 482-494 (1965).
18. Driebeek, N. J., An algorithm for the solution of mixed integer programming problems, Manage. Sci. 12, 576-587 (1966).
19. Echols, R. E., and L. Cooper, Solution of integer linear programming problems by direct search, J. Assoc. Comput. Mach. 15, 75-84 (1968).
20. Elmaghraby, S. E., and M. K. Wig, On the Treatment of Cutting Stock Problems as Diophantine Programs, Department of Industrial Engineering, North Carolina State University, Raleigh, 1970.
21. Elshafei, A. N., Facilities Location: Formulations, Methods of Solution, Applications and Some Computational Experience, Memo No. 276, Institute of National Planning, Cairo, Egypt, 1972.
22. Forrest, J. J. H., J. P. H. Hirst, and J. A. Tomlin, Practical solution of large mixed integer programming problems with UMPIRE, Manage. Sci. 20, 736-773 (1974).
23. Fortet, R., L'algebre de Boole et se application en recherche operationelle, Cah. Cent. Etud. Rech. Oper. 4, 215-280 (1959).
24. Garfinkel, R. S., and G. L. Nemhauser, Optimal political districting by implicit enumeration techniques, Manage. Sci. 16, B495-B508 (1970).
25. Garfinkel, R. S., and G. L. Nemhauser, Integer Programming, Wiley, New York, 1972.
26. Geoffrion, A. M., Integer programming by implicit enumeration and Balas' method, SIAM Rev. 7, 178-190 (1967).
27. Geoffrion, A. M., An improved implicit enumeration approach for integer programming, Oper. Res. 17, 437-454 (1969).
28. Geoffrion, A. M., and R. E. Marsten, Integer programming: A framework and state-of-the-art survey, Manage. Sci. 18, 465-491 (1972).
29. Gilmore, P. C., and R. E. Gomory, A linear programming approach to the cutting stock problem, Oper. Res. 9, 849-859 (1961).

30. Gilmore, P. C., and R. E. Gomory, A linear programming approach to the cutting stock problem—Part II, Oper. Res. 11, 863-888 (1963).
31. Glover, F., A bound escalation method for the solution of integer linear programs, Cah. Cent. Etud. Rech. Oper. 6, 131-168 (1965).
32. Glover, F., A multiphase-dual algorithm for the zero-one integer programming problem, Oper. Res. 13, 879-919 (1965).
33. Glover, F., A new foundation for a simplified primal integer programming algorithm, Oper. Res. 16, 727-740 (1968).
34. Glover, F., Surrogate constraints, Oper. Res. 16, 741-749 (1968).
35. Glover, F., Convexity cut and cut search, Oper. Res. 21, 123-134 (1973).
36. Glover, F., and R. E. Woolsey, Aggregating Diophantine Equations, Report No. 70-4, University of Colorado, Boulder, 1970.
37. Gomory, R. E., Outline of an algorithm for integer solutions to linear programs, Bull. Am. Math. Soc. 64, 275-278 (1958).
38. Gomory, R. E., All-Integer Integer Programming Algorithm, RC-189, IBM, Yorktown Heights, New York, 1960; also in Industrial Scheduling (J. F. Muth and G. L. Thompson, eds.), Prentice-Hall, Englewood Cliffs, New Jersey, 1963, pp. 193-206.
39. Gomory, R. E., An Algorithm for the Mixed Integer Problem, RM-2597, RAND Corp., Santa Monica, California, 1960.
40. Gomory, R. E., On the relation between integer and non-integer solutions to linear programs, Proc. Nat. Acad. Sci. U.S. 53, 260-265 (1965).
41. Greenberg, H., Integer Programming, Academic, New York, 1971.
42. Hammer, P. L., and S. Rudeanu, Boolean Methods in Operations Research and Related Areas, Springer, Berlin, 1968.
43. Healey, W. C., Jr., Multiple choice programming, Oper. Res. 12, 122-138 (1964).
44. Hillier, F. S., Derivation of probabilistic information for the evaluation of risky investments, Manage. Sci. 10, 443-457 (1963).
45. Hillier, F. S., Efficient heuristic procedures for integer linear programming with an interior, Oper. Res. 17, 600-637 (1969).
46. Hohn, F., Some mathematical aspects of switching, Am. Math. Mon. 62, 75-90 (1955).
47. Hu, T. C., Integer Programming and Network Flows, Addison-Wesley, Reading, Massachusetts, 1969.
48. Kunzi, H. P., and W. Oettli, Integer quadratic programming, in Recent Advances in Mathematical Programming (R. L. Graves and P. Wolfe, eds.), McGraw-Hill, New York, 1963, pp. 303-308.
49. Land, A. H., and A. G. Doig, An automatic method for solving discrete programming problems, Econometrica 28, 497-520 (1960).
50. Laughhunn, D. J., Quadratic binary programming with applications to capital-budgeting problems, Oper. Res. 18, 454-461 (1970).
51. Lawler, E. L., and M. D. Bell, A method for solving discrete optimization problems, Oper. Res. 14, 1098-1112 (1966).
52. Lemke, C. E., and K. Spielberg, Direct search zero-one and mixed integer programming, Oper. Res. 15, 892-914 (1967).
53. Manne, A. S., On the job-shop scheduling problem, Oper. Res. 8, 219-223 (1960).

54. Manne, A. S., Plant location under economies of scale-decentralization and computation, Manage. Sci. 11, 213-235 (1964).
55. Markowitz, H. M., and A. S. Manne, On the solution of discrete programming problems, Econometrica 25 84-110 (1957).
56. Mitten, L. G., Branch-and-bound methods: General formulation and properties, Oper. Res. 18, 24-34 (1970).
57. Petersen, C. C., Computational experience with variants of the Balas algorithm applied to the selection of R and D projects, Manage. Sci. 13, 736-750 (1967).
58. Peterson, D. E., and D. Laughhunn, Capital expenditure programming and some alternative approaches to risk, Manage. Sci. 17, 320-336 (1971).
59. Pritsker, A. A., L. J. Watters, and P. M. Wolfe, Multiproject scheduling with limited resources: A 0-1 programming approach, Manage. Sci. 16, 93-108 (1969).
60. Roth, J. P., Algebraic topological methods for the synthesis of switching systems—I, Trans. Am. Math. Soc. 88, 301-326 (1958).
61. Saaty, T. L., On nonlinear optimization in integers, Nav. Res. Logistics Q. 15, 1-22 (1968).
62. Salkin, H., Integer Programming, Addison-Wesley, Reading, Massachusetts, 1975.
63. Salveson, M. E., The assembly line balancing problem, J. Ind. Eng. 6, 18-25 (1955).
64. Shapiro, J. F., Group theoretic algorithms for the integer programming problem—II: Extension to a general algorithm, Oper. Res. 16, 928-947 (1968).
65. Shaw, M., Review of computational experience in solving large mixed integer programming problems, in Applications of Mathematical Programming Techniques (E. M. L. Beale, ed.), English Universities Press, London, 1970, pp. 406-412.
66. Taha, H. A., Sequencing by implicit ranking and zero-one polynomial programming, AIIE Trans. 2, 157-162 (1971).
67. Taha, H. A., On the Solution of Zero-One Linear Programs by Ranking the Extreme Points, Technical Report No. 71-2, Department of Industrial Engineering, University of Arkansas, Fayetteville, 1971.
68. Taha, H. A., A Balasian-based algorithm for zero-one polynomial programming, Manage. Sci. 18, B328-B343 (1972).
69. Taha, H. A., Integer Programming: Theory, Applications and Computations, Academic, New York, 1975.
70. Tomlin, J. A., Branch and bound methods for integer and nonconvex programming, in Integer and Nonlinear Programming (J. Abadie, ed.), American Elsevier, New York, 1970, pp. 437-450.
71. Trauth, C. A., and R. E. Woolsey, Integer linear programming: A study in computational efficiency, Manage. Sci. 15, 481-493 (1969).
72. Tuan, N. P., A flexible tree search method for integer programming problems, Oper. Res. 19, 115-119 (1971).
73. Valenta, J. R., Capital Equipment Decisions: A Model for Optimal System Interfacing, M.S. Thesis, MIT, 1969.
74. Wagner, H. M., An integer linear programming model for machine scheduling, Nav. Res. Logistics Q. 6, 131-140 (1959).
75. Watters, L. J., Reduction of integer polynomial problems to zero-one linear programming problems, Oper. Res. 14, 1171-1174 (1967).

76. Weingartner, H. M., Mathematical Programming and the Analysis of Capital Budgeting Problems, Prentice-Hall, Englewood Cliffs, New Jersey, 1963.
77. Wolsey, L. A., Extensions of the group theoretic approach in integer programming, Manage. Sci. 18, 74-83 (1971).
78. Wolsey, L. A., Group-theoretic results in mixed integer programming, Oper. Res. 19, 1691-1697 (1971).
79. Young, R. D., A simplified primal (all-integer) integer programming algorithm, Oper. Res. 16, 750-782 (1968).
80. Young, R. D., Hypercylindrically deduced cuts in 0-1 integer programming, Oper. Res. 19, 1393-1405 (1971).

3
GAME THEORY

WILLIAM F. LUCAS

School of Operation Research
and Industrial Engineering
Cornell University
Ithaca, New York

INTRODUCTION

The Nature of Game Theory

Game theory is a collection of mathematical models formulated to study decision making in situations involving conflict and cooperation. It recognizes that conflict arises naturally when various participants have different preferences, and that such problems can be studied quantitatively. Game theory attempts to abstract out those ingredients which are common and essential to many different competitive situations and to study them by means of the scientific approach. It is concerned with finding optimal solutions, stable outcomes, or equitable allocations when various decision makers have conflicting objectives in mind. These models describe how one should proceed in order to arrive at the best possible outcome in light of the options open to one's opponents. Game theory thus attempts to provide a normative guide to rational behavior when acting in a group whose members aim for different goals. In general, a game consists of players who must choose from a list of alternatives which will in turn bring about certain outcomes over which the participants may have different preferences. Chance and random events may also influence the final payoffs.

Related Subjects

Game theory is concerned with two or more participants involved in strategic encounters (games of skill). It thus extends beyond the classical theories of probability (games of chance) and decision theory (games against nature) which are frequently sufficient to solve situations involving merely one player and random elements. It is usually considered a distinct subject from gaming or simulation, even though interaction between these subjects often proves beneficial. Game theory also gains insights from conducting game theoretical types of experiments. And since such encounters are laden with various concepts of value and worth, game theory makes frequent use of utility theory which in its modern form was originally developed as a supplement to the game models.

Types of Games

It is obvious that the full scope of game theory is extremely broad and most ambitious, and that one must restrict himself to special cases if he is to obtain usable results. There are consequently many logical classifications of competitive situations. Some that have proved helpful are: the number of participants, the number of moves or choices, the constant-sum and general-sum cases, the cooperative and noncooperative situations, the various states of information available to the players, the different restrictions which may be placed on side payments, single play games or multistage games in which similar games are played repeatedly, as well as more dynamical situations in which choices are made continuously in time.

Forms of a Game

In 1944, John von Neumann and Oskar Morgenstern described three abstract models or forms for studying games: extensive form, normal form, and characteristic function form. Most of the models used in the mathematical theory of games are of one of these basic types, or a rather direct variation, extension, or generalization of them. A brief technical description of each form will be presented first, and then each type will be described in more detail, along with illustrations and applications.

An n-person game in extensive form can be described by a "rooted tree," that is, a connected graph in the plane with no loops and with a distinguished vertex or node. The distinguished or initial vertex (root) and each additional interior vertex (point, node, locus) corresponds either to one of the n players or to the "chance player." The arcs (edges, lines) ascending from a vertex correspond to the alternate choices this player can make if a play of a game leads to this point. To each terminal vertex there is associated an n-dimensional vector in which the components represent the payoffs to the respective players. For each node belonging to the chance player there is a probability distribution over the ascending arcs. The state of a player's information at any stage can be described by certain types of subsets of the set of all his vertices which are called information sets. The extensive form usually gives the most complete description of a game and is used to model many game-like situations. However, it has not yet proved to be the most suitable form for handling the computational problems involved in actually solving games, except for very simple cases.

An n-person game in normal form is characterized by n sets of integers N_1, N_2, ..., N_n (the players' strategy spaces) and a payoff function F from the Cartesian product $N_1 \times N_2 \times \cdots \times N_n$ into n-dimensional Euclidean space. The vector $F(i_1, i_2, \ldots, i_n)$ describes the respective payoffs to the n players when they each make the single choices $i_1, i_2, \ldots, i_n$. One can also incorporate chance into the game by including a set N_0 of chance moves and a probability distribution over such moves. One then works with expected payoffs in the obvious way. The subject of two-person "zero-sum" games in normal form (the matrix games) is well known, and it has had significant application to many fields. The n-person games can also be viewed as an n-dimensional matrix in which the "rows" in some one dimension correspond to a particular player's choices. The components of the matrix are n-dimensional payoff vectors. Games in normal form are used to study two-person games and n-person "noncooperative" games. In the latter case, the theory of "equilibrium points" is the main solution concept.

An n-person game in characteristic function form is determined by a real valued characteristic function v defined on the set 2^N of all subsets of a finite set N. That is, v assigns the real number v(S) to each subset S of N. N is the set of n players denoted by 1, 2, ..., n, and v(S) represents the value (wealth, power) which the coalition S can achieve when its players cooperate. Most models for n-person "cooperative" games are analyzed in this form or in some variation of it.

A game in extensive form can be reduced to a game in normal form by using the game tree to introduce a set of "pure strategies" for each player. Each pure strategy is a choice by the given player of how he will play through the entire game tree in the event of all possible choices by the other players. One then considers "mixed strategies" which are probability distributions over the set of pure strategies. In the case of the two-person zero-sum games, determining "optimal" mixed strategies is mathematically much simpler in this normal form. The optimal mixed strategies in the reduced game (in normal form) can be interpreted in the original extensive form of the game. As for determining optimal strategies, the game in normal form maintains many of the essential strategic features of the original game in extensive form. Historically, there has been little work done on deriving solution techniques directly for games in extensive form.

A game in normal form can be further reduced to a game in characteristic function form. If the game is "constant-sum," then v(S) is the "value" of the coalition S in the resulting two-person constant-sum game between S and its complement N - S. If the game is not constant-sum, then v(S) may be taken as the "maximum value" of S in the resulting two-person noncooperative general-sum game between S and N - S, or it may be determined in some other manner. In the former cases the determination of v(S) is rather conservative, because it assumes N - S will cooperate whenever S does. Furthermore, this reduction removes many of the detailed features of the original game: the nature and the timing of the moves, the nature of the information, and the specific payoffs. One can make several objections to it, especially in the nonconstant-sum case, and any results determined for the game in characteristic function form may have very limited use when interpreted in the normal or extensive form. However, the characteristic function seems most appropriate for studying coalition formation which is an essential feature in the cooperative games. From another point of view, one can begin by assuming that he has a game in characteristic function form independent of the other forms. The value v(S) can be given or derived somehow as the inherent wealth or power of the coalition S when its members cooperate.

Historical Highlights

Although some game theory concepts have arisen over the past couple of centuries, modern game theory dates from 1944 with the publication of the monumental work Theory of Games and Economic Behavior [25] by von Neumann and Morgenstern. Some particular two-person zero-sum matrix games were analyzed in the early part of the twentieth century, and the major theoretical result for these games (the famous Minimax Theorem) was proved by von Neumann in 1928 (see English translation in Ref. 24, 1959). Although a principal theorem was given by Zermelo in 1913 [29], the first complete definition of a game in extensive form also appeared in Theory of Games and Economic Behavior. However, a slightly different definition due to Kuhn [24, 1953] is the one most frequently used today. The first game theoretic model for n-person cooperative games was also presented by von Neumann and Morgenstern in 1944 when they presented a theory of solutions (stable sets) based on the characteristic function formulation. Several other solution concepts have since been defined for the cooperative theory. In 1950 Nash [14] generalized the "saddle point" concept and the minimax theorem for matrix games to the existence of equilibrium points for n-person general-sum noncooperative games in normal form.

An excellent survey of modern game theory until 1957 is presented in the book by Luce and Raiffa [11]. Many books contain an exposition on the matrix games. A current introduction to several major models in the n-person theory appears in Chapters 8, 9, and 10 of the text by Owen [16]. Shapley and Shubik have been preparing an extensive work on the multiperson games and their applications, and some chapters have already appeared as RAND Corporation reports [21]. Elementary expositions of the n-person theory have been published recently by Davis [4] and Rapoport [18]. A detailed bibliography on game theory until 1959 was compiled by D. M. and G. L. Thompson in Ref. 24, 1959. The best volume of references and short abstracts of game theory literature to 1968 appears currently only in Russian [26]. Several of the major mathematical papers in the field have appeared in five issues of the Annals of Mathematics Studies [24]. The introductions to these studies also provide excellent survey material on the subject. A new set of volumes on game theory has recently begun [28]. There is now an International Journal of Game Theory, and articles on this subject appear in the new journal Mathematics of Operations Research. A popular book On the Game of Business by McDonald [12] views several real cases from the world of business from a game theoretic point of view.

THE NORMAL FORM

Introduction

The simplest game theoretical model is the theory of games in normal form, also referred to as the strategic form or as the matrix (or polymatrix) games. This is simply a list of pure strategies for each one of the players along with a description of the resulting payoffs to the players for any possible choice of strategies by these participants. For the special situation of only two players and when one's loss corresponds to the other's gain, such games can be described by a real valued matrix. This simplest case will be described in some detail. Extensions to general-sum games and those with more than two players are mentioned later in less detail.

Matrix Games

A two-person zero-sum finite matrix game with real payoffs is characterized by an m by n matrix A with real components a_{ij}. Each number (or row) i in $S_I = \{1, 2, \ldots, m\}$ corresponds to a pure strategy for player I, the row player. Each number (or column) j in $S_{II} = \{1, 2, \ldots, n\}$ corresponds to a pure strategy for player II, the column player. So I has m pure strategies or choices, and II has n pure strategies. The number a_{ij} is the payoff from player II to player I if I plays his i-th strategy and II plays his j-th strategy. If $a_{ij} < 0$, then consider the payoff to be $|a_{ij}|$ from I to II. A play of the game consists of I choosing a row and II choosing a column, both in ignorance of the other's choice.

If player I were to pick the i-th row, then the worst that could happen to him is that he would obtain $\min_{1 \le j \le n} a_{ij}$. A conservative player I might play a row that corresponds to a maximum of these row minima. This is called a maximin strategy for I, and playing this strategy guarantees that he will obtain at least the lower value of the game:

$$v_I = \max_{1 \le i \le m} \min_{1 \le j \le n} a_{ij}$$

Similarly define a minimax strategy for player II to be a column which corresponds to a minimum of the column maxima. Playing such a strategy can assure II of losing no more than the upper value of the game:

$$v_{II} = \min_{1 \le j \le n} \max_{1 \le i \le m} a_{ij}$$

These concepts can be illustrated by the following array

	II's strategies							
I's strategies	1	2	...	j	...	n	row minima
1	a_{11}	a_{12}	...	a_{1j}	...	a_{1n}	$\min a_{1j}$
2	a_{21}	a_{22}	...	a_{2j}	...	a_{2n}	$\min a_{2j}$
⋮			...				⋮
i	a_{i1}	a_{i2}	...	a_{ij}	...	a_{in}	$\min a_{ij}$
⋮			...				⋮
m	a_{m1}	a_{m2}	...	a_{mj}	...	a_{mn}	$\min a_{mj}$
column maxima	$\max a_{i1}$	$\max a_{i2}$	...	$\max a_{ij}$	...	$\max a_{in}$	v_I \ v_{II}

It is easy to prove that $v_I \leq v_{II}$, that is, maximin $\leq$ minimax. As a result of this we can consider two cases: one when $v_I = v_{II}$ and the other when $v_I < v_{II}$.

Case 1: The Saddle Point Case. If $v_I = v_{II} = v$, then player I can assure himself of obtaining at least the amount v from II by playing a maximin strategy, and II can hold his losses down to at most v by playing a minimax strategy. Thus v is called the value of the game, and a maximin strategy i_0 and a minimax strategy j_0 are called optimal strategies for I and II, respectively. The number v and an optimal strategy for each of the players is called a solution of the game. The game is said to have a saddle point at (i_0, j_0) with value $a_{i_0 j_0} = v$. In this case, prior knowledge of the opponent's strategy choice is of no value to the opponent.

One can show that if player I has two optimal strategies i_1 and i_2, and player II has two optimal strategies j_1 and j_2, then $a_{i_1 j_1} = a_{i_1 j_2} = a_{i_2 j_1} = a_{i_2 j_2}$.

Case 2: The Mixed Strategy Case. Although the most that I can be sure of winning is v_I, it seems that he could "expect" in addition some of the gap $v_{II} - v_I > 0$ if he were to bring in probability considerations and use expected values. A mixed strategy for player I is a probability distribution over his pure strategy set, and it can be described by a probability vector $x = (x_1, x_2, \ldots, x_m)$ where $x_1 + x_2 + \cdots + x_m = 1$ and each $x_i \geq 0$. To play a mixed strategy means to use a random device to pick which pure strategy will be played, the i-th pure strategy to be chosen with probability x_i. Similarly, a mixed strategy for player II is a probability vector $y = (y_1, y_2, \ldots, y_n)$ where $y_1 + y_2 + \cdots + y_n = 1$ and each $y_j \geq 0$. Note that a pure strategy can be considered as a special case of a mixed strategy, and each player has an infinite set of mixed strategies.

If player I were to play the mixed strategy x and player II plays his pure strategy j, then I would "expect" to obtain

$$E(x, j) = x_1 a_{1j} + x_2 a_{2j} + \cdots + x_m a_{mj} = xA_j \tag{1}$$

which is the inner (dot, scalar) product of x and the j-th column of matrix A which is denoted A_j. Likewise, if II plays the mixed strategy y and I plays the pure strategy i, then I expects to obtain

$$E(i, y) = a_{i1} y_1 + a_{i2} y_2 + \cdots + a_{in} y_n = A^i y^T \tag{2}$$

which is the inner product of the i-th row of A and y. A^i is the i-th row of A and y^T is the transpose of the vector (matrix) y. If I and II play the mixed strategies x and y, respectively, then the expected win to I (or loss to II) is

$$E(x, y) = \sum_{j=1}^{n} \sum_{i=1}^{m} x_i a_{ij} y_j = xAy^T$$

The problem for player I is to find a probability vector x which maximizes the smallest component in the vector xA, i.e., that maximizes the minimum of the numbers of the form (1) as j varies from 1 to n. Let e_t be the t-dimensional row vector with all t components equal to 1. Then I's goal is to find the vector x that maximizes the number v subject to the constraints

$$xA \geq ve_n \quad (\max v)$$

$$xe_m^T = 1 \tag{3}$$

$$x \geq 0$$

Similarly, II's goal is to find a vector y that minimizes the number V subject to

$$Ay^T \leq Ve_m^T \quad (\min V)$$

$$e_n y^T = 1 \tag{4}$$

$$y \geq 0$$

Note that $v = ve_n y^T \leq xAy^T \leq xVe_m^T = xe_m^T V = V$.

The two optimization problems in (3) and (4) are a special case of a pair of dual linear programs in the theory of linear programming. So a game can be solved by solving a pair of dual linear programming problems. The Minimax Theorem, or Fundamental Theorem of two-person zero-sum matrix games, by von Neumann (1928) (or equivalently, the Duality Theorem in linear programming) states that there exist mixed strategies x_0 and y_0 for I and II, respectively, which solve the problems in (3) and (4), and that $v = V$. In other words, there exist probability vectors x_0 and y_0 and a number v such that

$$x_0 A \geq ve_n$$

$$Ay_0^T \leq ve_m^T$$

The triple (x_0, y_0, v) is called a solution of the game, x_0 and y_0 are said to be optimal strategies, and v is the value of the game.

Examples. (i) Consider the following matrix game.

$$\begin{pmatrix} -1 & 1 & 0 & 1 \\ 4 & 2 & 8 & 2 \\ 4 & 0 & 1 & 3 \\ 2 & 1 & 5 & 2 \end{pmatrix}$$

One can compute the row minima, column maxima, lower and upper values as follows:

$$\begin{pmatrix} & & & \\ & & & \\ & & & \\ & & & \end{pmatrix} \begin{matrix} -1 \\ 2 = v_I \\ 0 \\ 1 \end{matrix}$$

$$\begin{matrix} 4 & 2 & 8 & 3 \\ & \| & & \\ & v_{II} & & \end{matrix}$$

Since $v_I = v_{II} = 2 = a_{22} = v$, this game has a saddle point at (2, 2) and a value of 2.

(ii) Two athletes tie for the championship and are awarded with a single trophy worth \$6000. They agree to make sealed bids in multiples of \$1000 with the trophy going to the higher bidder and with the winner's bid being paid to the lower bidder. In case of a tie they will flip a coin for the trophy. If we use the expected value in the case of a tie, the game matrix in thousands of dollars is as follows:

	0	1	2	3	4	5	6
0	3	1	2	3	4	5	6
1	5	3	2	3	4	5	6
2	4	4	3	3	4	5	6
3	3	3	3	3	4	5	6
4	2	2	2	2	3	5	6
5	1	1	1	1	1	3	6
6	0	0	0	0	0	0	3

Each player has two optimal pure strategies; that is, to bid \$2000 or \$3000, and the value of the game is \$3000. The \$2000 bid "majorizes (dominates)" the \$3000 one.

(iii) A brewery worker can buy his beer before he leaves for his summer vacation through the brewery at a discount rate of \$3 per case, whereas it costs \$5 per case at the resort. He has no idea whether there will be cool or warm weather. If it is cool he will drink 10 cases during his vacation, and if it is warm he will drink 15 cases. Any left over is wasted since he drinks only at the brewery when not on vacation and there the beer is free. This game "against nature" has the following matrix.

	Cool	Warm
Buy 10 cases	-\$30	-\$55
Buy 15 cases	-\$45	-\$45

There is a saddle point at (Buy 15 cases, Warm) and the value is -\$45.

(iv) In matching pennies each player can pick heads H or tails T. Assume player I wins if the coins match and II wins otherwise. The game matrix is

	H	T
H	1	-1
T	-1	1

This game has no saddle point; $v_I = -1$ and $v_{II} = 1$. If I uses the mixed strategy $x = (\frac{1}{2}, \frac{1}{2})$ he will expect $E(x, j) = 0$ against either strategy j by II. Player II should use the same strategy. It is easy to check that these strategies are optimal and that the value of the game is 0. A player can effect this mixed strategy by flipping his coin, but he must not let the other know his choice in advance.

(v) The well-known children's game of scissors, paper, stone has the matrix

$$\begin{array}{c} \\ Sc \\ P \\ St \end{array}\begin{array}{c} \begin{array}{ccc} Sc & P & St \end{array} \\ \begin{pmatrix} 0 & 1 & -1 \\ -1 & 0 & 1 \\ 1 & -1 & 0 \end{pmatrix} \end{array}$$

The optimal strategies are (1/3, 1/3, 1/3) and the value is 0.

Solving Matrix Games

There are several methods for solving matrix games. A survey of some of these methods is given in Appendix 6 of Luce and Raiffa [11]. It was shown above that the problem of solving a game can be reduced to solving a pair of dual linear programs. This section illustrates how the "primal simplex method" for linear programming can be used to solve matrix games. There will be no detailed discussion of the simplex method in general, but only an illustration of how it works on a couple of specific examples. Those unfamiliar with linear programming can merely note that one proceeds repeatedly from one system of linear equations to another equivalent form until he reaches a system from which the desired solution appears obvious.

In this section we will assume that the game matrix $A = [a_{ij}]$ has $a_{ij} > 0$ for all i and j. This is necessary because we will divide through some inequalities by values which are assumed to be positive. One can prove that if he adds a number a to each entry in the matrix $[a_{ij}]$ to get the matrix $[a_{ij} + a]$, then these games both have the same optimal strategies and their values differ by a. This shows that one can add a constant to each entry in a matrix to obtain a new matrix $A > 0$, and thus v and $V > 0$; and solving this new game yields a solution for the original one by subtracting this constant from its value. One can also multiply all the entries in a game by a positive constant, for example, to clear it of fractions, since one can prove that the games $[a_{ij}]$ and $[ca_{ij}]$ for $c > 0$ have the same optimal strategies and that the value of the latter is c times the value of the former.

In the section entitled Matrix Games we saw that solving a game was equivalent to solving the dual linear programs in (3) and (4) for x, y and v (= V). In (3) one can divided through by v and define the m-dimensional vector

$$u = \frac{1}{v} x \tag{5}$$

to get

$$\begin{aligned} &uA \geq e_n \\ &ue_m^T = \frac{1}{v} \quad (\min \frac{1}{v}) \\ &u \geq 0 \end{aligned} \tag{6}$$

Similarly, one can let

$$w = \frac{1}{V} y \tag{7}$$

in (4) and divide by V to get

$$Aw^T \leq e_m^T$$

$$e_n w^T = \frac{1}{V} \quad (\max \frac{1}{V}) \tag{8}$$

$$w \geq 0$$

The "primal" problem in (8) can be expressed by the following "simplex table":

w_1	w_2	...	w_n	w_{n+1}	w_{n+2}	...	w_{n+m}	=	
a_{11}	a_{12}	...	a_{1n}	1	0	...	0	1	
a_{21}	a_{22}	...	a_{2n}	0	1	...	0	1	
		...				...		...	(min z)
a_{m1}	a_{m2}	...	a_{mn}	0	0	...	1	1	
-1	-1	...	-1	0	0	...	0	0	= -z

where all $w_k \geq 0$ and where

$$z = -1/V \tag{9}$$

The $w_{n+1}, w_{n+2}, \ldots, w_{n+m}$ are the "slack" variables. In matrix form this table is

w		
A	I_m	e_m^T
$-e_n$	o_m	z

where I_m is the m by m identity matrix and o_m is the m-dimensional vector of all zeros. One can solve the primal problem given in this table for optimal w and z. The final simplex table will also give the value of the optimal u for the dual problem in (6). From (5), (7), and (9) one can then compute the x, y, and v which will solve the game A. This method will now be illustrated by some examples.

Examples. If one adds 2 to the matrix for subsection (v) in the section entitled Examples, he obtains

$$A = \begin{pmatrix} 2 & 3 & 1 \\ 1 & 2 & 3 \\ 3 & 1 & 2 \end{pmatrix}$$

which has the table

w_1	w_2	w_3	w_4	w_5	w_6	
2	3	1	1	0	0	1
1	2	3	0	1	0	1
(3)	1	2	0	0	1	1
-1	-1	-1	0	0	0	0 = -z

One can arbitrarily pick the "pivot" element in the first column. The numbers in the last column (but not the last row) are divided by the respective positive elements in this first column, and the minimum occurs for the third row. Therefore, the circled number 3 in the third row is the pivot, and "pivoting" on it gives the second equivalent table

0	(7/3)	-1/3	1	0	-2/3	1/3
0	5/3	7/3	0	1	-1/3	2/3
1	1/3	2/3	0	0	1/3	1/3
0	-2/3	-1/3	0	0	1/3	1/3

The minimum in the last row occurs in the second column, and the minimum of the numbers in the last column divided by the positive numbers in the second column occurs in the first row. The next pivot is the circled number 7/3, and pivoting on it gives the third table

0	1	-1/7	3/7	0	-2/7	1/7
0	0	(18/7)	-5/7	1	1/7	3/7
1	0	5/7	-1/7	0	3/7	2/7
0	0	-3/7	2/7	0	1/7	3/7

The minimum in the last row occurs in the third column, and the minimum of the numbers in the last column divided by the positive numbers in the third column occurs in the first row. The next pivot is the circled number 7/3, and pivoting on it on it gives the fourth table

w_1	w_2	w_3				
0	1	0	7/18	1/18	-5/18	1/6
0	0	1	-5/18	7/18	1/18	1/6
1	0	0	1/18	-5/18	7/18	1/6
0	0	0	1/6	1/6	1/6	1/2

Since the last row is nonnegative, we have an optimal solution which is

$$w = (1/6,\ 1/6,\ 1/6), \qquad z = -1/2$$

The optimal solution u for the dual problem can be read off the last row under the slack variables, that is,

$$u = (1/6,\ 1/6,\ 1/6)$$

The optimal solution for the game A is

$$x = (1/3,\ 1/3,\ 1/3) = y, \qquad v = 2$$

The game

$$A = \begin{pmatrix} 7 & 5 & 6 \\ 0 & 9 & 4 \\ 14 & 1 & 8 \end{pmatrix}$$

is solved by the simplex method in the following tables. From the third table one determines

$$w = (0,\ 1/17,\ 2/17), \qquad u = (0,\ 7/68,\ 5/68), \qquad z = -3/17$$

The optimal solution is

$$y = (0,\ 1/3,\ 2/3), \qquad x = (0,\ 7/12,\ 5/12), \qquad v = 17/3$$

7	5	6	1	0	0	1
0	9	4	0	1	0	1
14	1	8	0	0	1	1
-1	-1	-1	0	0	0	0

-7/2	17/4	0	1	0	-3/4	1/4
-7	17/2	0	0	1	-1/2	1/2
7/4	1/8	1	0	0	1/8	1/8
3/4	-7/8	0	0	0	1/8	1/8

0	0	0	1	-1/2	-1/2	0
-14/17	1	0	0	2/17	-1/17	1/17
63/34	0	1	0	-1/68	9/68	2/17
1/34	0	0	0	7/68	5/68	3/17

Applications

The finite matrix games, along with their generalizations to the cases of infinitely many pure strategies and repeated play of such games, have proved useful in a great number of applications. These include allocation of resources (games of partitioning), duel theory (games of timing), search problems, as well as statistical decision theory. The theory of two-person, zero-sum games in normal form with finitely many strategies is mathematically equivalent to the theory of linear programming and consequently has as many uses. Two extremely simple examples will be given to illustrate such applications.

Examples. (i) A simple search game. Assume that there are n points on a line labeled (in order) 1, 2, ..., n. Player I picks one point at which to hide. Player II picks points, one at a time, at which to look for I. After each choice of a point, II is told whether he has found I, whether he has chosen a point too high (i.e., to the "right" of I), or whether he has chosen a point too low (i.e., to the "left" of I). The payoff to player I (from II) is the number of choices II must make in order to find I.

For player I there are n strategies: 1, 2, ..., n; i.e., the n places at which he can hide. A strategy for player II is a search plan or hunting strategy. This can be represented by an n-tuple with a "1" in the position of his first look, with a "2" to both the right and the left of the "1" which tells where his second look will be, with a "3" on both sides of each "2" (so that the "1" falls in the middle of the four "3's") which tells where his third look will be, etc. For example, if n = 7, then the hunting strategy

(2, 3, 1, 3, 4, 2, 3)

says that II's first look is in position 3, that if he is too high then his second look is in position 1, and if he is too low (on his first look) then his second look is in position 6, etc.

This game has been solved by S. M. Johnson for $n \leq 11$ in Ref. 24, 1964. The solution for the case when n = 3 follows. When n = 3, player I can pick position 1, 2, or 3 in which to hide. Player II has five hunting strategies:

(1, 2, 3) (1, 3, 2) (2, 1, 2) (2, 3, 1) (3, 2, 1)

The game matrix is

$$\begin{array}{cc} & \text{II} \\ \text{I} & \begin{array}{c} 1 \\ 2 \\ 3 \end{array} \begin{pmatrix} 1 & 1 & 2 & 2 & 3 \\ 2 & 3 & 1 & 3 & 2 \\ 3 & 2 & 2 & 1 & 1 \end{pmatrix} \end{array}$$

Note that the payoffs in each column give the same triple as the corresponding strategy for II. This game is solved in the following simplex tables. The optimal strategy for I is (.4, .2, .4) and for II it is (0, .2, .6, .2, 0). The value is 9/5.

1	1	2	2	3	1	0	0	1
2	3	1	3	2	0	1	0	1
3	2	2	1	1	0	0	1	1
-1	-1	-1	-1	-1	0	0	0	0
1/2	1/2	1	1	3/2	1/2	0	0	1/2
3/2	5/2	0	2	1/2	-1/2	1	0	1/2
2	1	0	-1	-2	-1	0	1	0
-1/2	-1/2	0	0	1/2	1/2	0	0	1/2
-1/2	0	1	3/2	5/2	3/2	0	-1/2	1/2
-7/2	0	0	9/2	11/2	2	1	-5/2	1/2
2	1	0	-1	-2	-1	0	1	0
1/2	0	0	-1/2	-1/2	0	0	1/2	1/2
2/3	0	1	0	2/3	5/6	-1/2	1/3	1/3
-7/9	0	0	1	11/9	4/9	2/9	-5/9	1/9
11/9	1	0	0	-7/9	-5/9	2/9	4/9	1/9
1/9	0	0	0	1/9	2/9	1/9	2/9	5/9

(ii). A duel. Consider a finite duel in which each player has a gun which contains one bullet. Player I has a silent gun, i.e., II does not know when I has fired; and player II has a noisy gun, i.e., I knows if II has fired. At the start the players are 10 steps apart. Assume that both players have the same accuracy function: after a player's k-th step his probability of hitting his opponent is k/5 where k = 0, 1, 2, 3, 4, 5. Any player who hits his opponent collects one unit from this opponent, and the duel then terminates.

If player I shoots after his i-th step and player II shoots after his j-th step, then the payoffs (in expected values) from II to I are:

$$a_{ij} = \frac{i}{5} - (1 - \frac{i}{5})\frac{j}{5} \qquad \text{when } i < j$$

$$a_{ij} = 0 \qquad \text{when } i = j$$

$$a_{ij} = (1 - \frac{j}{5}) - \frac{j}{5} \qquad \text{when } i > j$$

The game matrix (multiplied by 25) is

i \ j		0	1	2	3	4	5
		II					
	0	0	-5	-10	-15	-20	-25
	1	25	0	-3	-7	-11	-15
	2	25	15	0	1	-2	-5
I	3	25	15	5	0	7	5
	4	25	15	5	-5	0	15
	5	25	15	5	-5	-15	0

The third row (i = 2) "dominates" the first and second row, and the fourth row dominates the sixth row. These dominated rows can be deleted without losing all of the optimal solutions for this game. Similarly, the first two columns can be deleted because they dominate the third column. It is sufficient to solve the game:

i \ j	2	3	4	5
2	0	1	-2	-5
3	5	0	7	5
4	5	-5	0	15

Adding 5 to each entry in the matrix gives the matrix

$$B = \begin{pmatrix} 5 & 6 & 3 & 0 \\ 10 & 5 & 12 & 10 \\ 10 & 0 & 5 & 20 \end{pmatrix}$$

The latter game B is solved in the simplex tables

5	6	3	0	1	0	0	1
10	5	12	10	0	1	0	1
10	0	5	20	0	0	1	1
-1	-1	-1	-1	0	0	0	0
5	6	3	0	1	0	0	1
5	5	19/2	0	0	1	-1/2	1/2
1/2	0	1/4	1	0	0	1/20	1/20
-1/2	-1	-3/4	0	0	0	1/20	1/20

-1	0	-42/5	0	1	-6/5	3/5	4/10
1	1	19/10	0	0	1/5	-1/10	1/10
1/2	0	1/4	1	0	0	1/20	1/20
1/2	0	23/20	0	0	1/5	-1/20	3/20
-5/3	0	-14	0	5/3	-2	1	2/3
5/6	1	1/2	0	1/6	0	0	1/6
7/12	0	19/20	1	-1/12	1/10	0	1/60
5/12	0	9/20	0	1/12	1/10	0	11/60

From the last table one sees that the game B has value $v_B = 60/11$ and optimal strategies (1/12, 1/10, 0)(60/11) = (5/11, 6/11, 0) and (0, 1/6, 0, 1/60)(60/11) = (0, 10/11, 0, 1/11) for I and II, respectively. It follows that the original game matrix has value $v = v_B - 5 = 5/11$ and optimal strategies (0, 0, 5/11, 6/11, 0, 0) and (0, 0, 0, 10/11, 0, 1/11) for I and II, respectively. Since the original game matrix was obtained by multiplying by 25, one gets that the value of the duel is 1/55. Player I shoots with probabilities 5/11 and 6/11 after steps 2 and 3. Player II shoots with probability 10/11 after step 3, and with probability 1/11 he walks right up to I where he has a sure shot (if I has not already killed II).

In many duels and search problems the players have an infinite number of pure strategies; for example, all the points on an interval or all the continuous curves in the plane. Examples of such duels are given in Refs. 5 and 6.

Two-Person, General-Sum Games

A two-person general-sum game in normal form with a finite number of strategies is characterized by an m by n payoff matrix in which each entry is a pair of numbers corresponding to the payoffs to the two players. Such games can likewise be described by a pair of real valued matrices A and B, and hence are referred to as the bimatrix games. The rows and columns respectively correspond to the pure strategies for the players I and II. One can also consider mixed strategies for each player which are probability distributions over his set of pure strategies.

The best known example of general-sum games is the two-person prisoner's dilemma, due originally to A. W. Tucker in 1950 and since studied in great detail by Rapoport [19] and many others. Two men, I and II, are charged with a joint crime, and held separately by the police. Each is told that

(a) If one confesses and the other does not, then the former will receive a reward of one unit and the latter will be fined two units.
(b) If both confess, each will be fined one unit.

The prisoners know however that

(c) If neither confesses, both will go free.

The resulting payoff matrix follows.

		II	
		Confess	Not Confess
I	Confess	(-1, -1)	(1, -2)
	Not confess	(-2, 1)	(0, 0)

The first component in each payoff vector is for player I whereas II obtains the second component. For each man, the strategy "Confess" dominates the strategy "Not confess." If played noncooperatively, both players are likely to confess, resulting in each being penalized one unit. Technically, the payoff (-1, -1) is the unique "equilibrium" point in this game. On the other hand, if the players are allowed to play cooperatively, i.e., to communicate (perhaps via lawyers) and form binding agreements, then it is likely that they will not confess and both will go free. The latter outcome, which appears on the southeast edge of the "payoff space," is clearly the best solution from a global perspective.

Repeated play of the prisoner's dilemma game provides a miniature model of what happens in arms races, price wars, run-away advertising campaigns, and the "tragedy of the commons," in which each individual citizen uses more than his share of a renewable resource, thus causing its ultimate depletion. One obtains an interesting three-person game if he adds the State (or world community) to the above prisoner's dilemma, where the State's payoff depends upon the nature of the fines assessed.

When a game is general-sum or involves more than two players, then it is essential to distinguish whether this game is played cooperatively or noncooperatively. A cooperative game of two players involves elements of both cooperation and competition. The players normally agree to restrict the outcomes to "Pareto optimal" ones, i.e., payoff vectors on the "northeast" boundary of the set of all potential payoffs. They then must bargain over which payoff on this boundary is a distribution of wealth acceptable to both parties. This latter settlement is noncooperative in the sense that increasing one player's payoff decreases the other's outcome. Several methods for determining an equitable payoff vector from the Pareto optimal ones have been proposed. Chapter 6 in the book by Luce and Raiffa [11] is still an excellent discussion of many such cooperative, or bargaining solution, concepts. When there are more than two players in a cooperative game, then the theory changes drastically, since coalition formation then becomes a major aspect of the problem. Such multiperson cooperative games will be discussed in the section entitled The Characteristic Function Form.

n-Person Noncooperative Games

For noncooperative games in normal form, the main solution concept is that of an equilibrium strategy. Assume that each of the n players ($n \geq 2$) has chosen a particular strategy (pure or mixed). This strategy n-tuple is in equilibrium if no player can improve his payoff by he alone switching his strategy choice. That is, a unilateral deviation by just one player cannot improve his situation, if the other players continue to play these particular strategies. The strategy pair (Confess, Confess) is an equilibrium outcome in the prisoner's dilemma game. The optimal strategies in a two-person, zero-sum game are precisely the equilibrium outcomes.

Nash [14, 15] proved that every game with a finite list of pure strategies for each player must have at least one equilibrium n-tuple in mixed strategies. The problem of actually determining such equilibrium strategies corresponds to optimizing a multilinear function with inequality constraints. Precise algorithms exist for the case of two players, and recent algorithms for approximating fixed point values are often used currently for larger values of n.

Equilibrium points are important in applications of game theory to problems in fields such as economics and operations research where they may correspond to prices in an economic market, optimal bidding strategies, or the best allocation of vehicles over a road network.

One can read more about the theory of equilibrium points for noncooperative games in the books by Burger [2] and by Parthasarathy and Raghavan [17].

THE CHARACTERISTIC FUNCTION FORM

Introduction

The first detailed game theoretical model for multiperson cooperative games was presented in 1944 by von Neumann and Morgenstern and is called the cooperative, coalitional, or characteristic function form of a game. This model, which comprises the greatest part of their text [25], is employed in the study of situations involving three or more participants acting in a cooperative mode. In such cooperative behavior the agents are allowed to get together in coalitions and to undertake joint action for the purpose of mutual gain.

The estimated worth of any potential coalition is a crucial aspect of such models, and this value is expressed as a numerical measure by what is called the characteristic function. The term <u>von Neumann-Morgenstern solution</u> or simply <u>solution</u> in this context refers to the final solution concept or end product for their coalition formulation of cooperative games. A brief intuitive and verbal description of their model is presented first, and then a more precise technical definition is given.

An Overview of the Model

There are four essential concepts or fundamental definitions which make up the basic cooperative model. First, a multiperson <u>game</u> is characterized by merely assigning a real number to each possible coalition of the n participants, where each number represents the value, wealth, or power achievable by this coalition when its members cooperate. Next, one describes a set of n-dimensional payoff vectors called <u>imputations</u>. This presolution set represents all reasonable or realizable ways of distributing the available wealth among the n parties. Then, one introduces a preference relation between certain pairs of imputations. One imputation is said to <u>dominate</u> another if there is some coalition in which each of its members prefers the former payoff to the latter one, and if this coalition as a whole is not obtaining more in the former imputation than it can effectively realize in the game through its own efforts. Finally, a <u>von Neumann-Morgenstern solution</u> is any subset of the imputation space which possesses a certain internal and external consistency; namely, no imputation in a solution dominates another one in this solution, and any imputation not in the solution is dominated by another one within the solution. So a solution when taken as a whole is preferred to precisely those imputations outside of this set. That is, a solution set is the complement of its "dominion."

The Basic Model

An n-person game (in characteristic function form) consists of a pair (N, v) where $N = \{1, 2, \ldots, n\}$ is a set of n players labeled by 1, 2, ..., n, and where v is a characteristic function which assigns a real number v(S) to each nonempty subset (or coalition) S of N. (Note that n again represents the number of players in the game, rather than the number of strategies for player II as was the case in the section entitled The Normal Form.) One normally takes the integer $n \geq 3$, and assumes that v is superadditive, i.e., $v(S \cup T) \geq v(S) + v(T)$ for disjoint subsets S and T, but this latter assumption is not necessary for much of the theory. The set of imputations, denoted by A, consists of all vectors x such that $x_1 + x_2 + \cdots + x_n = v(N)$ and $x_i \geq v(\{i\})$ for all i in N, where each x_i is a real number and represents a payoff to player i. These two relations are referred to as Pareto optimality and individual rationality, respectively. An imputation x dominates an imputation y whenever there is some coalition S such that $x_i > y_i$ for all i in S and $\Sigma_{i \in S} x_i \leq v(S)$. When this latter condition is satisfied, one says that x is effective for S. A subset V of A is called a von Neumann-Morgenstern solution if it satisfies two conditions: no imputation x in V dominates any imputation in V, and every imputation y not in V is dominated by at least one imputation in V. These conditions are called internal stability and external stability, respectively, and now one frequently refers to such a solution as a stable set.

Example. An illustrative example, known as the three-person "constant-sum" game, is given by $N = \{1, 2, 3\}$, $v(\{1, 2, 3\}) = v(\{2, 3\}) = v(\{1, 3\}) = v(\{1, 2\}) = 1$, and $v(\{i\}) = 0$ for i = 1, 2, and 3. Intuitively, any coalition of two or three players has the power to distribute the total value 1 among the three players, e.g., a decision by majority rule, whereas an individual player by himself is assured of only his value 0. The imputation space consists of the triangular set or simplex $A = \{(x_1, x_2, x_3): x_1 + x_2 + x_3 = 1 \text{ and } x_i \geq 0 \text{ for } i = 1, 2, 3\}$. One solution, which has only three imputations as well as the "symmetry" of the game, is the set $V = \{(0, \frac{1}{2}, \frac{1}{2}), (\frac{1}{2}, 0, \frac{1}{2}), (\frac{1}{2}, \frac{1}{2}, 0)\}$. Any other solution for this game consists of all those imputations which give a fixed amount $x_i = c_i$, with $0 \leq c_i < \frac{1}{2}$, to one player i, and distributes nonnegative amounts x_j and x_k to the other two players so that $x_j + x_k = 1 - c_i$, and each of these latter sets is referred to as a discriminatory solution.

Nature of the Theory

To develop a mathematical theory for a given game model like the one above, one searches for answers to questions such as those about the existence, uniqueness, nature, mathematical properties, structure, and various representations and characterizations for the ensuing solution concepts. From the applied point of view, one is concerned with the computability and interpretation of solution sets, as well as with whether they provide any new insights into theory, or in practice when applied to situations in the real world.

It was finally demonstrated by Lucas [8, 9] that not every game need have a solution. However, all such known examples have ten or more players, and it appears that the nonexistence of solutions is rare. On the other hand, some solution sets have been characterized for several special but large classes of games. A good sample of the research in this direction is contained in several papers in four of the five volumes devoted to game theory in the Annals of Mathematics Studies edited by Tucker and others [24]. It turns out that many games have a great abundance of different solution sets. This lack of uniqueness, plus the great multiplicity of imputation in most individual solutions, leaves a great arbitrariness or ambiguity in determining any ultimate payoff vector for many games. Furthermore, some solutions are known to have a very complex or elaborate structure as is known from work of Shapley, and it is unlikely that one will be able to give a plausible intuitive or behavioralistic interpretation to all such irregular sets, or to design algorithmic methods for constructing them. It appears in general that solution theory is a deep and difficult mathematical subject and that many rich and fascinating structures are possible. A very readable introduction to solution theory as well as the most complete bibliography on the subject will appear in Chapter 6 of a forthcoming book by Shapley and Shubik. A preliminary version by the authors [21] has appeared in report form. A more technical survey of some important recent results on this subject was written by Lucas [10].

Several generalizations, variations, and extensions of the classical von Neumann-Morgenstern theory of solutions have also been investigated. These include in part: the games without side payments surveyed by R. J. Aumann in Ref. 22 and included in Shapley and Shubik [21], the games in partition function form presented by Thrall and Lucas [23], more dynamical approaches to solution concepts as discussed in the work of Weber [27], the game models which make use of an infinite number of players, and models built upon other abstract mathematical structures. In addition, several different types of models and alternate solution concepts along the lines of the classical formulation for games in coalition form have since been introduced. The core, the value theories, the bargaining sets, and the nucleolus for cooperative games are discussed in the next section.

Other Solution Concepts

Several other solution concepts have been developed since the original von Neumann-Morgenstern solutions (or stable sets). Four of these alternate solution ideas, which are well known and have use in applications, will be introduced in this section. These models make use of the same definition for a game (N, v) and for the set of imputations A as those given above.

The Core

The concept of the core of a game was defined explicitly by D. B. Gillies in Ref. 24 and Shapley in 1953. The core of the game (N, v) is the set

$$C = \{x \in A: \sum_{i \in S} x_i \geq v(S) \text{ for all } S \subset N\}$$

No coalition can overturn an agreed upon imputation in the core, since no element in A can dominate an imputation in C. For superadditive games, the core is precisely those imputations which are maximal with respect to the relation of dominance.

Consider the three-person game with $v(\{1, 2, 3\}) = v(\{1, 2\}) = v(\{1, 3\}) = 0$ and $v(\{2, 3\}) = v(\{1\}) = v(\{2\}) = v(\{3\}) = 0$. For this example, C consists of the one imputation (1, 0, 0) which gives the full amount to player 1. In this "veto-power" game, C treats 1 as though he were a "dictator."

A main problem with the core is that for many games it turns out to be the empty set. This is true, for example, in the three-person, constant-sum game in the Example on page 381. In applications of game theory to political science, many games have an empty core, whereas many examples of games which arise in economics do have a nonempty core, and the core then proves to be an important analytical tool. However, the example in this section illustrates that a nonempty core may sometimes prove to be "too small" and not to contain the outcomes which would likely arise from actually playing the game.

The Shapley Value

In 1953 Shapley [24] introduced a solution concept which determines a unique imputation ϕ for each game (N, v). This outcome is now called the <u>Shapley value</u> and is given by the formula

$$\phi_i = \sum \frac{(s - 1)!(n - s)!}{n!} [v(S) - v(S - \{i\})]$$

for each $i \in N$, where s is the number of players in coalition S and the summation is taken over all subsets S of N which contain the player i. The value ϕ is the unique imputation which satisfies four axioms stated by Shapley which are called efficiency, the dummy axiom (i.e., powerless players receive nothing), symmetry, and additivity. These axioms are properties which would be desirable of any equitable allocation scheme. The Shapley value can thus be viewed as a fair-division solution concept. It has been used in economic problems to set fair rates for services and in political science to assign power equitably. It also has a probabilistic interpretation as the average incremental gain caused by player i joining coalition S - i to obtain S. Furthermore, ϕ also arises from certain fair bargaining schemes, i.e., where players split equitably any gains from all potential ways of building up from n separate players to the grand coalition N.

The Nucleolus

Another solution concept which always gives a unique imputation for any game (N, v) is the nucleolus which was introduced by Schmeidler [20]. First, define the <u>excess</u> of any nonempty subset S of N with respect to the imputation $x \in A$ as $e(x, S) = v(S) - \Sigma_{i \in S} x_i$. One can set $e(x, \emptyset) = 0$ for the empty set $\emptyset$, and note that $e(x, N) = 0$. The excess represents the "group attitude" of the coalition S toward its payoff at imputation x. In the core where each coalition S is satisfied, $e(x, S) \leq 0$. If S gets less than its value v(S) at some imputation x, then the corresponding excess is negative and S has incentive to form and ask for more. The <u>nucleolus</u> is the imputation which minimizes the largest excess, i.e., it simultaneously minimizes the largest coalitional "complaint." In the event that this maximum excess attains a minimum at various imputations, then one compares the next largest excess, and so on, in order to arrive at the unique nucleolus. In other words, for each x one arranges the excesses e(x, S) in nonincreasing order. The nucleolus is then the one particular x which gives a minimum in the "lexicographical" ordering of these arrangements.

The nucleolus always exists and is unique for each game. It is an element in the core when the latter is nonempty, and it is also a member of many, but not all, of the bargaining sets mentioned below. It can be computed by solving a sequence of linear programming problems or equivalently one very large program, but such computations are lengthy unless n is small. In applications, the nucleolus can also be used to set rates for services and appears to be a good approach for setting tax rates. It has been proposed as a scheme for setting fines or other charges for firms that pollute the environment.

Bargaining Sets

In 1964 R. J. Aumann and M. Maschler published in Ref. 24 some dozen different solution concepts referred to as bargaining sets. Their models imitate rather closely what actually takes place in real bargaining situations, and are similar to what happens in game theoretical experiments. They also consider more explicitly how the set N of players fragments into <u>coalition structures</u>, i.e., into partitions $P = \{P_1, P_2, \ldots, P_m\}$ of N, where the P_i are disjoint coalitions whose union is N. They also enlarge the set of imputations to include more vectors than given by A. Every such extended imputation x along with a corresponding coalition structure P will be called a <u>payoff</u>. A payoff (x, P) is <u>stable</u>, or is an element of a <u>bargaining set</u>, if it can be "defended" against various other payoffs which will be proposed instead.

The crucial concepts in their models are the relations of objection and counterobjection. These technical definitions are rather lengthy, and thus only intuitive statements about them are presented here. An <u>objection</u> by a coalition K to a disjoint coalition L at a payoff (x, P) is a second payoff (y, Q) in which each player in K obtains more and in which each player they must use to achieve this new payoff does at least as well. A <u>counterobjection</u> by L to K in this objection is a third payoff (z, R) in which the players in L obtain at least their original amounts in x and in which any "partners" in R whose cooperation L needs are not worse off than they were in x or y. A payoff is <u>stable</u> if for each objection to it there exists a corresponding counterobjection. A <u>bargaining set</u> is the set of those payoffs which are stable. Since there are different ways to define payoff, objection, and counterobjection, there are several different solution concepts which are called bargaining sets.

So a payoff is not unstable merely because some group can object to it. Many such objections really carry little weight, since they can easily be countered. A payoff is unstable, however, if there is any objection to it which cannot be countered.

An Application to Politics

A game (N, v) in characteristic function form is called a <u>simple</u> game if v(S) = 0 or 1 for each coalition S contained in the player set N. A simple game is said to be <u>monotone</u> if $S \subset T$ implies $v(S) \leq v(T)$. Monotone simple games can be used to model many voting situations: a <u>losing</u> coalition S has value v(S) = 0 and a <u>winning</u> coalition S is assigned value v(S) = 1. A particular class of such voting games is the <u>weighted voting</u> games in which each player i in N casts a ballot with weight w_i and in which a coalition S wins whenever its total weight $\Sigma_{i \in S} w_i$ equals or exceeds some given number q called the quota.

Value-type solution concepts, such as the Shapley value, can be used as indices to measure the voting powers of the various players. One such value concept, proposed by the lawyer Banzhaf [1], has actually been accepted in several court rulings in the United States as a reasonable measure of power for determining equity in certain weighted voting systems.

As an illustration, consider a corporation with five stockholders denoted by 1, 2, 3, 4, and 5 who own 40, 25, 20, 10, and 5% of the stock, respectively. Assume that a simple majority is necessary to pass any legislation. The minimal sized winning coalitions are $\{1, 2\}$, $\{1, 3\}$, and $\{2, 3, 4\}$. The Shapley value for this game is the vector $\phi = (5, 3, 3, 1, 0)/12$, and it can be defended as a reasonable measure of the respective voting powers for the players. Note that player 5 is powerless; if he alone switches his vote, he will never change the outcome on any ballot.

An Application to Economics

In a pure exchange economy each trader brings a bundle of commodities to a market, and these participants proceed to redistribute their goods. Presumably the goal of each such economic agent is to depart with a commodity bundle which he prefers over his initial one. Assume furthermore that each such trader has a utility function which expresses his preferences over the possible bundles which he can secure, and assume that each one attempts to maximize his resulting utility. This situation can be treated as a multiperson cooperative game in which the characteristic function value v(S) of each coalition S is the maximum total utility to the players in S if they achieve an optimal trade with those goods available within their coalition. To simplify our model, assume that the value to coalition S is the sum of the individual utilities of the players in S for the bundles they each receive.

As a simple illustration, consider a friendly coffee break with three participants denoted by 1, 2, and 3. Each participant brings certain ingredients and each has a different preference for what he wishes to drink. Let the commodities be coffee, tea, cream, and sugar. Player 1 has two units of coffee, and prefers to drink tea with sugar. Player 2 has one unit of tea, and wishes to drink coffee with cream. Player 3 has two units of cream and three units of sugar, and desires to drink coffee with cream and sugar. One can describe a typical commodity vector as $y = (y_1, y_2, y_3, y_4)$, and denote the initial bundles for the three players by $w^1 = (2, 0, 0, 0)$, $w^2 = (0, 1, 0, 0)$, and $w^3 = (0, 0, 2, 3)$. And their respective utility functions will be assumed to be $u^1(y) = \max\{y_2, y_4\}$, $u^2(y) = \max\{y_1, y_3\}$, and $u^3(y) = \max\{y_1, y_3, y_4\}$. Then the total utility available to a coalition $S \subset \{1, 2, 3\}$ is $\max \Sigma_{i \in S} u^i(y^i)$ where y^i is player i's bundle and where this maximum is subject to the constraint, or conservation law, $\Sigma_{i \in S} y^i = \Sigma_{i \in S} w^i$. Under these assumptions the characteristic function for this game is $v(\{1\}) = v(\{2\}) = v(\{3\}) = v(\{1, 2\}) = v(\{2, 3\}) = 0$, $v(\{1, 3\}) = 2$, and $v(\{1, 2, 3\}) = 3$. The set of imputations is $A = \{(x_1, x_2, x_3): x_1 + x_2 + x_3 = 3;\ x_1, x_2, \text{ and } x_3 \geq 0\}$, and the core and unique stable set is $C = V = \{x \in A: x_1 + x_3 \geq 2\}$. No distribution x in C can be improved upon by some coalition which decides to go it alone.

GAMES IN EXTENSIVE FORM

Graph Theory

A few very simple ideas from graph theory will be mentioned briefly, since a game in extensive (or tree) form is usually defined in terms of such concepts. A graph consists of a set of points, called vertices or nodes, along with a set of (unordered) pairs of distinct vertices, called edges or arcs. The edges can be viewed as lines connecting the corresponding pairs of vertices. The degree of a vertex is the number of distinct edges in which it appears, i.e., the number of lines emanating from this vertex. A graph has a cycle if one can "pass" through a sequence of vertices and return to the original one by means of distinct edges. Such a closed path will not retrace any of its edges. A graph is connected if one can pass from any vertex to another one by means of a sequence of adjacent edges.

Definitions

A general n-person game in extensive form is a topological tree (i.e., a finite connected graph with no cycles and with no vertices of degree two) with the following specifications. There is one distinguished vertex corresponding to the starting point, and referred to as the root. Each nonterminal vertex is a choice point and is labeled with one of the n players 1, 2, ..., n; or by the "chance player" denoted by 0. Each edge "leading out" of a vertex describes a possible move by the corresponding player if this point in the game is reached. Each vertex labeled by 0 has a probability distribution over its moves. To each terminal vertex there is assigned an n-dimensional payoff vector whose components describe the outcomes to the respective players when the game ends there. The set of all vertices for a particular player i is partitioned into information sets. When it is one of his turns to make a choice, i is aware of the information set he is in, but he does not know the precise vertex within this set. For each vertex in a given information set there must be the same number of potential moves, and one normally assumes that in playing a game no such vertex can come "after" another one in the same information set.

When engaged in a particular game, each player is faced with the problem of how to best play the game in order to maximize his expected payoff. A player's complete plan for playing a game is called a strategy. There are several different ways in which one can specify such a coherent plan of action. A pure strategy for player i is a rule for picking a particular move at each of his information sets. It is a function which assigns to each information set a particular one of his available choices. A mixed strategy for i is a global randomization over his pures; that is, a probability distribution over his set of all pure strategies. A behavioral strategy for i is an overall plan for local randomization; that is, a behavioral strategy consists of a class of probability distributions such that one particular distribution is assigned to the set of moves at each one of his information sets. Certain combinations of "partial" pure strategies called "signaling strategies" and "associated" behavioral strategies are called composite strategies.

The main solution concept for games in extensive form is the same Nash equilibrium point which was introduced in the section entitled The Normal Form; i.e., a collection of n optimal strategies (one for each player) such that a unilateral change in strategy by only one player cannot possibly improve his expected payoff. The principal computational problem is to determine such equilibrium points for a given game.

Information and Existence Theorems

Most of the major results about extensive games are of a theoretical nature. These theorems guarantee the existence of optimal strategies which will achieve an equilibrium. In 1912 Zermelo [29] demonstrated the existence of an optimal pure strategy for two-person zero-sum games with perfect information; that is, games such as checkers or chess in which all information sets contain a single vertex. "Perfect information" implies that a player can always reconstruct the entire past history of the game along the unique path from the initial vertex (root) up to his current choice vertex. Kuhn [7; 24, 1953] extended these results about optimal pure strategies to the n-person general-sum games with perfect information. In 1928 von Neumann showed the existence of optimal mixed strategies for any two-person zero-sum game; this is the Minimax Theorem or Fundamental Theorem of Game Theory mentioned before. Nash [14, 15] extended the equilibrium concept beyond the two-person zero-sum games, and he proved that there are optimal mixed strategies in the n-person general-sum case. Kuhn also showed the existence of optimal behavioral strategies for games with "perfect recall" such as in poker. A game has perfect recall if each player is "aware" at each of his moves of precisely what moves he chose prior to it, but may not know all the prior choices made by the other players. G. L. Thompson [24, 1953] proved that composite strategies suffice to obtain an equilibrium point for an arbitrary game, and he illustrated his theory for simplified models of bridge in which partners are treated as single players who alternately "forget" certain information.

Examples. The following examples will illustrate some of the definitions and theorems presented above.

(i) Consider a three-person game with perfect information in which the players are denoted by I, II, and III (Fig. 1). Player I begins by choosing either the upper

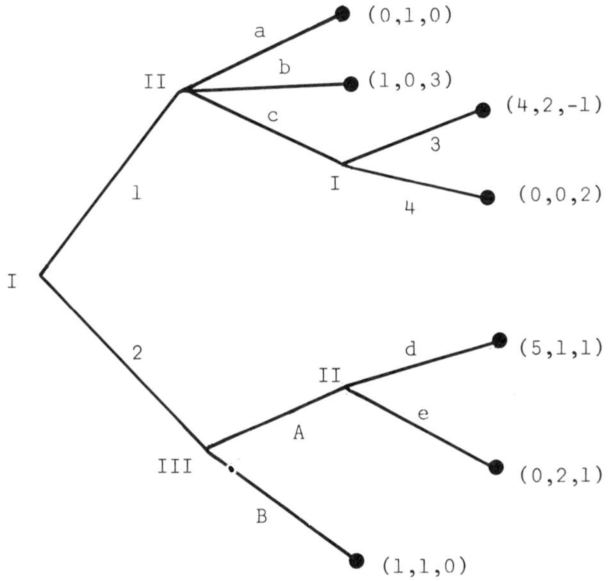

FIGURE 1.

branch labeled 1 or the lower branch labeled 2. In the former case, II then picks a, b, or c. Choice a or b will end the game with the payoff vectors (0, 1, 0) and (1, 0, 3), respectively, whereas choice c leaves the final move up to player I who then picks 3 or 4 to terminate at outcomes (4, 2, -1) or (0, 0, 2). If I had initially chosen the lower branch 2, then player III chooses between moves A and B. Move B ends the game, whereas choice A leaves the final move to player II who picks either d or e. The equilibrium payoff (4, 2, -1) is obtained when players I, II, and III play their pure strategies (1, 3), (c, e) and A, respectively.

The proof that a (finite) game with perfect information has a pure strategy optimal solution follows from a simple backwards induction; i.e., the game tree can be repeatedly "pruned" back from the terminal vertices to the root. In our example, the branch

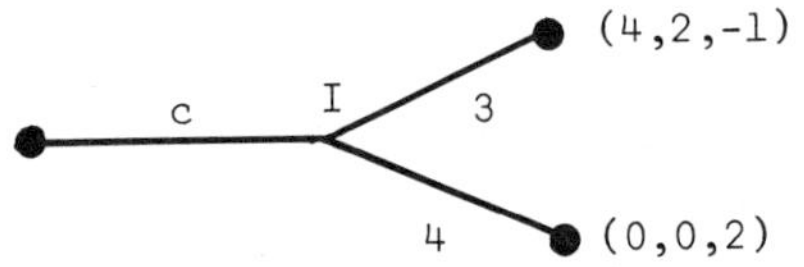

can be replaced by

since a rational player I will surely use alternative 3 instead of 4. One can work back in this way from the terminal notes to the initial one.

(ii) The well-known two-person game of matching pennies can be described as in Fig. 2. Player I picks either heads H or tails T, and player II has the same choices H and T. Player I wins if the coins match and II wins if they do not match. The information set which encloses the two vertices for II indicates that he is unaware of I's choice; e.g., I and II may in practice move simultaneously. Optimal mixed (or behavioral) strategies for this game are $(\frac{1}{2}, \frac{1}{2})$ and $(\frac{1}{2}, \frac{1}{2})$; i.e., each plays H and T with equal probability. The value of the game is 0, which indicates that it is a fair game. In this game, player II is not sure of where he is because of an unknown move by I, and not because he forgot his own choice at some prior vertex.

(iii) The two simple examples of Fig. 3 indicate how one can forget what he had done or known at an earlier stage of a game. One who is unable to so reconstruct his past history of play when at one of his information sets does not have perfect recall. The payoffs are not listed in these examples. In the example on the left, I forgets what action he took at an earlier stage, and in the example on the right, II forgets what what he knew at the previous move.

(iv) The example of Fig. 4, given by Kuhn, indicates how behavioral strategies need not suffice to achieve the best solution in a game without perfect recall. Chance, denoted by 0, deals a high card H to one player and a low card L to the other, with each deal being equally probable. The holder of the high card then receives 1 unit from the other player, and also has the choice of whether to stop the game s or to continue it c. In the latter case, player I has the option of leaving the cards the way they are, ℓ, or of trading cards with his opponent, t. However, at this stage of the game, I has forgotten who originally held the high card.

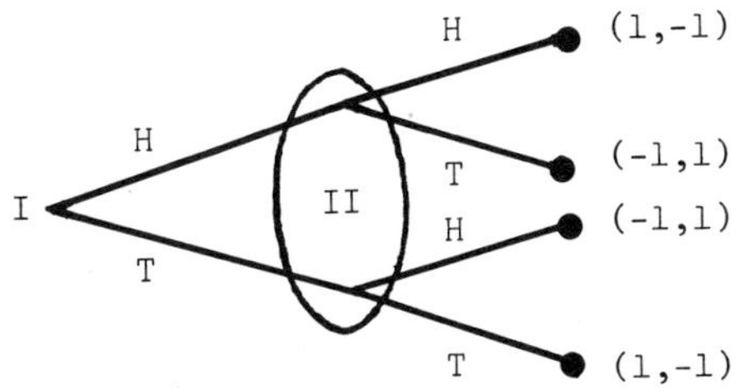

FIGURE 2.

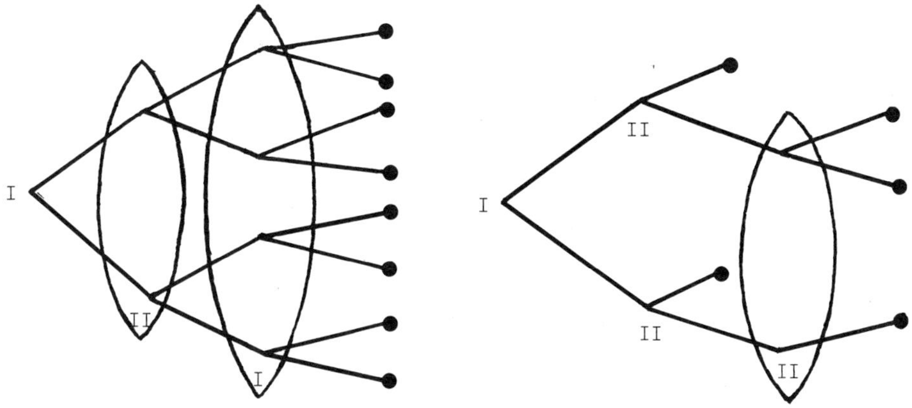

FIGURE 3.

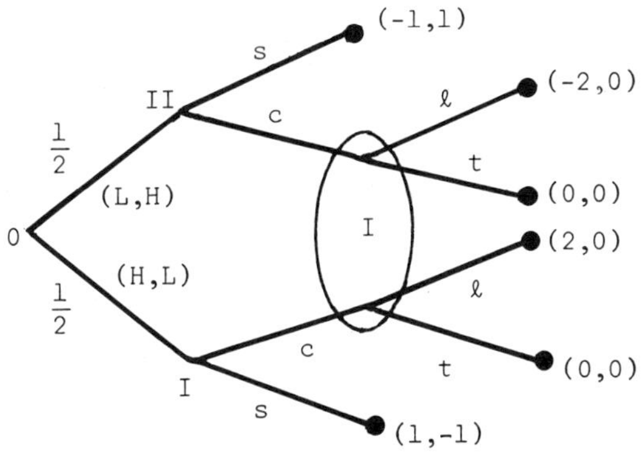

FIGURE 4.

The optimal mixed strategy for player I is to mix his pure strategies (s, t) and (c, ℓ) with equal probability, and II should play both s and c with probability 1/2. An expected payoff to player I of 1/4 will result. One can show, however, that if player I were restricted to a behavioral strategy, i.e., one probability distribution over c and s and another over ℓ and t, then the best expected value which he can assure himself is 0.

It is true that each mixed strategy induces a behavioral strategy in a natural way, and each behavioral strategy induces a particular mixed strategy which in turn reinduces that same behavioral strategy. However, if a game lacks perfect recall, then not all mixed strategies can be induced from behavioral strategies.

Applications

When a game arises in applications, one frequently begins his analysis by describing the game in extensive form. This gives a rather complete picture of the problem, including the detailed structure or rules of the game, the sequence of all possible moves, the state of each player's information, and the payoffs. Except for very elementary games, however, one does not normally pursue his investigation in the extensive form. This is because the mathematical problem of actually determining optimal strategies for this form in an efficient manner has not been resolved.

Instead, one reduces the game to its normal form by considering the full list of pure strategies for each player and the resulting payoffs when these strategies are employed. In theory it is a much simpler problem to compute optimal strategies in the normal form. As mentioned in the case of the two-person zero-sum games, it is equivalent to solving a pair of dual linear programs. Good algorithms also exist in the case of two-person, general-sum games. For larger numbers of players it is usually difficult to determine optimal solutions, and this is still an active area of research for games in normal form. Furthermore, in practice one usually gets an enormous increase in the number of parameters necessary to describe a mixed strategy in the resulting normal form compared to the number of variables necessary to determine a behavioral strategy in the original game tree. For example, Kuhn has described a trivial poker game in which the simplex of mixed strategies has dimension 8191 while the "cube" of behavioral strategies has dimension 13. This astronomical increase in the number of variables to be determined actually occurs in some important real-world problems and often forces the analyst to abandon the game theoretic approach.

In addition, even if the reduction to normal form results in a problem which is computationally feasible, there is still a major concern about whether this approach is philosophically sound. It is true that for games with perfect recall an optimal mixed strategy induces a behavioral strategy which will achieve the same expected payoff. It is questionable, at least in some people's minds, whether one would or should always play throughout the game according to this latter strategy.

Some of the many applications of the games in extensive form have been to areas of pure mathematics, including topics in logic as well as in the very foundations of mathematics. Certain problems in the area of combinatorial games relate to questions on computational complexity. Some of the developments in this direction are indicated in the recent book by Conway [3]. It has been shown that certain game trees of <u>infinite</u> length and perfect information need not have a pure strategy optimal solution if one assumes the axiom of choice. This has given rise in turn to alternate assumptions and to questions about the relationship between such fundamental assumptions, which are most basic to much modern mathematics. A paper by Mycielski [13] gives an indication of some results and problems in this latter direction.

REFERENCES

1. Banzhaf, J. F., One man, 3.312 votes: A mathematical analysis of the electoral college, Villanova Law Rev. 13, 303-332 (Winter 1968).
2. Burger, E., Introduction to the Theory of Games, Prentice-Hall, Englewood Cliffs, New Jersey, 1963.
3. Conway, J. H., On Numbers and Games, Academic, New York, 1976.
4. Davis, M. D., Game Theory: A Nontechnical Introduction, Basic Books, New York, 1970.
5. Dresher, M., Games of Strategy: Theory and Applications, Prentice-Hall, Englewood Cliffs, New Jersey, 1961.
6. Karlin, S., Mathematical Methods and Theory in Games, Programming, and Economics, Vols. I and II, Addison-Wesley, Reading, Massachusetts, 1959.
7. Kuhn, H. W., Extensive games, Proc. Natl. Acad. Sci. U.S. 36, 570-576 (1950).
8. Lucas, W. F., A game with no solution, Bull. Am. Math. Soc. 74, 237-239 (1968).
9. Lucas, W. F., The proof that a game may not have a solution, Trans. Am. Math. Soc. 137, 219-229 (1969).
10. Lucas, W. F., Some recent developments in n-person game theory, SIAM Rev. 13, 491-523 (1971).
11. Luce, R., and H. Raiffa, Games and Decisions: Introduction and Critical Survey, Wiley, New York, 1957.
12. McDonald, J., The Game of Business, Doubleday, Garden City, New York, 1975.
13. Mycielski, J., On the axiom of determinateness, Fundam. Math. 53, 205-224 (1963/64).
14. Nash, J. F., Jr., Equilibrium points in n-person games, Proc. Natl. Acad. Sci. U. S. 36, 48-49 (1950).
15. Nash, J. F., Jr., Noncooperative games, Ann. Math. 54, 286-295 (1951).
16. Owen, G., Game Theory, Saunders, Philadelphia, 1968.
17. Parthasarathy, T., and T. E. S. Raghavan, Some Topics in Two-Person Games, American Elsevier, New York, 1970.
18. Rapoport, A., N-Person Game Theory: Concepts and Applications, University of Michigan Press, Ann Arbor, 1970.
19. Rapoport, A., and A. M. Chammah, Prisoner's Dilemma: A Study of Conflict and Cooperation, The University of Michigan Press, Ann Arbor, 1965.
20. Schmeidler, D., The nucleolus of a characteristic function game, SIAM J. Appl. Math. 17, 1163-1170 (1969).
21. Shapley, L. S., and M. Shubik, Game Theory in Economics—Chapter 6: Characteristic Function, Core, and Stable Set (R-904-NSF/6), RAND Corp., Santa Monica, California, July 1973.
22. Shubik, M. (ed.), Essays in Mathematical Economics: In Honor of Oskar Morgenstern, Princeton University Press, Princeton, New Jersey, 1967.
23. Thrall, R. M., and W. F. Lucas, n-Person games in partition function form, Nav. Res. Logist. Q. 10, 281-298 (1963).

24. Tucker, A. W., et al. (eds.), Contributions to the Theory of Games, Vols. 1, 2, 3, 4; and Advances in Game Theory (Annals of Mathematics Studies, Nos. 24, 28, 39, 40, and 52), Princeton University Press, Princeton, New Jersey, 1950, 1953, 1957, 1959, and 1964.
25. von Neumann, J., and O. Morgenstern, Theory of Games and Economic Behavior, Princeton University Press, Princeton, New Jersey, 1944; 2nd ed., 1947; 3rd ed., 1953.
26. Vorob'ev, N. N. (ed.), Game Theory: Annotated Index of Publications to 1968, Science Publisher, Leningrad, 1976 (in Russian).
27. Weber, R. J., Bargaining Solutions and Stationary Sets in n-Person Games (Technical Report No. 223), Department of Operations Research, Cornell University, Ithaca, New York, July 1974.
28. Zauberman, A. (ed.), Differential Games and Other Game Theoretic Topics in Soviet Literature: A Survey (Studies in Game Theory and Mathematical Economics), New York University Press, New York, 1975.
29. Zermelo, E., Über eine Anwendung der Mengenlehre auf die Theorie des Schachspiels, in Proceedings of the Fifth International Congress of Mathematics, Vol. II, Cambridge, England, 1913, pp. 501-504.

4
GOAL PROGRAMMING

SANG M. LEE

Department of Management
University of Nebraska
Lincoln, Nebraska

DECISION ANALYSIS WITH MULTIPLE OBJECTIVES

The traditional concept of "economic man" not only postulates that he is rational, but also suggests that he is the optimizer who strives to allocate scarce resources in the most economic manner. It is assumed that he possesses knowledge of relevant aspects of the decision environment, a stable system of preference, and ability to analyze alternative courses of action. However, recent developments in the theory of the firm have raised a serious question as to whether such assumptions regarding economic man can be applied to the decision maker in any realistic sense. For an individual to be perfectly rational in decision analysis, he must be capable of attaching a definite preference to each possible outcome of the alternative courses of action. Furthermore, he should be able to specify the exact outcomes by employing scientific analysis. According to broad empirical investigation, however, there is no evidence that any one individual is capable of performing such exact analysis for a complex decision problem. Moreover, there is considerable doubt that the individual's value system is identical to that of the firm.

The primary goal of economic man as an optimizer is assumed to be maximization of profits. If this were the situation, decision analysis would not be such a difficult task. In reality, the decision maker may have only a vague idea as to what is the best outcome for the organization in a global sense. Furthermore, he often is incapable of identifying the optimal choice due to either his lack of analytical ability or the complexity of the organizational environment. There is an abundance of evidence which suggests that the practice of decision making is affected by the epistemological assumptions of the individual who makes the decision [33]. Indeed, scientific methodology and rational choice are not always directly applicable to decision analysis. The decision maker constantly is concerned with his environment, and always relates possible decision outcomes and their consequences to its unique conditions. Stated differently, the decision maker does not only consider isolated economic payoffs, but is extremely conscious of the implications of the decision for his surroundings. This concern with the environmental context of the decision results in modifications which further remove him from the classical concept of economic man.

The concept of the individual profit maximizer in classical economic theory does not provide either a descriptive or normative model for the decision maker in an organization. To be sure, profit is difficult to define and even when defined is extremely difficult to measure. Another problem is that there exists a considerable variety of opinion about the objectives of an organization.

In today's complex organizational environment, the decision maker is regarded as one who attempts to achieve a set of objectives to the fullest possible extent in an environment of conflicting interests, incomplete information, and limited resources [34]. The soundness of decision making is measured by the degree of organizational objectives achieved by the decision. Therefore, recognition of organizational objectives provides the foundation for decision making. Decisions are also constrained by environmental factors such as government regulations, welfare of the public, and long-run effects of the decision on environmental conditions (i.e., pollution, quality of life, use of nonrenewable resources, etc.). In order to determine the best course of action, therefore, a comprehensive analysis of multiple and often conflicting organizational objectives and environmental factors must be undertaken. Indeed, the most difficult problem in decision analysis is the treatment of multiple conflicting objectives [33]. The issue becomes one of value trade-offs in the complex socioeconomic structure of conflicting interests. Regardless of the type of problem on hand, it is extremely difficult to answer questions such as what should be done now, what can be deferred, what alternatives are to be explored, and what should be the priority structure for the objectives.

Important developments in the field of management have clearly indicated that organizations, as well as society in general, have become so fragmented into various interest and value groups that there is no longer one predominant objective for any organization. Consequently, one of the most important and difficult aspects of any decision problem is to achieve an equilibrium among multiple and conflicting interests and objectives of various components of the organization. Several recent studies concerning the future of the industrialized society [31] have echoed the same theme. When the society is based on enormous technological development and change, stability of the system must be obtained by achieving a delicate balance among such multiple objectives as food production, industrial output, pollution control, population growth, use of natural resources, international cooperation for economic stability, and civil rights and equal opportunity provisions. There is obviously a need for continuous research in the analysis of multiple conflicting objectives.

GOAL PROGRAMMING FOR MULTIPLE-OBJECTIVE DECISIONS

One of the most promising techniques for multiple objective decision analysis is goal programming. Goal programming is a powerful tool which draws upon the highly developed and tested technique of linear programming, but provides a simultaneous solution to a complex system of competing objectives. Goal programming can handle decision problems having a single goal with multiple subgoals [22]. The technique was originally introduced by Charnes and Cooper [4-6, 8], and further developed by Ijiri [18] and Lee [22-24].

Often goals set by management compete for scarce resources. Furthermore, these goals may be incommensurable. Thus there is a need to establish a hierarchy of importance among these conflicting goals so that low order goals are satisfied or have reached the point beyond which no further improvements are desirable. If the decision maker can provide an ordinal ranking of goals in terms of their contributions or importance to the organization and if all relationships of the model are linear, the problem can be solved by goal programming.

In goal programming, instead of attempting to maximize or minimize the objective criterion directly, as in linear programming, the deviations between goals and what can be achieved within the given set of constraints are minimized. In the simplex algorithm of linear programming such deviations are called slack variables. These variables take on a new significance in goal programming. The deviational variable is represented in two dimensions, both positive and negative deviations from each subgoal or goal. Then the objective function becomes the minimization of these deviations based on the relative importance or priority assigned to them.

The solution of any linear programming problem is based on the cardinal value such as profit or cost. The distinguishing characteristic of goal programming is that it allows for an ordinal solution. The decision maker may be unable to obtain information about the value or cost of a goal or a subgoal, but often can determine its upper or lower limits. Usually, the decision maker can determine the priority of the desired attainment of each goal or subgoal and rank the priorities in an ordinal sequence. Obviously, it is not possible to achieve every goal to the extent desired. Thus, with or without goal programming, the decision maker attaches a certain priority to the achievement of a particular goal. The true value of goal programming, therefore, is its contribution to the solution of decision problems involving multiple and conflicting goals according to the decision maker's priority structure.

Goal programming has been applied to a wide range of planning, resource allocation, policy analysis, and functional management problems. The first application was made by Charnes et al. [10] for advertising media planning. However, the first real-world application was in the area of manpower planning by Charnes et al. [9]. Subsequently, goal programming was applied to aggregate production planning [22, 28], transportation logistics [27], academic resource planning [25], hospital administration [22], marketing planning [29], capital budgeting [22], portfolio selection [26], municipal economic planning [22], and resource allocation for environmental protection [7].

THE GOAL PROGRAMMING MODEL

Goal programming is a linear mathematical model in which the optimum attainment of multiple goals is sought within the given decision environment. The decision environment determines the basic components of the model, namely, the decision variables, constraints, and the objective function.

Let us now consider the goal programming model through a simple illustration. First, goal programming involving a simple goal with multiple subgoals will be discussed, followed by an analysis of multiple goals.

Single Goal with Multiple Subgoals

Example 1. A furniture manufacturer produces two kinds of products, desks and tables. The unit profit of a desk is \$80 and of a table is \$40. The goal of the plant manager is to earn a total profit of exactly \$640 in the next week.

We can interpret the profit goal in terms of subgoals, which are sales volumes of desks and tables. Then a goal programming model can be formulated as follows:

Minimize

$$Z = d_1^- + d_1^+$$

subject to

$$\$80X_1 + \$40X_2 + d_1^- - d_1^+ = \$640$$

$$X_1, X_2, d_1^-, d_1^+ \geq 0$$

where

X_1 = number of desks sold

X_2 = number of tables sold

d_1^- = underachievement of the profit goal of \$640

d_1^+ = overachievement of the profit goal of \$640

If the profit goal is not completely achieved, then obviously the slack in the profit goal will be expressed by d_1^-, which represents the underachievement of the goal (or negative deviation from the goal). On the other hand, if the solution shows a profit in excess of \$640, the d_1^+ will show some value. If the profit goal of \$640 is exactly achieved, both d_1^- and d_1^+ will be zero. It should be noted that d_1^- and d_1^+ are complementary. If d_1^- takes a nonzero value, d_1^+ will be zero, and vice versa. Since at least one of the deviational variables will always be zero, $d_1^- \times d_1^+ = 0$. In the above example, there are an infinite number of combinations of X_1 and X_2 that will achieve the goal. The solution will be any linear combination of X_1 and X_2 between the two points ($X_1 = 8$, $X_2 = 0$) and ($X_1 = 0$, $X_2 = 16$). This straight line is exactly the isoprofit function when total profit is \$640.

In the above example we did not have any model constraints. Now let us suppose that in addition to the profit goal constraint considered in Example 1, the following two constraints are imposed. The marketing department reports that the maximum number of desks that can be sold in a week is six. The maximum number of tables that can be sold is eight.

Now the new goal programming model can be presented in the following way:

Minimize

$$Z = d_1^- + d_1^+$$

subject to

$$\$80X_1 + \$40X_2 + d_1^- - d_1^+ = \$640$$

$$X_1 \leq 6$$

$$X_2 \leq 8$$

$$X_1, X_2, d_1^-, d_1^+ \geq 0$$

The solution to the above problem can be easily calculated on the back of an envelope. Thc solution is $X_1 = 6$ and $X_2 = 4$. With this solution the deviational variables d_1^- and d_1^+ will both be zero. The plant manager's profit goal can be achieved under the new constraints imposed on the subgoals.

Analysis of Multiple Goals

The model illustrated above can be extended to handle cases of multiple goals. Let us assume that these goals are conflicting and incommensurable.

Example 2. Let us consider the furniture manufacturer case illustrated in Example 1. Now the manager desires to achieve a weekly profit as close to \$640 as possible. He also desires to achieve sales volume for desks and tables close to six and to four respectively. The manager's decision problem can be formulated as a goal programming model as follows:

Minimize

$$Z = d_1^- + d_2^- + d_3^- + d_1^+$$

subject to

$$\$80X_1 + \$40X_2 + d_1^- - d_1^+ = 640$$

$$X_1 + d_2^- = 6$$

$$X_2 + d_3^- = 4$$

where d_2^- and d_3^- represent underachievements of sales volume for desks and tables, respectively. It should be noted that d_2^+ and d_3^+ are not included in the second and third constraints, since the sales goals are given as the maximum possible sales volume. The solution to this problem can be found by a simple examination of the problem: If $X_1 = 6$, and $X_2 = 4$, all goals will be completely attained. Therefore, $d_1^- = d_2^- = d_3^- = d_1^+ = 0$.

Ranking and Weighting of Multiple Goals

In Example 2 we had a case in which all goals are achieved simultaneously within the given constraints. However, in a real decision environment this is rarely the case. Quite often, most goals are competitive in terms of need for scarce resources. In the presence of incompatible multiple goals the manager needs to exercise his judgment about the importance of the individual goals. In other words, the most important goal must be achieved to the extent desired before the next goal is considered.

Goals of the decision maker may simply be meeting a certain set of constraints. For example, the manager may set a goal concerning a stable employment level in the plant, which is simply a part of the production constraint. Or the goal may be an entirely separate function from the constraints of the system. If that is the case, the goal constraint must be generated in the model. The decision maker must analyze the system and investigate whether all of his goals are expressed in the goal programming model. When all constraints and goals are completely identified in the model, the decision maker must analyze each goal in terms of whether over- or underachievement of the goal is satisfactory or not. Based on this analysis he can assign deviational variables to the regular and/or goal constraints. If overachievement is acceptable, positive deviation from the goal can be eliminated from the objective function. On the other hand, if underachievement of a certain goal is acceptable, negative deviation should not be included in the objective function. If the exact achievement of the goal is desired, both negative and positive deviations must be represented in the objective function.

In order to achieve the ordinal solution—that is, to achieve the goals according to their importance—negative and/or positive deviations about the goal must be ranked according to the "preemptive" priority factors. In this way the low-order goals are considered only after high-order goals are achieved as desired. If there are goals in several ranks of importance, the preemptive priority factor p_k ($k = 1, 2, \ldots, k$) should be assigned to the negative and/or positive deviational variables. The preemptive priority factors have the relationship of $p_1 >>> p_2 >>> p_3 \cdots p_k >>> p_{k+1}$, where >>> means "very much greater than." The priority relationship implies that multiplication by n, however large it may be, cannot make the lower-level goal as important as the higher goal. It is, of course, possible to refine goals even further by means of decomposing (subdividing) the deviational variables. To do this, additional constraints and additional priority factors may be required.

One more step to be considered in the goal programming model formulation is the weighting of deviational variables at the same priority level. For example, if the sales goal involves two different products, there will be two deviational variables with the same priority factor. The criterion to be used in determining the differential weights of deviational variables is the minimization of the opportunity cost or regret. This implies that the coefficient of regret, which is always positive, should be assigned to the individual deviational variable with the identical p_k factor. The coefficient of regret simply represents the relative amount of unsatisfactory deviation from the goal. Therefore, deviational variables on the same priority level must be commensurable, although deviations that are on different priority levels need not be commensurable.

Example 3. Consider the following modified case of the illustration given in the previous examples. Production of either a desk or a table requires 1 hr of production capacity in the plant. The plant has a normal maximum production capacity of 10 hr/week. Because of the limited sales capacity, the maximum number of desks and tables that can be sold are six and eight per week, respectively. The unit profit for a desk is \$80 and for a table \$40.

The plant manager has set the following goals, arranged in order of importance.

1. He wants to avoid any underutilization of production capacity (providing job security to the plant employees).
2. He wants to sell as many desks and tables as possible. Since the unit profit from the sale of a desk is twice the amount of profit from a table, he has twice as much desire to achieve the sales goal for desks as for tables.
3. He wants to minimize overtime operation of the plant as much as possible.

In the above example, the plant manager is to make a decision that will achieve his goals as closely as possible with the minimum sacrifice. Since overtime operation is allowed in this example, production of desks and tables may take more than the normal production capacity of 10 hr. Therefore, the operational capacity can be expressed as

$$X_1 + X_2 + d_1^- - d_1^+ = 10$$

where X_1 = number of desks to be produced
X_2 = number of tables to be produced
d_1^- = idle (underutilization of) production capacity
d_1^+ = overtime operation

Accordingly, the sales capacity constraints can be written as

$$X_1 + d_2^- = 6$$

$$X_2 + d_3^- = 8$$

where d_2^- = underachievement of sales goal for desks
d_3^- = underachievement of sales goal for tables

It should be noted that d_2^+ and d_3^+ are not in the equation, since the sales goals given are the maximum possible sales volume.

In addition to the variables and constraints stated above, the following preemptive priority factors are to be defined:

P_1: The highest priority, assigned by management to the underutilization of production capacity (i.e., d_1^-).

P_2: The second priority factor, assigned to the underutilization of sales capacity (i.e., d_2^- and d_3^-). However, management puts twice the importance on d_2^- as that on d_3^- in accordance with respective profit figures for desks and tables.

P_3: The lowest priority factor, assigned to overtime in the production capacity (i.e., d_1^+).

Now the complete model can be formulated. The objective is the minimization of deviation from goals. The deviant variable associated with the highest preemptive priority must be minimized to the fullest possible extent. When no further improvement is desirable or possible in the highest goal, then the deviations associated with the next highest priority factor will be minimized. The model can be expressed as:

Minimize

$$Z = P_1 d_1^- + 2P_2 d_2^- + P_2 d_3^- + P_3 d_1^+$$

subject to

$$X_1 + X_2 + d_1^- - d_1^+ = 10$$

$$X_1 + d_2^- = 6$$

$$X_2 + d_3^- = 8$$

$$X_1,\ X_2,\ d_1^-,\ d_2^-,\ d_3^-,\ d_1^+ \geq 0$$

From the simple investigation of the model we can derive the following optimal solution: $X_1 = 6$, $X_2 = 8$, $d_1^- = d_2^- = d_3^- = 0$, $d_1^+ = 4$. The first two goals are completely attained, but the third goal is only partially achieved, since the overtime operation could not be minimized to zero. This result is due to the direct conflict between the second (sales) goal and the third (minimization of overtime) goal. This kind of result reflects the everyday problem experienced in business when there are several conflicting goals.

Now, the general goal programming model can be presented as below [22]:

Minimize

$$Z = \sum_{k=1}^{K} \sum_{i=1}^{m} P_k (w_i^- d_i^- + w_i^+ d_i^+)$$

subject to

$$\sum_{j=1}^{n} a_{ij} X_j + d_i^- - d_i^+ = b_i \qquad (i = 1, m)$$

$$X_j,\ d_i^-,\ d_i^+ \geq 0$$

In this model, P_k is the preemptive priority factor assigned to goal k; w_i^- and w_i^+ are numerical weights assigned to the deviations of goal i at a given priority level; d_i^- and d_i^+ are the negative and positive deviations, respectively; a_{ij} is the technological coefficient of X_j in goal i; and b_i is the right-hand-side value of goal i.

THE MODIFIED SIMPLEX METHOD OF GOAL PROGRAMMING

In general, the goal programming model is a linear mathematical model in which the optimum attainment of objectives is sought within the given decision environment. The decision environment determines the basic components of the model, namely the constraints (system and goal), decision variables, and the objective function. A goal programming model can perform three types of analysis: (1) it determines the input (resource) requirements to achieve a set of goals, (2) it determines the degree of attainment of defined goals with given resources, and (3) it provides the optimum solution under the varying inputs and priority structures of goals.

The system constraints represent the absolute restrictions imposed by the decision environment on the model. For example, there are only 7 days in a week (time constraint), the production capacity in a short run is limited to certain available hours (manpower constraint), the production should be limited to demand and storage capacity (physical constraint); these are good illustrations of the system constraints. The system constraints must be satisfied before any of the goal constraints can be considered.

The goal constraints represent those functions that present desired levels of certain measurements. For example, desired level of pollution control, desired level of profit, desired diversification of investments among various available alternatives, desired market share for each product, and the like are good illustrations of goal constraints. In order to achieve the ordinal solution, negative and/or positive deviations about the goal must be minimized based on the preemptive priority weights assigned to them. Thus the low-order goals are sought only after the higher order goals are fully attained as desired ($P_k >>> P_{k+1}$). When there are multiple goals at a given priority level, differential weights (w_i) are assigned based on the numerical opportunity costs. The detailed discussion of model formulation, application areas, and limitations of goal programming can be found in Ref. 22.

The most widely used solution method of goal programming is the modified simplex method (see Ref. 22 for a detailed discussion). For those who are not familiar with goal programming, the following simple example seems appropriate.

An Example Problem

A textile company produces two types of linen materials, a strong upholstery material and a regular dress material. The upholstery material is produced according to direct orders from furniture manufacturers. The dress material, on the other hand, is distributed to retail fabric stores. The average production rates for the upholstery and for the dress material are identical: 1000 yards/hr. By running two shifts, the operational capacity of the plant is 80 hr/week.

The marketing department reports that the estimated maximum sales for the following week are 70,000 yards of the upholstery material and 45,000 yards of the dress material. According to the accounting department, the approximate profit from a yard of upholstery material is \$2.50 and from a yard of dress material \$1.50.

The president of the firm has the following four goals, listed in order of importance:

P_1: Avoid any underutilization of production capacity (i.e., maintain stable employment level at normal capacity).
P_2: Limit overtime operation of the plant to 10 hr.
P_3: Achieve the sales goal of 70,000 yards of upholstery material and 45,000 yards of dress material.
P_4: Minimize the overtime operation of the plant as much as possible.

A goal programming model for the problem can be formulated as follows:

Minimize

$$Z = P_1 d_1^- + P_2 d_{11}^+ + 5P_3 d_2^- + 3P_3 d_3 + P_4 d_1^+$$

subject to

$$X_1 + X_2 + d_1^- \qquad - d_1^+ = 80$$
$$X_1 \qquad + d_2^- \qquad = 70$$
$$X_2 \qquad + d_3^- \qquad = 45$$
$$d_{11}^- + d_1^+ - d_{11}^+ = 10$$
$$X_j,\ d_i^-,\ d_i^+ \geq 0$$

where X_1 = number of hours used for producing the upholstery material
X_2 = number of hours used for producing the dress material
d_1^- = underutilization of production capacity, below 80 hr
d_1^+ = overtime operation of the plant, over 80 hr
d_2^- = underachievement of sales goal of the upholstery material
d_3^- = underachievement of sales goal of the dress material
d_{11}^- = negative deviation of overtime operation from allowed 10 hr
d_{11}^+ = overtime operation in excess of 10 hr

In the above model, the overtime operation goal constraint (last constraint) can be easily substituted by $X_1 + X_2 + d_4^- - d_4^+ = 90$.

Before the solution is presented for the problem, several points deserve our attention. First, in goal programming the purpose of the model is to minimize the unattained portions of goals as much as possible. This is achieved by minimizing the deviational variables through the use of preemptive priority factors and differential weights. There is no profit maximization or cost minimization per se in the objective function. Therefore the preemptive factors and differential weights represent C_j values. Second, it should be remembered that preemptive priority factors are ordinal weights and they are not commensurable. Consequently, Z_j or $Z_j - C_j$

TABLE 1

C_j	v	rhs	X_1	X_2	P_1 d_1^-	$5P_3$ d_2^-	$3P_3$ d_3^-	d_{11}^-	P_4 d_1^+	P_2 d_{11}^+
P_1	d_1^-	80	1	1	1				-1	
$5P_3$	d_2^-	70	(1)			1				
$3P_3$	d_3^-	45		1			1			
	d_{11}^-	10						1	1	-1
$Z_j - C_j$	P_4	0							-1	
	P_3	485	5	3						
	P_2	0								-1
	P_1	80	1	1					-1	

becomes a $k \times n$ size matrix, where k represents the number of priority levels and n is the number of variables. Third, since the simplex criterion $(Z_j - C_j)$ is expressed as a matrix rather than as a row, a new procedure must be devised for identifying the pivot column. Since $P_k >>> P_{k+1}$, the selection procedure of the pivot column must be initiated from P_k and move gradually to the lower priority levels.

Table 1 presents the initial tableau of the goal programming problem. The initial solution is at the origin $(X_1 = X_2 = 0)$. Consequently, d_i^- will be the basic variables. The simplex criterion $(Z_j - C_j)$ is a 4×8 matrix because there are four priority factors and eight variables (two decision and six deviational) in the model. It should be apparent that the selection of the pivot column is based on the per-unit contribution rate of each variable in achieving the goals. When the first goal is completely attained, then the pivot column selection criterion moves on to the second goal, and so on. This is why the preemptive priority factors are listed from the lowest to the highest so that the pivot column can be easily identified at the bottom of the tableau. To make the simplex tableau relatively simple, the Z_j matrix is omitted.

In goal programming, the Z_j values $(P_4 = 0,\ P_3 = 485,\ P_2 = 0$, and $P_1 = 80)$ in the right-hand-side column represent the unattained portion of each goal (U_k). The pivot column is the one which has the largest positive $Z_j - C_j$ at the P_1 level, as there exists an unattained portion of this highest goal $(P_1 = 80)$. There are two identical positive $Z_j - C_j$ values in the X_1 and X_2 columns. In order to break this tie, we check the subsequently next-lower priority levels. Thus X_1 is the pivot column. The pivot row is the one that has the minimum nonnegative value when we divide the right-hand-side values by the coefficients in the pivot column.

By utilizing the regular simplex procedure (see Dantzig Simplex Method in Vol. 7), the first tableau is revised to obtain the second tableau and subsequently the optimum solution as shown in Table 2. The solution is optimal in the sense that it enables the decision maker to attain his goals as closely as possible within the given decision environment and priority structure. The optimum solution is $X_1 = 70,\ X_2 = 20,\ d_1^+ = 10,\ d_3^- = 25$. In the final simplex tableau, since the third goal is not completely attained, there is a positive value in $Z_j - C_j$ at the P_3 level. This value is found as (3) in the d_{11}^+ column. Obviously, we can attain the third goal to a greater extent if we introduce d_{11}^+ in the solution. However, there is a negative value (-1) at a higher priority level (P_2) in the same column. This implies that if we introduce d_{11}^+ into the solution, the third goal would be improved but only at the expense of achieving the second goal. Thus we cannot introduce d_{11}^+ into the solution. The same logic applies to the d_{11}^- column, where there is a positive $Z_j - C_j$ value at the P_4 level. The rule is that if there is a positive $Z_j - C_j$ at a lower priority level, the variable in that column cannot be chosen as the entering variable as long as there is a negative $Z_j - C_j$ at a higher priority level.

From an analysis of the final simplex tableau, one can perform a sensitivity analysis to determine the conflict among the goals and their tradeoffs. For a detailed discussion of the modified simplex method and sensitivity analysis, see Chapters 5 and 7 of Ref. 22. The complete solution procedure of the modified simplex for the problem is shown in Fig. 1.

For any management science technique to be a truly valuable tool for decision analysis, it must accommodate itself to a computer-based solution. The complexity of real-world problems usually compels the use of computers. An efficient computer

TABLE 2

C_j	v	rhs	X_1	X_2	P_1 d_1^-	$5P_3$ d_2^-	$3P_3$ d_3^-	d_{11}^-	P_4 d_1^+	P_2 d_{11}^+
P_1	d_1^-	10		(1)	1	-1			-1	
	X_1	70	1			1				
$3P_3$	d_3^-	45		1			1			
	d_{11}^-	10						1	1	-1
$Z_j - C_j$	P_4	0							-1	
	P_3	135		3		-5				
	P_2	0								-1
	P_1	10		1		-1			-1	
	X_2	10		1	1	-1			-1	
	X_1	70	1			1				
$3P_3$	d_3^-	35			-1	1	1		1	
	d_{11}^-	10						1	(1)	-1
$Z_j - C_j$	P_4	0							-1	
	P_3	105			-3	-2			3	
	P_2	0								-1
	P_1	0			-1					
	X_2	20		1	1	-1		1		-1
	X_1	70	1			1				
$3P_3$	d_3^-	25			-1	1	1	-1		1
P_4	d_1^+	10						1	1	-1
$Z_j - C_j$	P_4	10						1		-1
	P_3	75			-3	-2		-3		3
	P_2	0								-1
	P_1	0			-1					

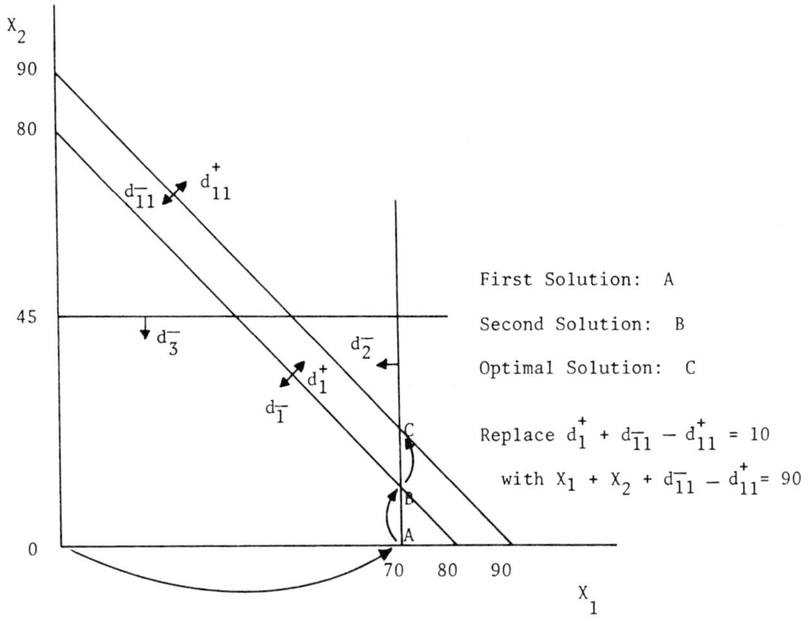

FIGURE 1.

program has been developed by Lee [22]. This program has been widely applied to real-world problems. For complete information about the goal programming program, either consult Ref. 22 or contact the author.

NEW DEVELOPMENTS IN GOAL PROGRAMMING

Goal programming is a powerful and flexible technique that can be applied to a variety of decision problems involving multiple objectives. It should, however, be pointed out that goal programming is by no means a panacea for contemporary decision problems. The fact is that goal programming is applicable only under certain specified assumptions and conditions. Most goal programming applications have thus far been limited to well-defined deterministic problems. Furthermore, the primary analysis has been limited to the identification of an optimal solution that optimizes goal attainment to the extent possible within specified constraints. In order to develop goal programming as a universal technique for modern decision analysis, many refinements and further research are necessary. In this section, several areas where research is being conducted will be discussed.

Sensitivity Analysis

An analysis of the effects of parameter changes after determining the optimal solution is a very important part of any solution process. This procedure is broadly defined as the postoptimal sensitivity analysis. Because there usually exists some degree of uncertainty in real-world problems concerning the model parameters—i.e., priority factor, technological coefficients, and goal levels or available resources—sensitivity analysis should be an important part of the goal programming solution. If the optimal solution is relatively sensitive to changes in certain parameters, special efforts should be directed to forecasting the future values of these parameters. By the same token, if the optimal solution has very little sensitivity to changes in certain parameters, it might be a waste of time and effort to try to estimate the values of parameters more accurately.

The dual solution procedure of goal programming has not been fully explored, although there have been a couple of working papers on the subject. Consequently, the usual sensitivity analysis employed in linear programming cannot be applied to goal programming. However, from the final modified simplex tableau, one can obtain enough data to perform a sensitivity analysis for changes in C_j (priority factor), b_i (goal levels or resources), and a_{ij} (technological coefficient). For detailed information, see Chap. 7 of Ref. 22.

Goal Programming under Uncertainty

One of the frequent problems in the practical application of goal programming is the difficulty of determining proper values of model parameters, especially the right-hand-side values (b_i) and technological coefficients (a_{ij}). If the values of certain parameters are based on random events that are difficult to predict, the problem becomes goal programming under the condition of uncertainty.

Some very limited work has been done to date toward offering approaches to goal programming under uncertainty. Charnes, Cooper, and Niehaus [9] have developed a manpower planning model which considers the effect of Markov processes from period to period. Contini [11] has suggested a form of chance-constrained goal programming using normally distributed random variables associated with the deviational variables. Lee [22, Chap. 7] presented the piece-wise approximation and variance-covariance approaches as possible ways to handle goal programming problems under uncertainty. It appears that the well-known methods of chance-constrained programming and separable programming can be adapted to goal programming. There is a need for further research in this area of goal programming.

Integer Goal Programming

Since the introduction of the concept, many researchers have further studied goal programming during the past several years [5, 7, 8, 22, 27, 29]. However, these studies are based on the conventional divisibility requirement of linear programs. In many practical decision problems with multiple conflicting objectives, the decision variables make sense only if they assume nonfractional or discrete values [12]. The decision variables in this situation might be people, crews composed of various personnel and equipments, assembly lines, indivisible investment alternatives, construction projects, or pieces of equipment. A simple rounding of values of the decision variables in the optimum solution obtained by the regular goal programming algorithm to the nearest integers can be easily accomplished. However, the rounding procedure frequently yields either infeasible or a less than optimal solution. Thus there is a need to develop integer goal programming techniques. In this section we shall discuss three integer goal programming algorithms: the cutting plane method for all-integer goal programming problems, the branch and bound method for all-integer or mixed-integer goal programming problems, and the implicit enumeration method for zero-one goal programming problems [24].

Cutting Plane Method

The cutting plane method of integer goal programming is adapted from Gomory's [16, 17] methodology for the general integer linear program. The basic solution procedure can be summarized as follows:

Step 1: Solve the model by the ordinary modified simplex method of goal programming.

Step 2: Examine the optimal solution. If all the basic variables have integer values, the integer optimal solution is derived—stop. If one or more basic variables have fractional or continuous values, go to Step 3.

Step 3: Develop a cutting plane and find a new optimal solution by the modified simplex method. Go to Step 2.

The primary difference between the Gomory approach and the cutting plane goal programming method lies in the treatment of the multidimensional priority weights. Perhaps the best way to introduce the method is through a simple illustration. Consider the following problem:

Minimize

$$Z = P_1 d_1^+ + P_2 d_2^+ + P_3 d_3^-$$

subject to

$$X_1 + \frac{1}{10} X_2 + d_1^- - d_1^+ = 4$$

$$X_2 + d_2^- - d_2^+ = 2$$

$$4X_1 + X_2 + d_3^- - d_3^+ = 20$$

X_j = 0 or nonnegative integers

TABLE 3

C_j							P_3	P_1	P_2	
	v	rhs	X_1	X_2	d_1^-	d_2^-	d_3^-	d_1^+	d_2^+	d_3^+
	X_1	$3\frac{4}{5}$	1		1	-1/10		-1	1/10	
	X_2	2		1		1			-1	
P_3	d_3^-	$2\frac{4}{5}$			-4	-3/5	1	4	3/5	-1
$Z_j - C_j$	P_3	$2\frac{4}{5}$			-4	-3/5		4	3/5	-1
	P_2	0							-1	
	P_1	0						-1		

Table 3 presents the optimal solution by the modified simplex method. The optimal noninteger solution is $X_1 = 3\frac{4}{5}$, $X_2 = 2$, $d_3^- = 2\frac{4}{5}$, $U_1 = 0$ (underattainment of priority 1 goal), $U_2 = 0$, and $U = 2\frac{4}{5}$. Since the model requires integer solutions for all the basic variables, this solution is not acceptable. Therefore, a cutting plane must be developed and a further solution procedure is required. In accordance with the Gomory approach, a cutting plane is to be generated from a solution vector whose solution variable has the largest fractional part. This procedure assures a new constraint which tends to "cut" as deeply as possible. In the example, the solution vector for X_1 is selected to generate a cutting plane. Again, based on the Gomory procedure (see Cutting Plane Methods in Vol. 7), the following cutting plane can be obtained from the X vector:

$$-\frac{9}{10}d_2^- - \frac{1}{10}d_2^+ \leq -\frac{4}{5}$$

The above constraint can be transformed to a goal constraint $-\frac{9}{10}d_2^- - \frac{1}{10}d_2^+ + \hat{d}_4^- = -\frac{4}{5}$. By utilizing the dual modified simplex method of goal programming, a new feasible solution can be obtained and the cutting plane procedure continued. Of course, in order to satisfy the cutting plane constraint, $\hat{d}_4^+$ should be minimized with P_1 priority factor and other real priority goals should be relegated one level lower accordingly. Although the dual modified simplex method works well, the complicated pivoting procedure makes it quite cumbersome. Therefore, a new approach will be used by taking advantage of the flexibility of goal programming.

The cutting plane we derived above can be easily converted to $\frac{9}{10}d_2^- + \frac{1}{10}d_2^+ \geq \frac{4}{5}$. Now we can transform this to the following goal constraint.

$$\frac{9}{10}d_2^- + \frac{1}{10}d_2^+ + \hat{d}_4^- - \hat{d}_4^+ = \frac{4}{5}$$

In order to satisfy the inequality constraint, the negative deviation ($\hat{d}_4^-$) should be minimized. Since the right-hand-side value is positive, the solution will be feasible. Thus there is no need for the dual modified simplex method. After assigning P_1 to $\hat{d}_4^-$ and relegating the real priority goals by one level lower, the regular modified simplex method can be applied to derive a new solution. As shown in Table 4(a), $\hat{d}_4^-$ is inserted as the basic variable. It should be simple to identify d_2^- as the pivot column and $\hat{d}_4^-$ as the pivot row. The new solution derived in Table 4(b) also indicates a noninteger solution.

A new cutting plane is generated from the X vector as follows:

$$\frac{1}{9}\hat{d}_4^- + \frac{1}{9}d_2^+ + \frac{8}{9}\hat{d}_4^+ + \hat{d}_5^- - \hat{d}_5^+ = \frac{8}{9}$$

By following the same procedure described above, a new solution is derived in Table 5(b). The new solution is an all-integer optional solution. The solution is $X_1 = 4$, $X_2 = 0$, $d_2^- = 2$, $d_3^- = 4$, $\hat{d}_4^+ = 1$, $U_1 = 0$ (cutting plane priority level), $U_2 = 0$, $U_3 = 0$, and $U_4 = 4$.

The integer optimal solution can be compared with the fractional optimal solution in terms of the degree of goal attainment as below:

Fractional Optimal	Integer Optimal
$U_1 = 0$, $U_2 = 0$, $U_3 = 2\frac{4}{5}$	$U_1 = 0$, $U_2 = 0$, $U_3 = 4$

TABLE 4

	C_j							P_4	P_1	P_2	P_3		
		v	rhs	X_1	X_2	d_1^-	d_2^-	d_3^-	d_4^-	d_1^+	d_2^+	d_3^+	$\hat{d}_4^+$
(a)		X_1	$3\frac{4}{5}$	1		1	-1/10			-1	1/10		
		X_2	2		1		1				1		
	P_4	d_3^-	$2\frac{4}{5}$			-4	-3/5	1		4	3/5	-1	
	P_1	$\hat{d}_4^-$	4/5				(9/10)		1		1/10		-1
	$Z_j - C_j$	P_4	$2\frac{4}{5}$			-4	-3/5			4	3/5	-1	
		P_3	0								-1		
		P_2	0							-1			
		P_1	4/5				9/10				1/10		-1
(b)		X_1	35/9	1		1			1/9	-1	1/9		-1/9
		X_2	10/9		1				-10/9		-10/9		10/9
	P_4	d_3^-	10/3			-4		1	2/3	4	2/3	-1	-2/3
		d_2^-	8/9				1		10/9		1/9		-10/9
	$Z_j - C_j$	P_4	10/3			-4			2/3	4	2/3	-1	-2/3
		P_3	0								-1		
		P_2	0							-1			
		P_1	0						-1				

As expected, the integer optimal solution is inferior to the fractional optimal as the additional cutting plane constraints further restricted the convex set. The integer solution yielded $1\frac{1}{4}$ less goal attainment for the third priority goal. This difference represents the cost of indivisibility.

The general mixed-integer linear programming approach of Gomory can be adapted for the goal programming problem in the same manner as the procedure discussed thus far. The computational experience of the author, however, generally indicates that the branch and bound method of goal programming is more efficient for the mixed-integer problem than the cutting plane method.

TABLE 5

	C_j	v	rhs	X_1	X_2	d_1^-	d_2^-	$\overset{P_4}{d_3^-}$	$\overset{P_1}{\hat{d}_4^-}$	$\overset{P_1}{\hat{d}_5^-}$	$\overset{P_2}{d_1^+}$	$\overset{P_3}{d_2^+}$	d_3^+	$\hat{d}_4^+$	$\hat{d}_5^+$
(a)		X_1	35/9	1		1			1/9		-1	1/9		-1/9	
		X_2	10/9		1				-10/9			-10/9		10/9	
	P_4	d_3^-	10/3			-4		1	2/3		4	2/3	-1	-2/3	
		d_2^-	8/9				1		10/9			1/9		-10/9	
	P_1	$\hat{d}_5^-$	8/9						1/9	1		1/9		(8/9)	-1
	$Z_j - C_j$	P_4	10/3			-4			2/3		4	2/3	-1	-2/3	
		P_3	0									-1			
		P_2	0								-1				
		P_1	8/9						-8/9			1/9		8/9	-1
(b)		X_1	4	1		1			1/8	1/8	-1	1/8			-1/8
		X_2	0		1				-5/4	-5/4		-5/4			5/4
	P_4	d_3^-	4			-4		1	3/4	3/4	4	3/4	-1		-3/4
		d_2^-	2				1		5/4	5/4		1/4			-5/4
		$\hat{d}_4^+$	1						1/8	9/8		1/8		1	-9/8
	$Z_j - C_j$	P_4	4			-4			3/4	3/4	4	3/4	-1		-3/4
		P_3	0									-1			
		P_2	0								-1				
		P_1	0						-1	-1					

The Branch and Bound Method

Integer programming problems quite frequently have upper and/or lower bounds for their decision variables. Since the bounded integer programming problem has a finite number of feasible solutions, an enumeration procedure for searching an optimum solution is a sensible approach. The branch and bound algorithm (see Branch and Bound Technique in Vol. 4), first suggested by Land and Doig [20, 21], can also be adapted to an integer goal programming problem.

The basic idea of the branch and bound method of goal programming can be summarized as follows [24]:

Step 1: Solve the model by the ordinary modified simplex method of goal programming.

Step 2: Examine the optimal solution. If the basic variables that have integer requirements are integer valued, the integer optimal solution is obtained—stop. If one or more basic variables do not satisfy integer requirements, go to Step 3.

Step 3: The set of feasible solutions is branched into subsets (subproblems). The purpose of branching is to eliminate continuous solutions that do not satisfy the integer requirements of the problem. The branching is achieved by introducing mutually exclusive constraints that are necessary to satisfy integer requirements while making sure no feasible integer solution is excluded.

Step 4: For each subset the optimal value of the objective function (degree of goal attainment, U_k) is determined as the lower bound. The optimal U_k of a feasible solution which satisfies integer requirements becomes the upper bound. Those subsets having lower bounds that exceed the current upper bound must be excluded from further analysis. A feasible solution having U_k which is as good as or better than the lower bound for any subset is to be found. If there exists such a solution, it is optimal—stop. If such a solution does not exist, a subset with the best lower bound is selected and go to Step 3.

Implicit Enumeration Method

The implicit enumeration method (see Bivalent Programming by Implicit Enumeration in Vol. 2) is developed to solve the goal programming problem which requires either zero or one values for the decision variables. The method is basically a combination of Balas' [1-3] additive algorithm and Glover's [15] backtracking procedure. The solution combinations are evaluated by introducing one decision variable at a time. When no further variable can be added to improve the solution beyond the current optimal solution, the solution is called as "fathomed." Then a backtracking technique is instituted to evaluate other combinations in a systematic fashion. The optimal zero-one solution is the upper bound solution when all possible combinations are evaluated. For a detailed explanation of the technique, see Ref. 24.

Decomposition Goal Programming

The decomposition algorithm was originally developed as a computational device for solving a large model by decomposing it into smaller models. However, subsequent developments and the very underlying philosophy of decomposition have gradually shifted to the process of decision making in a decentralized organization. There are basically two different approaches to the decomposition technique, although there have been numerous decomposition models proposed during the past 10 years. The first is the price directive method developed by Dantzig and Wolfe [13]. The second is the resource directive method developed by Karnai and Liptak [19].

Most decomposition techniques have several important limitations. One predominant limitation is their computational inefficiency. The speed at which the value of the objective function converges to the optimal solution is very slow indeed. Another, and perhaps more important, limitation rests upon their single goal assumption of either maximizing profit or minimizing cost. Rucfli [32] and Freeland [14] employed the notion of goal programming to avoid this limitation, but the attempt was incomplete as their models did not employ the preemptive priority factors for multiple goals. Recently, Lee and Rho [30] have developed decomposition goal programming algorithms based on the preemptive priority factors which generalize the model and enable it to effectively handle the multiple conflicting objectives of the organization.

Interactive Goal Programming

One of the primary difficulties in the application of goal programming is the concept of ordinal solution based on preemptive priority weights which are assigned to multiple conflicting objectives. It is especially difficult to analyze the trade-offs among the objectives when their priority structure changes in time due to the changing decision environment. The best approach to analyze such a problem appears to be an interactive mode where the decision maker and the goal programming model interact via a computer terminal [24]. The same interactive approach can be equally effective for the changing goal levels (b_i) and technological coefficients (a_{ij}), addition or deletion of constraints, and addition or deletion of decision variables. Thus the interactive approach can provide an instant analysis of the effect of changes in model parameters as well as a complete sensitivity analysis of the optimal solution. The interactive goal programming approach provides a systematic process in which the decision maker seeks the most satisfactory solution. The process allows the decision maker to reformulate the model and systematically compare the solutions in terms of their achievement of multiple objectives.

The interactive system of Lee [22] is installed on IBM 370-158 and consists of a control program (CP) and the conversation monitor system (CMS). The system allows the user to create and edit files (programs, problems, etc.) and to execute programs conversationally. The interactive goal programming is based on the modified simplex method of Lee [22, Chap. 6]. Consequently, once the preliminary logon procedure is completed, the input format is exactly the same as the regular goal programming input required by Lee's program.

The interactive approach of goal programming developed by Lee [24] for the sensitivity analysis purpose can be applied to integer and decomposition programming. The interactive integer goal programming requires only slight changes in the program which can be easily accommodated by the file editing procedure. Thus the interactive integer approach can be used not only for deriving an integer or decomposition solution, but it can be utilized for sensitivity analysis for the model.

CONCLUSION

Virtually all models developed for managerial decision analysis have neglected the unique organizational environment, bureaucratic decision process, and multiple conflicting nature of organizational objectives. In reality, however, these are important factors that greatly influence the decision process. In this article, the goal programming approach is discussed as a tool for the optimization of multiple objectives while permitting an explicit consideration of the existing decision environment.

Developing and solving the goal programming model reveals those goals which cannot be achieved under the desired policy and, hence, where trade-offs due to limited resources must be made. Furthermore, the model solution allows the decision maker to review critically the priority structure.

The goal programming approach is not the ultimate solution for all managerial decision problems. It requires that the decision maker be capable of defining, quantifying, and ordering objectives. The technique simply provides the best solution under the given constraints and priority structure of goals. Although there have been numerous important developments in goal programming, as presented in this article, there are future research needs concerning the identification, definition, and ranking of goals. The development of a systematic methodology to generate such knowledge is very much needed.

REFERENCES

1. Balas, E., An additive algorithm for solving linear programs with zero-one variables, Oper. Res. 13, 517-546 (1965).
2. Balas, E., Discrete programming by the filter method, Oper. Res. 15, 915-957 (1967).
3. Balas, E., Intersection cuts—A new type of cutting plane for integer programming, Oper. Res. 19, 19-39 (1971).
4. Charnes, A., and W. W. Cooper, Management Models and Industrial Applications of Linear Programming, Wiley, New York, 1961, 2 vols.
5. Charnes, A., and W. W. Cooper, Goal programming and constrained regression—A comment, Omega 3(4), 403-409 (1975).
6. Charnes, A., W. W. Cooper, and R. O. Ferguson, Optimal estimation of executive compensation by linear programming, Manage. Sci. 1(2), 138-151 (1955).
7. Charnes, A., W. W. Cooper, J. Harrald, K. Karwan, and W. A. Wallace, A Goal Interval Programming Model for Resource Allocation in a Marine Environmental Protection Program, Rensselaer Polytechnic Institute, School of Management Research Report, September 1975.
8. Charnes, A., W. W. Cooper, D. Klingman, and R. J. Niehaus, Explicit solutions in convex goal programming, Manage. Sci. 22(4), 438-448 (1975).
9. Charnes, A., W. W. Cooper, and R. J. Niehaus, Studies in Manpower Planning, U.S. Navy Office of Civilian Manpower Management, Washington, D.C., 1972.
10. Charnes, A., et al., Note on the application of a goal programming model for media planning, Manage. Sci. 14(8), 431-436 (1968).
11. Contini, B., A stochastic approach to goal programming, Oper. Res. 16(3), 576-586 (1968).
12. Dantzig, G. B., D. R. Fulkerson, and S. M. Johnson, Solution of a large-scale travelling salesman problem, Oper. Res. 2, 393-410 (1954).
13. Dantzig, G. B., and P. Wolfe, The decomposition algorithm for linear programs, Econometrica 29(4), 767-778 (1961).
14. Freeland, J. R., Conceptual Models of the Resource Allocation Decision Process in Hierarchical Decentralized Organization, Ph.D. Dissertation, Georgia Institute of Technology, 1973.

15. Glover, F., Multi-phase dual algorithm for the zero-one integer programming problems, Oper. Res. 13(6), 879-919 (1965).
16. Gomory, R. E., An algorithm for integer solutions to linear programs, Bull. Am. Math. Soc. 64, 275-278 (1958).
17. Gomory, R. E., An Algorithm for the Mixed Integer Problem, Rand Corporation P-1885, Santa Monica, California, 1960.
18. Ijiri, Y., Management Goals and Accounting for Control, Rand McNally, Chicago, 1956.
19. Kornai, J., and T. Liptak, Two-level planning, Econometrica 33(1), (1965).
20. Land, A. H., and A. Doig, An automatic method of solving discrete programming problems, Econometrica 28, 497-520 (1960).
21. Lawler, E. L., and D. E. Wood, Branch and bound methods: A survey, Oper. Res. 14, 699-719 (1966).
22. Lee, S. M., Goal Programming for Decision Analysis, Auerbach, Philadelphia, 1972.
23. Lee, S. M., Goal programming for decision analysis with multiple objectives, Sloan Manage. Rev. 14(2), 11-24 (1973).
24. Lee, S. M., Interactive and Integer Goal Programming, paper presented at the Joint ORSA/TIMS Meeting, Las Vegas, 1975.
25. Lee, S. M., and E. R. Clayton, A goal programming model for academic resource allocation, Manage. Sci. 17(8), 395-408 (1972).
26. Lee, S. M., and A. J. Lerro, Optimizing the portfolio selection for mutual funds, J. Finance 28(8), 1087-1101 (1972).
27. Lee, S. M., and L. J. Moore, Optimizing transportation problems with multiple objectives, AIIE Trans. 5(4), 333-338 (1973).
28. Lee, S. M., and J. J. Moore, A practical approach to production scheduling, Prod. Inventory Manage. 15(1), 79-92 (1974).
29. Lee, S. M., and R. Nicely, Goal programming for marketing decisions: A case study, J. Marketing 38(1), 24-32 (1974).
30. Lee, S. M., and B. H. Rho, Decomposition Goal Programming, Research Report, Virginia Polytechnic Institute, 1976.
31. Meadows, D. H., et al., The Limits to Growth, New American Library, Washington, D.C., 1972.
32. Ruefli, T., A generalized goal decomposition model, Manage. Sci. 17(9), 505-518 (1971).
33. Schubik, M., Approaches to the study of decision-making relevant to the firm, in The Making of Decisions: A Reader in Administrative Behavior (W. J. Gore and J. W. Dyson, eds.), Free Press of Glencoe, London, 1964.
34. Simon, H. A., The New Science of Management Decision, Harper, New York, 1960.

5
MULTICRITERIA DECISION MAKING

JAMES S. DYER* and RAKESH K. SARIN

Graduate School of Management
University of California
Los Angeles, California

INTRODUCTION

The process of decision making involves several interrelated stages. In the first stage, the problem is recognized and bounded. That is, the decision maker must first become aware of a problem and then determine what elements and issues will be relevant in his problem-solving efforts. In the second stage, alternative solutions are generated and evaluated. This requires an understanding or definition of the system that is being analyzed and its environment so that the problem solver can predict the effects of implementing each alternative solution. Finally, these outcomes must be evaluated so that an alternative can be selected. The third stage is the implementation of the solution. These stages are not isolated from one another, and usually there will be feedback among the stages.

Here we focus on the second stage with emphasis on alternatives that generate multiple effects to be evaluated by the problem solver. For example, a decision among alternative job opportunities may require the consideration of such effects (outcomes, attributes, or criteria) as salary, duties and responsibilities, and location. Similarly, the choice of an automobile may be based on the multiple outcomes of cost, horsepower, and appearance; a national evaluation plan may be evaluated on predictions of its effects on economic growth, employment, and inflation. Other decisions that often involve the consideration of multiple outcomes are resource allocations in education, health, and criminal justice, the formulation of a corporate marketing strategy, and personnel selection. Typically, decisions involving multiple criteria are complex, major decisions involving a significant allocation of resources. They are often among the most difficult and the most important decisions that an individual or an organization will make.

**Current affiliation:* Department of Management, University of Texas, Austin, Texas

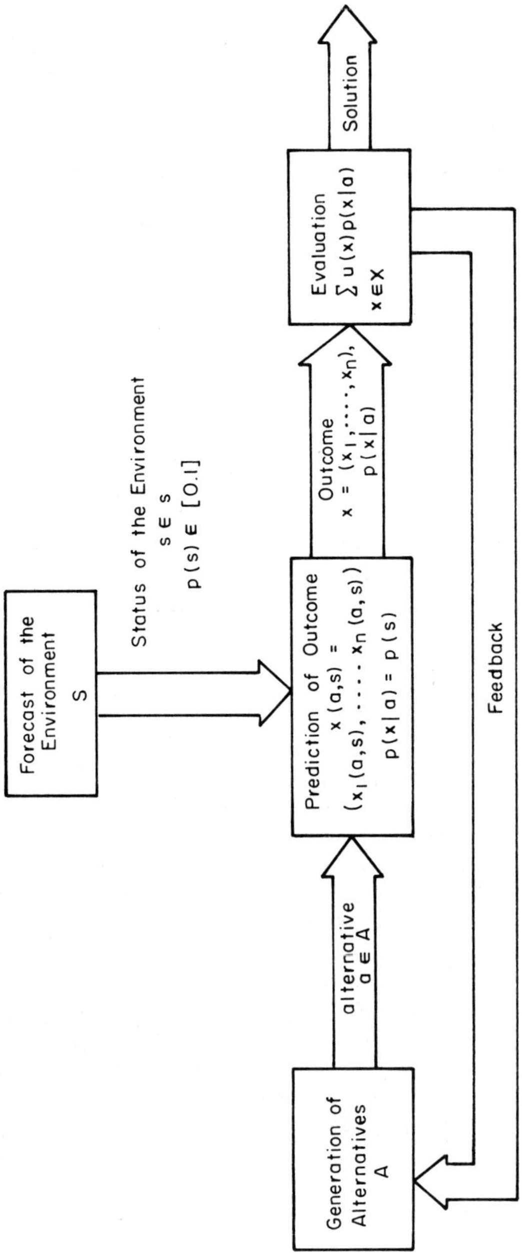

FIG. 1. The second stage of the decision-making process.

In the past decade or so, a large body of literature dealing with multicriteria decision making has appeared. Contributors to this literature include mathematicians, economists, management scientists, and psychologists. In this article we provide a summary of some of the major contributions in this area of research. However, methods for multicriteria decision making have been a topic of interest to scholars for a much longer period of time, as evidenced by this letter written by Benjamin Franklin to Joseph Priestly in 1772:

Dear Sir,

In the affair of so much importance to you, wherein you ask my advice, I cannot, for want of sufficient premises, advise you what to determine, but if you please I will tell you how. When those difficult cases occur, they are difficult, chiefly because while we have them under consideration, all the reasons pro and con are not present to the mind at the same time; but sometimes one set present themselves, and at other times another, the first being out of sight. Hence the various purposes or informations that alternatively prevail, and the uncertainty that perplexes us. To get over this, my way is to divide half a sheet of paper by a line into two columns; writing over the one Pro, and over the other Con. Then, during three or four days consideration, I put down under the different heads short hints of the different motives, that at different times occur to me, for or against the measure. When I have thus got them all together in one view, I endeavor to estimate their respective weights; and where I find two, one on each side, that seem equal, I strike them both out. If I find a reason pro equal to some two reasons con, I strike out the three. If I judge some two reasons con, equal to three reasons pro, I strike out the five; and thus proceeding I find at length where the balance lies; and if, after a day or two of further consideration, nothing new that is of importance occurs on either side, I come to a determination accordingly. And, though the weight of the reasons cannot be taken with the precision of algebraic quantities, yet when each is thus considered, separately and comparatively, and the whole lies before me, I think I can judge better, and am less liable to make a rash step, and in fact I have found great advantage from this kind of equation, and what may be called moral or prudential algebra.

Wishing sincerely that you may determine for the best, I am ever, my dear friend, yours most affectionately.

B. Franklin (signed) [46]

Problem Statement and Notation

Figure 1 provides some additional detail regarding the second stage of the decision-making process. We define A as the feasible set of decision alternatives, hereafter termed alternatives. In order to predict the outcome of choosing an alternative, it may be necessary to know the state of the environment $s \in S$ at some future time. When the probability of the occurrence of a state is one ($p(s) = 1.0$), we say that the decision is made under the condition of certainty. Otherwise, when each of several states may occur with a finite probability, we say that the decision is made under the condition of risk. Naturally, we assume that

$$\sum_{s \in S} p(s) = 1.0$$

For expositional simplicity, we have introduced notation that implies that S is a finite set. In general, this development could be extended so that p(s) is a probability measure.

Given an alternative $a \in A$ and a state of the environments', we can predict the outcome x(a, s). Here we emphasize the case where x(a, s) can be decomposed into multiple outcomes, attributes, or criteria, and write $x(a, s) = (x_1(a, s), \ldots, x_n(a, s))$ to indicate that there are n attributes. Thus the outcomes of the alternatives are represented by the elements in an n-dimensional outcome space X. The set of values or scores that an alternative may possibly obtain on the i-th criterion is denoted as X_i. Therefore, the outcome space X is the Cartesian product of the X_i, denoted $X_1 \times \cdots \times X_n$. An example will illustrate the notation.

Suppose the relevant criteria in evaluating a national plan are economic growth, balance of payment, and the employment level. Economic growth is measured in terms of high, medium, and low. The balance of payments can either be positive (surplus) or negative (deficit). The employment level is measured on a five point scale, 1 to 5, with 1 denoting poor employment and 5 denoting excellent employment in the economy. Then X_1 = (high, medium, low), X_2 = (surplus, deficit), and X_3 = (1, 2, 3, 4, 5). The outcome space X consists of $3 \times 2 \times 5 = 30$ different outcomes. A possible outcome x is (medium, deficit, 4).

Under the condition of risk, the outcome x(a, s) occurs with the probability p(s), which we denote by the definition $p(x|a) = p(s)$. As indicated in Fig. 1, the evaluation of an alternative a is based on the outcomes x(a, s) and the probability of their occurrence.

In order to complete the evaluation, we require some information regarding the decision maker's utility function u, defined on the outcome space X. The major emphasis of the work on multiattribute decision making has been on questions involving u: on necessary conditions for its existence, on conditions for its decomposition into simple polynomials, on methods for its assessment, and on methods for obtaining sufficient information regarding u so that the evaluation can proceed without its explicit identification. Therefore, it is worthwhile to pause and reflect a moment on the meaning and interpretation of u before we proceed.

In general, we merely assume that the decision maker has preferences. That is, given $a, b \in A$, he can state that he prefers a to b (denoted $a \succ b$), he prefers b to a ($b \succ a$), or he is indifferent between the two alternatives b and a ($b \sim a$). We use the notation $a \succsim b$ to mean $a \succ b$ or $a \sim b$. Under conditions of certainty, each alternative a is associated with a unique outcome x(a), where the notation s is deleted since $p(s) = 1$. If the decision-maker's preferences are, in some sense, consistent and satisfy other conditions that we shall detail later, then we are assured that there exists some utility function u such that $a \succ b$ if and only if $u(x(a)) > u(x(b))$. Under conditions of risk, we need only an additional assumption or two, and we are assured that there exists u such that a is preferred to b if and only if

$$\sum_{x \in X} u(x)p(x|a) > \sum_{x \in X} u(x)p(x|b)$$

Once the existence of u has been established, we can address the problem of designing an efficient means of interacting with a decision maker in order to obtain sufficient information regarding u so that a feasible alternative that is preferred to any other alternative in A can be identified. Thus our focus is on methods for assisting the decision maker in choosing among alternatives, so that his final choice is consistent with his underlying preferences. Theories that provide guidance for this task are termed prescriptive or normative theories of decision making.

Descriptive or predictive theories of decision making focus on the psychology of thinking. The emphasis in these theories is to uncover how people do make decisions and to predict their actual choices. A manufacturer of toothpaste, for example, would be interested in how people do choose toothpastes, no matter how illogical their choice process may be.

The relevance of a descriptive or a prescriptive theory depends on the context and the objectives of the decision maker. However, a good understanding of how people do behave (make decisions) is essential before prescribing how they should make decisions. Descriptive theories may thus aid in building normative theories, and certainly are crucial for the application of these theories. In the training or education of decision makers to make better decisions, both descriptive and prescriptive theories are needed. Even though at present the research in these two fields seems to be progressing somewhat in parallel, a greater interaction between the two theories is inevitable in the future.

In this article, our scope is limited to the discussion and synthesis of the prescriptive theories of multicriteria decision making, and only occasional digressions to the descriptive theory will be made. This limitation is necessary in order to reduce the length and complexity of this survey. Reviews of descriptive decision theories are provided by Fischer [30], Huber [57, 58], and von Winterfeldt [113].

By avoiding the descriptive theories, we restrict the discussion to theories that include the assumption that the decision maker can provide tradeoffs between levels of the criteria. These approaches are termed compensatory tradeoff models. The noncompensatory models include the lexicographic ranking schemes reviewed by Fishburn [43], conjunctive/disjunctive constraints (e.g., see Coombs [13], Dawes [15], Einhorn [24]), elimination by aspects (Tversky [110, 111]), maximin (Wald [117]), and several others as reviewed in MacCrimmon [76].

We also limit our scope to the case of a single decision maker, or to a group of persons that may be viewed as a single decision maker for the purposes of the analysis. Theories that explicitly recognize multiple decision makers are discussed by Sen [96], Fishburn [42], and Keeney and Raiffa [66, Chap. 10].

The organization of this survey is as follows. The next section provides a discussion of methods for identifying the relevant criteria for an analysis. Then we review the prescriptive theories of multiattribute decision making under certainty and under risk. This dichotomy is a natural one, since the complexities introduced by risk require substantially different theories for these two cases. The final section provides a brief review of the applications that have been reported in the literature.

IDENTIFYING THE CRITERIA

We use the terms attributes, outcomes, criteria, and effects interchangeably throughout our discussion, even though some writers differentiate among them. The important question we address here is the identification of the criteria that are relevant for comparing alternatives. These criteria are simply the outcomes that are both affected by the choice of an alternative and affect the decision-maker's preference for the alternative. The task of recognizing the aspects of the system that will be affected by the choice of an alternative may be the most difficult and the most important task in the analysis. The development of the ability to predict outcomes and the evaluation scheme may have to be accomplished simultaneously or

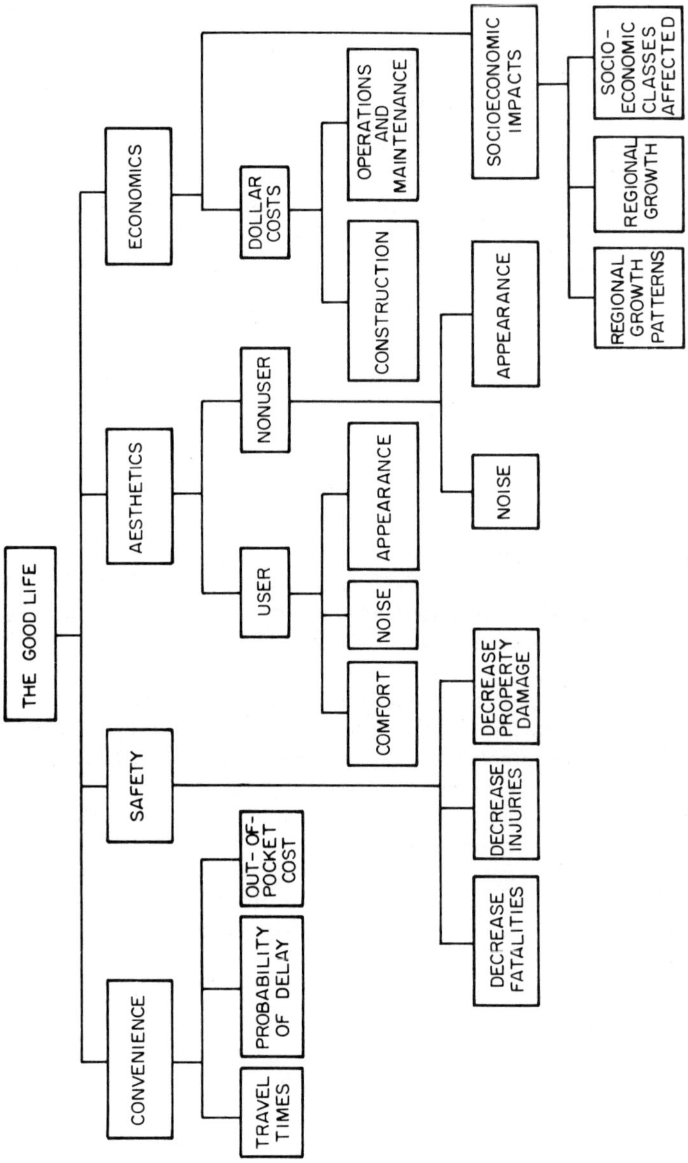

FIG. 2. A hierarchy of criteria for the urban transportation system problem.

iteratively, since the criteria may change as the decision maker learns more about the problem.

As an example, consider the imposition of a 55-mile-per-hour speed limit during the oil embargo in 1974. The stated objective of this law was to conserve fuel by forcing automobiles to travel slower. It also had the effect of communicating the seriousness of the situation to the general public. In addition, deaths and injuries from traffic accidents fell drastically during this period. On the negative side, this law also touched off a nationwide strike by truckers and curtailed the demand at businesses depending on motorists and tourists. It is not clear that Congress or the President were aware of all of these outcomes prior to the passage of the legislation.

Thus the criteria may include outcomes, such as the possibility of a strike by truckers, not directly related to the primary purpose of the alternative—conserving gasoline in this case. Nevertheless, these criteria must be included in the evaluation if they affect the preference of the decision maker.

In order to get started in identifying criteria, one may merely mull things over in an active fashion and identify many outcomes of choosing various alternatives. The decision maker can make a list of such outcomes, which may provide the basis for the determination of the final set of criteria.

As an alternative, the decision maker may begin with a general statement of his purpose, an overall objective, and refine this statement into more and more specific items. The result is a hierarchical representation of criteria that eventually ends at the lowest level with criteria whose associated values can be determined. This approach has been described by Manheim and Hall [78] and by Miller [81]. This important problem of identifying criteria is discussed in depth by Keeney and Raiffa [66, Chap. 2].

The example used by Manheim and Hall dealt with the establishment of an urban transportation system. They began with a "super goal" of "the good life," which seems particularly noncontroversial. They then suggested that any transportation system would be a means to that end if it were convenient, safe, aesthetically appealing, and economically attractive. These four rather vague subgoals were each decomposed further into more specific criteria. For example, safety was specified more sharply by the three criteria of a decrease in fatalities, a decrease in injuries, and a decrease in property damage. The complete hierarchy of criteria is shown in Fig. 2.

DECISIONS UNDER CERTAINTY

In this section it is assumed that the outcome $x(a) \in X$ of any alternative $a \in A$ is known with certainty. How can we choose the "best" or most preferred feasible alternative (or associated outcome)? We shall assume that for any outcome $x \in X$ there exists an associated feasible alternative. Thus we may speak of preferences for alternatives (e.g., $a \succ b$) or for their associated outcomes (e.g., $x(a) \succ x(b)$) without fear of ambiguity.

We begin with the statement of the conditions that must be satisfied in order to guarantee the existence of a utility function defined on the set of outcomes X. Next, we discuss when this utility function can be decomposed into an additive form in order to simplify its practical use. Finally, we focus on special solution strategies for the case where X is defined by a set of equality and inequality relationships.

Existence of a Value Function

A decision-maker's preferences for multiple outcomes ($X = X_1 \times \cdots \times X_n$) must satisfy the following conditions for any x', x'', $x''' \in X$ in order to guarantee the existence of an ordinal utility function (Luce et al. [73, p. 260]; see also Debreu [17] and Fishburn [36]).[1]

1. Transitivity: if $x' \succsim x''$ and $x'' \succsim x'''$, then $x' \succsim x'''$
2. Connectivity: $x' \succsim x''$ or $x'' \succsim x'$
3. Nonsatiety: if $x' \geq x''$ (for every i, $x_i' \geq x_i''$, and for some i, $x_i' > x_i''$), then $x' \succsim x''$
4. Continuity: if $x' \succsim x''$ and $x'' \succsim x'''$, then there is a real number λ such that $0 \leq \lambda \leq 1$ and $\lambda x' + (1 - \lambda)x''' \sim x''$

These four conditions are necessary and sufficient for the existence of a numerical function v such that for all x', $x'' \in X$,

$$v(x') \geq v(x'') \quad \text{if and only if } x' \succsim x'' \tag{1}$$

The function v is commonly called an ordinal utility function. However, we shall refer to it as a value function in order to distinguish it more clearly from the parallel development for the utility function in the risky case. v is unique up to a monotone increasing transformation; that is, if a value function v satisfies (1), any other value function constructed by a monotone increasing transformation of v will also represent the same preferences.

The conditions discussed above merely guarantee the existence of a value function, but do not suggest how such functions can be constructed. In fact, in many situations it may be practically impossible to construct these functions unless the preferences of the decision-maker also satisfy some other simplifying conditions. This is mainly because the provision of the amount of information needed for the construction of these functions would tax the capability of any human decision-maker. This leads us to the examination of some additional conditions which simplify the evaluation of multicriteria alternatives.

Additivity

The most common approach for evaluating the multicriteria alternatives under certainty is to use an additive representation. In the additive representation, a real value v is assigned to each outcome ($x_1, \ldots, x_n$) by

$$v(x_1, x_2, \ldots, x_n) = v_1(x_1) + v_2(x_2) + \cdots + v_n(x_n) \tag{2}$$

where v_i is a single attribute value function over X_i. Debreu [18] proved for $n \geq 3$ that if the substitution rate between X_i and X_j is independent of all other attributes for all pairs X_i and X_j, there exists a set of functions v_i, $i = 1, 2, \ldots, n$, such

[1]These conditions have not remained unchallenged. Fishburn [39] argues that the transitivity of statements of indifference may not hold. Other authors, including Aumann [2, 3] and Roy [87] do not assume connectivity.

that v in (2) satisfies (1). When this condition is satisfied, we say that X_i and X_j are preferentially independent of the remaining criteria.

Preferential independence plays such an important role in our development that we shall present an example. Suppose a decision maker uses the criteria çost, horsepower, and appearance to choose among alternative automobiles. Suppose he prefers the automobile described by the 3-tuple ($3500, 175 H.P., so-so) to one described by ($3400, 125 H.P., so-so). He will also prefer ($3500, 175 H.P., beautiful) to ($3400, 125 H.P., beautiful) if cost and horsepower are preferentially independent of appearance; that is, his preferences for automobiles that differ only in the criteria cost and horsepower will not depend on the common value of appearance.

Essentially, preferential independence implies that the indifference curves for any pair of criteria are unaffected by the fixed levels of the remaining criteria. Luce and Tukey [75] gave a proof of this proposition for n = 2, and Krantz [69] gave an algebraic proof for arbitrary n. Gorman's [50] results show that preference independence need not be verified for all pairs of X_i and X_j. Instead, for example, if X_i and X_j are preference independent of the remaining criteria for any fixed i and all $j \neq i$, then the additive representation in (2) will hold.

The representation in (2) facilitates the assessment of the value function v as the single dimensional value functions, the v_i's, can be constructed with a comparatively moderate information demand on the decision maker. However, the assessment techniques do require a series of tradeoff judgments between pairs of criteria while the remaining criteria are held constant, so the process may still be tedious and time-consuming. Keeney and Raiffa [66, Chap. 3] provide detailed descriptions and examples of the use of these procedures for constructing an additive value function.

So far we have assumed that the objective is simply to rank order the alternatives in A. Sometimes we may wish to order the differences in the strength of preference between pairs of alternatives (e.g., see Stevens [107]). We shall use the term measurable value function for a function that may be used to order the preference differences between the alternatives. Several alternative axiom systems have been proposed that imply the existence of a measurable value function (see Alt [1], Suppes and Winet [108], and Suppes and Zinnes [109]). A review of the state-of-the-art in difference measurement is provided by Krantz et al. [70, Chap. 4].

Krantz et al. [70, p. 492] and Dyer and Sarin [21] provide axioms for an additive measurable value function. The key condition, called difference independence by Dyer and Sarin, asserts that the preference difference between two alternatives that differ only in terms of one criterion does not depend on the common values of the other n - 1 criteria. When difference independence for one criterion is coupled with preferential independence for all pairs of criteria, the additive value function is also measurable.

Our automobile example may be helpful again in illustrating this concept. If the decision maker prefers a car with 150 H.P. to one with 125 H.P., ceteris paribus, and horsepower is difference independent of cost and appearance, then the preference difference between ($3500, 150 H.P., beautiful) and ($3500, 125 H.P., beautiful) must be equal to the preference difference between ($4500, 150 H.P., ugly) and ($4500, 125 H.P., ugly). Loosely speaking, his feelings about the value of the additional 25 H.P. do not depend on the cost and the appearance of the automobile.

This condition is important for assessment since it allows the construction of a value function v_i for each criterion i by direct rating or by other methods that involve the notion of strength of preference. Further, it is no longer necessary to simultaneously consider two or more criteria in the assessment procedure.

A commonly used approach for multicriteria decision making in practice is to set up a table with a row for each criterion and a column for each alternative. Each alternative is assigned a subjective rating on each criterion, and each criterion is assigned a numerical weighting to represent its importance. The weights are then multiplied by the ratings in their corresponding rows, and the products are summed down each column. The alternative associated with the column receiving the largest sum is ranked first, and so forth (see Edwards [23]). This approach is valid only if v is a measurable additive value function.

A special form of the additive model in (2) is

$$\sum_{i=1}^{n} \lambda_i x_i$$

which assumes $X_i \subseteq Re$ and where λ_i is the weight for the i-th criterion. This weighted linear model has been widely used in the descriptive approach for the assessment of value functions. In this approach a value function is inferred from previous choices or from the observed behavior of the decision maker. One strategy often encountered in the literature is to ask the decision maker to rate multicriteria alternatives on an interval scale (see Huber [57, 58] and Fischer and Edwards [31]). These ratings are treated as the dependent variable and the criteria as the independent variables in a multiple regression analysis of the responses. The resulting regression equation is then proposed as a value function for the decision maker.

Since these ratings are assumed to provide an interval scale where differences are also ordered, this approach implicitly assumes a measurable additive value function. Other methods that only require rankings of the alternatives are described by Green and Wind [51]. These latter methods do not require the assumption of difference independence.

Several researchers have found that the precise estimation of weights in the linear model may not be required in some decision situations and therefore all criteria can be equally weighted (e.g., see Dawes and Corrigan [16], Einhorn and Hogarth [25], and Wainer [116]). These researchers have also found that a linear model constructed using the data on the previous choices of the decision maker often outperforms the decision maker himself in the prediction of an external criterion of success; that is, the average of his responses is a better guide to the prediction. This seems to occur because the linear model is more consistent and not subject to random errors. A further discussion of man versus models of man has been provided in Libby [71, 72] and Goldberg [49].

Other special forms of the additive model are discussed in Krantz et al. [70], along with several interdependent cases (see Fishburn [40] and Srinivasan and Shocker [99] also). Dyer and Sarin [22] extend the results regarding measurable value functions to include multiplicative and other nonlinear forms, and describe several methods of assessment for these functions.

In some situations it may be convenient to use trade off schemes in evaluating the multicriteria alternatives. In these schemes, corresponding to every outcome an indifferent outcome is determined which has standard values on all other criteria except a chosen reference criterion. Thus tradeoffs are used to find $\hat{x}_1$ such that an outcome $(x_1, \ldots, x_n) \sim (\hat{x}_1, \bar{x}_2, \ldots, \bar{x}_n)$, where x_1 is selected as the reference criterion and $\bar{x}_2, \ldots, \bar{x}_n$ are conveniently chosen standard levels on the remaining criteria. In our automobile example, suppose cost is the reference criterion and the standard levels of horse power and appearance are, respectively, 120 H.P. and so-so. For each automobile described by the 3-tuple, e.g., ($8,000, 150 H.P., beautiful), the decision maker determines an indifferent outcome ($\hat{x}_1$, 120 H.P., so-so). Since all outcomes now differ only on the reference criterion, the decision maker would hopefully have less difficulty in ranking these outcomes. In the automobile example the ranking would be straightforward as less cost is always desirable and all outcomes differ only on cost. In general, however, the decision maker may have to consider the entire n-tuple in ranking these standardized outcomes.

A further discussion of tradeoff schemes and the estimation procedures is provided in Raiffa [85], von Winterfeldt and Fisher [114], MacCrimmon and Wehrung [77], Keeney and Raiffa [66], etc.

Mathematical Programming Approaches

So far we have assumed that all feasible alternatives are easily identifiable, or that corresponding to every x in X there exists an alternative a in A. In many situations a is a vector with m components and $A \subseteq R^m$ is a set defined by the constraints on the m components of a. These components of a are often termed variables in the optimization literature. For example, an alternative a could be the budget allocation to m different departments of an organization, but a is feasible only if the total allocation does not exceed the total budget. The set of all possible allocations that satisfy the budget constraint would be A. The optimization literature deals with precisely this situation when only one criterion needs to be maximized (or minimized) and certain assumptions (convexity, differentiability, etc.) for the constrained set of feasible alternatives A and the criterion function x(a) are satisfied.

Direct Assessment of v. In the multicriteria situation, $x_i(a)$ may be interpreted as a criterion function; that is, $x_i : A \rightarrow X_i \subseteq Re$. If a value function $v(x_1(a), \ldots, x_n(a))$ has been assessed, then the decision problem may be solved by determining the solution to the optimization problem

$$\max_{a \in A} v(x(a)) \tag{3}$$

If v is additive and each v_i is concave, then (3) may be solved by a standard convex programming algorithm. If v is not additive, (3) may not be a convex programming problem, and more general solution strategies will be required. Keefer [61] provides an example of an actual application illustrating the latter case.

Approximation of v. If v is additive and each v_i is concave, (3) is a separable nonlinear programming problem. When A is a convex polyhedron defined by a set of linear constraints, a common approach to solving (3) is to construct a piecewise linear approximation to each v_i, and then apply a standard linear programming algorithm (Wagner [115]).

Dyer [20] has noted that a special case of this strategy of piecewise linear approximation of each v_i is equivalent to solving the multicriteria optimization problem using goal programming (see Charnes and Cooper [10, 11], Ijiri [59], and

Kornbluth [68]). Goal programming approximates each v_i by asking the decision maker to identify a goal L_i for each criterion, $i = 1, \ldots, n$. Positive deviations from L_i are weighted by λ_i^+ and negative deviations from L_i are weighted by λ_i^-. The objective is to minimize the weighted sum of these deviations.

As shown by Charnes, Cooper, and Ferguson [12], this problem may be formulated as

minimize

$$\sum_{i=1}^{n} (\lambda_i^+ y_i^+ + \lambda_i^- y_i^-)$$

subject to

$$x_i(a) + y_i^- - y_i^+ = L_i; \qquad i = 1, \ldots, n$$

$$y_i^+ \geq 0, \quad y_i^- \geq 0; \qquad i = 1, \ldots, n$$

$$a \in A$$

and solved using conventional linear programming techniques. Although it is not readily apparent, goal programming implicitly assumes that v is a measurable additive value function (Dyer [20]).

Man-Machine Interaction. Rather than attempting to assess v in detail, we might try to obtain the minimal amount of information regarding v that is necessary to solve a multicriterion problem, and thus reduce the burden on the decision maker. One such scheme has been proposed by Geoffrion [47]. In this scheme it is assumed that A is a compact, convex subset of an m-dimensional Euclidean space defined by a family of differentiable constraints on the m components of a, and there exists an implicit ordinal preference function

$$v(x_1(a), \ldots, x_n(a))$$

where v is concave, increasing, and differentiable, and the x_i are differentiable, concave real valued criterion functions. It is further assumed that A and the x_i are known explicitly but v is only implicitly known; that is, v has not been assessed but the decision maker is able to provide some specific kinds of information about it. A large-step gradient ascent algorithm is employed to maximize v, and the decision maker is required to provide only local information about his preferences during the optimization process.

Geoffrion, Dyer, and Feinberg [48] discuss this approach in detail and illustrate it by an application to a resource allocation problem in an academic department. Dyer [19] describes a time-sharing computer program to conveniently implement it. Here we summarize the salient aspects of this approach.

Several standard methods in nonlinear programming consist of a direction-finding subproblem and a one-dimensional optimization problem for the choice of a "step-size" in that direction (see Zoutendijk [132]). The marginal substitution rates between the i-th criterion and a fixed reference criterion at any point a in A define the tangential hyperplane to the indifference surface passing through the point. Thus the responses to the (n - 1) questions, "With all other criteria held constant at the point x(a), how much would you be willing to decrease the value of criterion i to obtain an increase of Δx_1 in criterion 1?" provide an approximation of the gradient of v at a. This information is used to determine a feasible direction of movement along which v is initially increasing. A new feasible solution with a higher v value is obtained by asking the decision maker to identify his most preferred point in this direction. Thus, at each iteration of the algorithm, the decision maker provides (n - 1) marginal substitution rates and selects the value of step size such that $v(x_1(a), \ldots, x_n(a))$ is maximized. Note that if v is linear, then only one iteration would be needed in the above scheme to identify the optimal or most preferred solution. Some extensions of Geoffrion's method and descriptions of other applications are provided by Wehrung [120] and Oppenheimer [83].

In a similar spirit, Zionts and Wallenius [131] propose a method within the framework of the simplex algorithm for linear programming that sequentially solicits certain tradeoff information from the decision maker in the choice of the most preferred alternative. This approach utilizes the concept of an efficient solution as a key element of the interactive strategy. An alternative a is called efficient if there is no other alternative in A having a more preferred outcome for at least one criterion and equally preferred outcomes for the other criteria. The assumption is also made that v is strictly increasing in each x_i.

Initially, a weighted-sum problem is optimized to produce an efficient solution (the choice of positive weights can be arbitrary). Since v is only implicit, the general idea in this approach is to involve the decision maker in the choice of a nonbasic variable to enter the basis at each iteration of the simplex method. The procedure screens out those nonbasic variables which, when introduced into the basis, do not lead to an efficient adjacent extreme point solution. For a given nonbasic variable that would lead to an efficient solution, a set of tradeoffs, in which some criteria values are increased and others are reduced, is presented to the decision maker. These tradeoffs are the reduced costs for each of the criterion functions corresponding to the given nonbasic variable. The decision-maker's responses regarding whether he considers the tradeoffs desirable, undesirable, or neither are used to impute the new set of multipliers, or weights, and the associated efficient solution is obtained; the process is repeated until no further improvements are found.

Within the linear programming framework, the key difference between this approach and Geoffrion's approach is that the former requires several ordinal comparisons whereas the latter requires fewer (n - 1) but more difficult, indifference judgments. Wallenius [118] describes an evaluation of these two methodologies.

Several other interactive approaches based on the concept of "best compromise" have been proposed (Saska [92], Benayoun et al. [6], Belensen and Kapur [4]). The vector of most preferred feasible outcomes determined by optimizing each criterion function independently represents the "ideal point." In general this ideal point is not feasible. The interactive scheme is guided by the relative distance from the idea point in the Benayoun et al. procedure. The other schemes employ different strategies for exploring the feasible set in moving toward the "best compromise" solution.

All of the above procedures are developed within the continuous mathematical programming context. Similar procedures for discrete mathematical programming with multiple criterion functions have recently been proposed. In principle, Geoffrion's approach can be employed in conjunction with standard branch and bound procedures, but, unless some clever ways are found to prune the branches, the amount of information required of the decision maker would be prohibitive for any practical use. Zionts [129] has suggested an extension of the interactive strategy by Zionts and Wallenius for solving multicriteria problems with discrete alternatives.

Generation of Efficient Solutions. When the assessment or approximation of v is not practical, and the man-machine interactive strategies are not considered appropriate, the mathematical programming formulation of a multicriteria problem may be used to generate a subset of A that contains the optimal solution. If we simply assume that v is strictly increasing in each criterion x_i, i = 1, ..., n, the optimal solution will be an efficient solution.

In the linear programming framework, we assume that A is a polyhedral set defined by a set of linear constraints on the m components of a, and each x_i is also linear. If v is also assumed to be linear (but not explicitly known), then it would be sufficient to examine only the efficient extreme points of the polyhedral A. Two questions arise: How do we identify all of the efficient alternatives in A, and how do we select the most preferred alternatives from the set of efficient alternatives?

Several algorithms have been proposed to identify all the efficient alternatives in A (see Philip [84], Zeleny [128], and Evans and Steuer [26]). As described in Steuer [104], these algorithms involve three phases. In Phase I an initial feasible extreme point is identified as in single objective linear programs. An efficient extreme point is obtained in Phase II by solving the simple weighted-sums problem:

maximize

$$\sum_{i=1}^{n} \lambda_i x_i(a)$$

subject to

$$\lambda_i \geq 0; \quad i = 1, \ldots, n$$

$$a \in A$$

In Phase III the algorithms determine if a given efficient alternative is the only one that exists, and if not, the remaining efficient extreme points are found. Several different types of procedures have been proposed to obtain the remaining efficient extreme points in Phase III (see Zeleny [128], Yu and Zeleny [126], and Evans and Steuer [26]). Computer programs are also available to generate all efficient extreme points (Steuer [102, 103]).

Unfortunately, the set of efficient alternatives is often quite large. Any attempt to reduce the efficient set of alternatives requires information about the decision-maker's preferences, and thus some interactive approach is called for. Several schemes have been suggested (Yu [125], Zeleny [127], and Steuer [105]), but in these schemes the interaction with the decision maker is not dictated by the algorithm and no attempt is made to uncover the preference function of the decision maker.

Guided Trial and Error Procedures

In relatively more complex problem areas, much of the work has focused on models and methods that will predict the outcomes of an alternative in terms of multiple criteria. The decision maker chooses an alternative and uses this prediction of the outcome as a basis for a trial and error modification of this initial solution. Often these applications involve problems requiring the design of a system that can be represented visually on a cathode-ray tube (CRT) display. There is a huge literature in this area, so we shall summarize only a few specific approaches in order to provide a sampling of the work that has been done.

INTUVAL is a computer graphic program that utilizes computer graphics for the systematic evaluation and iterative modification of design proposals (Kamnitzer and Hoffman [60]). Although the concept is more general, INTUVAL has been applied to the problem of designing a freeway system in an urban area. The designer plots a freeway route using a light-pen on a graphic display of a map of the region. The program then evaluates the route on the criteria driving time, visual interest, safety, conservation, community patterns, and cost. These evaluations are presented in the form of bar graphs. The decision maker may then modify his design, and observe the changes in the criteria.

There are other interactive graphic procedures that assist in multicriterion design problems. Schneider, Miller, and Friedman [95] describe an application to the problem of locating and sizing park-ride lots. Weinzapfel and Handel [121] give examples of the use of the system IMAGE for assisting in determining the physical layout of new facilities.

The use of interactive graphics can also be applied to other problems that can be represented visually. The Geodata Analysis and Display System (GADS) has been made available to users studying urban growth (Cristiani et al. [14]), designing police beats (Carlson and Suttan [9]), designing school boundaries (Wytock and Grall [124]), and closing schools (Holloway and Montey [55]). In addition to the visual display of each alternative, the system will also provide tables showing the ratings of each alternative in terms of multiple criteria.

DECISIONS UNDER RISK

Recall from Fig. 1 that the outcome is determined by both the alternative a and the state s of the environment. In the previous section on decision making under certainty, we assumed that the state s was known with probability equal to 1, so the notation s was deleted for simplicity. In a risky decision situation there exists a probability distribution over the states; as a consequence, the outcome of an alternative $a \in A$ is characterized by a probability distribution over X.

We denote $\tilde{X}$ as the set of all simple probability distributions over X. We shall assume that for any $p \in \tilde{X}$ there exists an $a \in A$, and therefore p could be termed a <u>risky alternative</u>. The outcome of an alternative $p \in \tilde{X}$ might be represented by the <u>lottery</u> which assigns probabilities $p_1, p_2, \ldots, p_\ell$, adding to one, to the outcomes $x^1, x^2, \ldots, x^\ell \in X$, respectively. Recall that for simplicity we have assumed that s is finite.

For the choice among risky alternatives, the expected utility model is appropriate if certain assumptions about the decision-maker's preference relation $\succsim$ over $\tilde{X}$ are satisfied. The expected utility model states that for any $p, q \in X$,

$$p \succsim q \quad \text{if and only if} \quad \sum_{x \in X} p(x)\, u(x) \geq \sum_{x \in X} q(x)\, u(x)$$

where u is a real valued utility function on X. The assumptions that must be satisfied for expected utility to be an appropriate guide for the selection of a risky alternative were first stated by von Neumann and Morgenstern [112]. Since then, alternative sets of assumptions have been suggested for various cases of the model (e.g., see Marschak [79], Herstein and Milnor [53], Luce and Raiffa [74], and Fishburn [38]).

In the research on multicriteria decision making under risk, the emphasis has been on "independence conditions" which, if satisfied, would allow the decomposition of the utility function into easily assessed components. Thus the expected utility model could be estimated while placing relatively moderate information requirements on the decision maker.

Additivity

A major portion of the literature in multicriteria utility theory in risky choice situation deals with the additive utility function of the form

$$u(x_1, x_2, \ldots, x_m) = \sum_{i=1}^{n} k_i u_i(x_i)$$

where u and u_i are utility functions scaled from zero to one and $\sum_{i=1}^{n} k_i = 1$. Fishburn [33, 37, 38] has derived necessary and sufficient conditions for a utility function to be additive. Fishburn's condition is commonly known as the marginality condition. Simply stated, the marginality condition requires that the desirability of any lottery over X should depend only on the marginal probability distributions over X_i and not on their joint probability distribution. If the marginality condition is satisfied, then the attributes are called value independent, or sometimes just independent of one another.

As an illustrative example, consider a simplified economy in which a plan is evaluated on the amount of grain production, the balance of payments, and the employment level. The marginality condition would require a planner to be indifferent between the two plans shown in Table 1 since each outcome has the identical marginal probability of occurrence in each plan. It is quite possible that the planner may prefer Plan 1 to avoid a 1/3 chance of the poor outcome (95, -, 2) in Plan 2. If so, the marginality condition would be violated, so that no additive utility function would exist.

TABLE 1

	Plan 1			Plan 2		
Attributes/Probabilities	1/3	1/3	1/3	1/3	1/3	1/3
Food availability (million tons)	95	100	105	95	100	105
Balance of payment (surplus (+), deficit (-))	+	+	-	-	+	+
Economic growth (%)	3	2	4	2	3	4

If the marginality assumption is verified, then the n-dimensional utility function decomposes into n single dimension utility functions. Methods for assessing the single dimension utility functions are reviewed in Keeney and Raiffa [66, Chap. 4].

The Multiplicative Utility Function

A widely used independence condition is "utility independence." The concept of utility independence was first introduced by Keeney [62]. Loosely speaking, a criterion is said to be utility independent of the other criteria if the decision-maker's preferences for lotteries over this criterion do not depend on the fixed levels of the remaining criteria. For example, consider again the problem of planning in the simplified economy. It seems that a planner's preferences for uncertain employment levels may not depend on the fixed levels of the other two criteria. However, it is quite likely that his preferences for an uncertain amount of grain production may be influenced by the levels of the balance of payments and employment in the economy. A planner may prefer a 0.6 chance of 104 million tons and a 0.4 chance of 94 million tons of grain production to a 0.5 chance of 100 million tons and 0.5 chance of 98 million tons when the balance of payments is good, but his preferences may reverse for a poor balance of payments since a favorable balance of payments may facilitate grain imports. In this situation, employment is utility independent of the remaining criteria, but the amount of grain production is not.

If we denote $X_{\bar{i}} \equiv X_1 \times \cdots \times X_{i-1} \times X_{i+1} \times \cdots \times X_n$, then X_i is utility independent of $X_{\bar{i}}$ if and only if there exists $u(x_i, x_{\bar{i}})$ such that

$$u(x_i, x_{\bar{i}}) = g(x_{\bar{i}}) + h(x_{\bar{i}})\, u(x_i, x_{\bar{i}}^0); \qquad \text{for all } x_{\bar{i}}$$

where g and h are scalar valued functions and $x_{\bar{i}}^0$ represents one specific level of the $X_{\bar{i}}$. This result is due to Keeney [62]. Note that if X_i is utility independent of $X_{\bar{i}}$, then it does not necessarily follow that $X_{\bar{i}}$ is utility independent of X_i (see Raiffa [85] for an illustrative example). Moreover, if each X_i is utility independent of the respective $X_{\bar{i}}$, it does not follow that any $X_{\bar{i}}$ is utility independent of the corresponding X_i (Keeney [64]).

If the appropriate utility independence conditions are satisfied, the task of estimating the utility function is simplified for the simple reason that only conditional utility functions over each attribute need to be assessed. A conditional utility function is assessed on a single attribute while the other n - 1 attributes are held fixed; in general, it is conditional on the fixed values of these n - 1 attributes.

For two attributes, if X_1 is utility independent of X_2 but X_2 is not utility independent of X_1, Raiffa [85] and Keeney [63] give a procedure for estimating the utility function u that requires only three one-attribute conditional utility functions. Keeney [63] shows that if X_1 is utility independent of $X_2 \times X_3$ and X_2 is utility independent of $X_1 \times X_3$, then six one-attribute conditional utility functions need to be assessed in order to estimate u.

A nonadditive decomposition of the utility function that has been used in several applications in recent years is multiplicative or log-additive. Keeney [65] shows that if for some attribute i, X_i is utility independent of $X_{\bar{i}}$, and $X_i \times X_j$ are preferentially independent of the remaining attributes for all $j \neq i$, then the following holds: either

$$u(x) = \sum_{i=1}^{n} k_i u_i(x_i)$$

or

$$1 + ku(x) = \prod_{i=1}^{n} (1 + kk_i u_i(x_i))$$

where u and u_i are utility functions scaled from zero to one, the k_i are scaling constants such that $0 < k_i < 1$, and $k > -1$ is a nonzero scaling constant, satisfying $(1 + k) = \prod_{i=1}^{n} 1 + kk_i$. This result is important because it allows both additive and multiplicative representations of the utility function. Keeney [65] also gives a procedure to estimate the utility function.

It should be noted that value independence implies that any subset of attributes is utility independent of the rest. However, utility independence conditions alone do not imply value independence. Therefore, value independence is a sufficient condition for utility independence, but the utility independence conditions are necessary conditions for value independence.

Other independence conditions have been identified that lead to more complex nonadditive decompositions of the utility function. Examples are provided by Fishburn [41, 44] and Farquhar [27, 28]. These general conditions and several others are reviewed by Farquhar [29].

Recently a new direction of research in approximating the utility function has begun. Fishburn [45] assumes no independence condition and shows how $u(x_1, \ldots, x_n)$ can be approximated. Bell [5] uses interpolations of conditional utility functions in the approximate estimations of the utility functions. Kirkwood's [67] notion of parametrically-dependent preferences can be viewed as an approximation to utility independence condition.

When the multidimensional u can be decomposed into some well-known forms (e.g., additive or multiplicative), a complete assessment of the utility function may not be necessary to identify a most preferred alternative from a finite set of available alternatives. Sarin [89] shows that the information required from the decision maker might be reduced considerably if a sequential strategy is adopted using the framework of linear programming in the evaluation of alternatives (see Fishburn [32, Chap. 11] also).

When decision alternatives are few, a tradeoff approach to obtain $\hat{x}_i$ such that

$$(\hat{x}_i, \bar{x}_{\bar{i}}) \sim (x_i, x_{\bar{i}})$$

can be used. The $u(\hat{x}_i, \bar{x}_{\bar{i}})$ can then be estimated for computation of expected utility. In general, this approach does not require any independence conditions to be satisfied. Of course, if X_i is utility independent of $X_{\bar{i}}$, the assessment of u is greatly simplified since it does not depend on the value of $\bar{x}_{\bar{i}}$.

Even though the tradeoff approach and the decomposition approach are identical in theory, some authors, especially the Decision Analysis Group of Stanford Research Institute, have maintained that the tradeoff approach is more practical (see Matheson and Roths [80], Stanford Research Institute [100], and Howard et al. [56] for several applications of the tradeoff approach).

We shall conclude this section by citing a few references on the assessment procedures for the multicriteria expected utility model. The two components required in the expected utility model are the probability distributions and the utility functions. If sample data are inadequate to estimate the probability distributions, subjective estimates are needed. Winkler [122, 123], Savage [93], Spetzler and Stael von Holstein [98], and Hogarth [54] provide assessment methods and surveys for probability encoding. The assessment of utilities (and criteria weights/scaling constants) is discussed extensively in the book by Keeney and Raiffa [66], a survey article by Fishburn [35], and by several other authors (e.g., Raiffa [85] and Huber [57, 58]). Computer programs are also available to assist the decision maker (and analyst) in the estimation of the utility function (Schlaifer [94], Nair and Sicherman [82]).

If utilities are precisely estimated but only incomplete information on probabilities (e.g., ordinal rank and interval estimates) can be obtained, the approaches of Fishburn [32, 34] and Sarin [90] can be used for choosing among alternatives. Sarin [88] also discusses how a preferred alternative may be identified within the expected utility framework when both utilities and probabilities are incompletely known. Also see Roy [86] for related work.

APPLICATIONS

Several applications of multicriteria utility theory are discussed in the book by Keeney and Raiffa [66]. The most widely applied model is the additive utility model (see Fishburn [36], Slovic and Lichtenstein [97], Brown et al. [8], Halter et al. [52], and the publication by the Stanford Research Institute [101]). The applications of nonadditive utility models are largely restricted to the multiplicative functional form (Keeney and Raiffa [66], Bodily [7], Sarin et al. [91]). Edwards [23] gives several examples of the implicit use of the additive measurable value functions.

A great majority of the applications deal with decisions in which an alternative is chosen from a finite number of well-defined alternatives. Keefer [61] provides an interesting application where utility theory is used in conjunction with mathematical programming.

Within the mathematical programming framework, Charnes and Cooper [11], Geoffrion et al. [48], Steuer and Schuler [106], Wallenius et al. [119], and Zionts and Deshpande [130] discuss several applications. The number of applications using multicriteria mathematical programming techniques has been relatively small, but seems likely to grow in future years.

A promising area for the use and application of multicriteria decision making is in real-time decision contexts. This is an area where descriptive, normative, heuristic, and several other decision process oriented approaches would all be required to understand the decision making and educate the decision maker in making "better" decisions.

ACKNOWLEDGMENTS

It is a pleasure to acknowledge the helpful comments by K. Nair and L. Schwarz.

REFERENCES

1. Alt, F., Uber die Messbarkeit des Nutzens, Z. Nationalokonome 7, 161-169 (1936); reprinted (in English) in J. S. Chipman, L. Hurwicz, M. K. Richter, and H. F. Sonnenschein (eds.), Preferences, Utility and Demand, Harcourt Brace Jovanovich, New York, 1971.
2. Aumann, R. J., Utility theory without the completeness axiom, Econometrica 30, 445-462 (1962).
3. Aumann, R. J., Utility theory without the completeness axiom: A correction, Econometrica 32, 210-212 (1964).
4. Belensen, S. M., and K. C. Kapur, An algorithm for solving multicriterion linear programming problems with examples, Oper. Res. Q. 24, 65-77 (1973).
5. Bell, D. E., Joint Interpolation Independence, presented at the Conference on Multiple Criteria Problem Solving—Theory, Methodology, and Practice, The State University of New York at Buffalo, August 22-26, 1977.
6. Benayoun, R., J. de Montgolfier, J. Tergny, and O. Laritchev, Linear programming with multiple objective functions: Step method (STEM), Math. Programming 1, 366-375 (1971).
7. Bodily, S. E., The Utilization of Frozen Red Cells in Blood Banking Systems: A Decision Theoretic Approach, Technical Report 92, Operations Research Center, M.I.T., Cambridge, Massachusetts, 1974.
8. Brown, R. V., A. S. Kahr, and C. Peterson, Decision Analysis for the Manager, Holt, Rinehart, and Winston, New York, 1974.
9. Carlson, E. D., and J. A. Sutton, A Case Study of Non-Programmer Interactive Problem-Solving, IBM Research Report RJ 1382, IBM Research Laboratory, San Jose, California, April 1974.
10. Charnes, A., and W. W. Cooper, Management Models and Industrial Applications of Linear Programming, Wiley, New York, 1961.
11. Charnes, A., and W. W. Cooper, Goal programming and multiple objective optimizations, Part 1, Eur. J. Oper. Res. 1, 39-54 (1977).
12. Charnes, A., W. W. Cooper, and R. Ferguson, Optimal estimation of executive compensation by linear programming, Manage. Sci. 1(2), 138-151 (January 1955).
13. Coombs, C. H., A Theory of Data, Wiley, New York, 1964.
14. Cristiani, E. J., et al., An interactive system for aiding evaluation of local government policies, IEEE Trans. Syst., Man Cybern. SMC-3(2), 141-146 (March 1973).
15. Dawes, R. M., Social selection based on multidimensional criteria, J. Abnorm. Soc. Psychol. 68, 104-109 (1964).
16. Dawes, R. M., and B. Corrigan, Linear models in decision making, Psychol. Bull. 81, 95-106 (1974).
17. Debreu, G., Representation of a preference ordering by a numerical function, in Decision Processes (R. M. Thrall, C. H. Coombs, and R. L. Davis, eds.), Wiley, New York, 1954.

18. Debreu, G., Topological methods in cardinal utility theory, in Mathematical Methods in the Social Sciences, Stanford University Press, Stanford, California, 1960, pp. 16-26.
19. Dyer, J. S., A time-sharing computer program for the solution of the multiple criteria problem, Manage. Sci. 19, 1379-1383 (1973).
20. Dyer, J. S., On the Relationship between Multiattribute Utility and Goal Programming, Discussion Paper, Management Science Study Center, Graduate School of Management, UCLA, Los Angeles, California, October 1977.
21. Dyer, J. S., and R. K. Sarin, An Axiomatization of Cardinal Additive Conjoint Measurement Theory, WP265, Western Management Science Institute, University of California, Los Angeles, 1977.
22. Dyer, J. S., and R. K. Sarin, Measurable Multiattribute Value Functions, WP275, Western Management Science Institute, University of California, Los Angeles, 1977.
23. Edwards, W., How to use multiattribute utility measurement for social decision making, IEEE Trans. Syst., Man Cybern. SMC-7(5), (May 1977).
24. Einhorn, H. J., The use of nonlinear, noncompensatory models in decision making, Psychol. Bull. 73, 221-230 (1970).
25. Einhorn, H. J., and R. M. Hogarth, Unit weighing schemes for decision making, Organ. Behav. Human Performance 13, 171-192 (1975).
26. Evans, J. P., and R. E. Steuer, A revised simplex method for linear multiple objective programs, Math. Programming 5, 54-72 (1973).
27. Farquhar, P. H., A fractional hypercube decomposition theorem for multiattribute utility functions, Oper. Res. 23, 941-967 (1975).
28. Farquhar, P. H., Pyramid and semicube decompositions of multiattribute utility functions, Oper. Res. 24, 256-271 (1976).
29. Farquhar, P. H., A survey of multiattribute utility theory and applications, TIMS Stud. Manage. Sci. 6, 59-89 (1977).
30. Fischer, G. W., Convergent validation of decomposed multi-attribute utility assessment procedures for risky and riskless decisions, Organ. Behav. Human Performance 18, 295-315 (1977).
31. Fischer, G. W., and W. Edwards, Technological Aids for Inference, Evaluation and Decision Making: A Review of Research and Experience, Engineering Psychology Laboratory, University of Michigan, Ann Arbor, Michigan, 1973.
32. Fishburn, P. C., Decisions and Value Theory, Wiley, New York, 1964.
33. Fishburn, P. C., Independence in utility theory with whole product sets, Oper. Res. 13, 28-45 (1965).
34. Fishburn, P. C., Analysis of decisions with incomplete knowledge of probabilities, Oper. Res. 13, 217-237 (1965).
35. Fishburn, P. C., Methods of estimating additive utilities, Manage. Sci. 13, 435-453 (1967).
36. Fishburn, P. C., Utility theory, Manage. Sci. 14, 335-378 (1968).
37. Fishburn, P. C., A study of independence in multivariate utility theory, Econometrica 37, 107-121 (1969).
38. Fishburn, P. C., Utility Theory for Decision Making, Wiley, New York, 1970.
39. Fishburn, P. C., Intransitive indifferences in preference theory: A survey, Oper. Res. 18, 207-228 (1970).
40. Fishburn, P. C., Interdependent preferences on finite sets, J. Math. Psychol. 9, 225-236 (1972).

41. Fishburn, P. C., Bernoullian utilities for multiple-factor situations, in Multiple Criteria Decision Making (J. L. Cochrane and M. Zeleny, eds.), University of South Carolina Press, Columbia, South Carolina, 1973, pp. 47-61.
42. Fishburn, P. C., The Theory of Social Choice, Princeton University Press, Princeton, New Jersey, 1973.
43. Fishburn, P. C., Lexicographic orders, utilities and decision rules: A survey, Manage. Sci. 20, 1442-1471 (1974).
44. Fishburn, P. C., Von-Neumann-Morgenstern utility functions on two attributes, Oper. Res. 22, 35-45 (1974).
45. Fishburn, P. C., Approximations of two-attribute utility functions, Math. Oper. Res. 2, 30-44 (1977).
46. Franklin, B., Letter to Joseph Priestly, in The Benjamin Franklin Sampler, Fawcett, New York, 1956.
47. Geoffrion, A., Vector Maximal Decomposition Programming, Working Paper No. 164, University of California, Los Angeles, 1970.
48. Geoffrion, A., J. S. Dyer, and A. Feinberg, An interactive approach for multi-criterion optimization, with an application to the operation of an academic department, Manage. Sci. 19, 357-368 (1972).
49. Goldberg, L. R., Man versus model of man: Just how conflicting is that evidence?, Organ. Behav. Human Performance 16, 13-22 (1976).
50. Gorman, W. M., Symposium on aggregation: The structure of utility functions, Rev. Econ. Stud. 35, 367-390 (1968).
51. Green, P. E., and Y. Wind, Multiattribute Decisions in Marketing: A Measurement Approach, Dryden, Hillsdale, Illinois, 1973.
52. Halter, A. N., and G. W. Dean, Decisions Under Uncertainty with Research Applications, South-Western, Cincinnati, Ohio, 1971.
53. Herstein, I. N., and J. Milnor, An axiomatric approach to measurable utility, Econometrica 21, 291-297 (1953).
54. Hogarth, R. M., Cognitive processes and the assessment of subjective probability distributions, J. Am. Stat. Assoc. 70, 271-289 (1975).
55. Holloway, C. A., and P. E. Montey, Implementation of an Interactive Graphics Model for Design of School Boundaries, Research Paper No. 229, Graduate School of Business, Stanford University, Stanford, California, March 1976.
56. Howard, R. A., J. E. Matheson, and D. W. North, The decision to seed hurricanes, Science 176, 1191-1202 (1972).
57. Huber, G. P., Methods for quantifying subjective probabilities and multiattribute utilities, Decision Sci. 5, 430-458 (1974).
58. Huber, G. P., Multi-attribute utility models: A review of field and field-like studies, Manage. Sci. 20, 1393-1402 (1974).
59. Ijiri, Y., Management Goals and Accounting for Control, North Holland, Amsterdam, 1965.
60. Kamnitzer, P., and S. Hoffman, INTUVAL: An interactive computer graphic aid for design and decision making in urban planning, in Proceedings of ERDA II, 1970.
61. Keefer, D. L., Applying Multiobjective Decision Analysis to Resource Allocation Planning Problems, presented at the Conference on Multiple Criteria Problem Solving—Theory, Methodology, and Practice, The State University of New York at Buffalo, August 22-26, 1977.
62. Keeney, R. L., Multidimensional Utility Functions: Theory, Assessment, and Applications, Technical Report No. 43, Operations Research Center, M.I.T., Cambridge, Massachusetts, October 1969.

63. Keeney, R. L., Utility independence and preferences for multiattributed consequences, Oper. Res. 19, 875-893 (1971).
64. Keeney, R. L., Concepts of independence in multiattribute utility theory, in Multiple Criteria Decision Making (J. L. Cochrane and M. Zeleny, eds.), University of South Carolina Press, Columbia, South Carolina, 1973.
65. Keeney, R. L., Multiplicative utility functions, Oper. Res. 22, 22-34 (1974).
66. Keeney, R. L., and H. Raiffa, Decisions with Multiple Objectives: Preferences and Value Tradeoffs, Wiley, New York, 1976.
67. Kirkwood, C. W., Parametrically dependent preferences for multiattributed consequences, Oper. Res. 24, 92-103 (1976).
68. Kornbluth, J. S. H., A survey of goal programming, OMEGA 1, 193-205 (1973).
69. Krantz, D. H., Conjoint measurement: The Luce-Tukey axiomatization and some extensions, J. Math. Psychol. 1, 248-277 (1964).
70. Krantz, D. H., R. D. Luce, P. Suppes, and A. Tversky, Foundations of Measurement, Academic, New York, 1971.
71. Libby, R., Man versus model of man: Some conflicting evidence, Organ. Behav. Human Performance 16, 1-12 (1976).
72. Libby, R., Man versus model of man: The need for a nonlinear model, Organ. Behav. Human Performance 16, 23-26 (1976).
73. Luce, R. D., R. R. Bush, and E. Galanter, Handbook of Mathematical Psychology, Vol. 3, Wiley, New York, 1965.
74. Luce, R. D., and H. Raiffa, Games and Decisions, Wiley, New York, 1957.
75. Luce, R. D., and J. W. Tukey, Simultaneous conjoint measurement: A new type of fundamental measurement, J. Math. Psychol. 1, 1-27 (1964).
76. MacCrimmon, K. R., An overview of multiple objective decision making, in Multiple Criteria Decision Making (J. L. Cochrane and M. Zeleny, eds.), University of South Carolina Press, Columbia, South Carolina, 1973, pp. 18-44.
77. MacCrimmon, K. R., and D. A. Wehrung, Trade-Off Analysis: Indifference and Preferred Proportion, WP323, Faculty of Commerce and Business Administration, University of British Columbia, Vancouver, Canada, 1975.
78. Manheim, M., and F. Hall, Abstract Representation of Goals, P-67-24, Department of Civil Engineering, M.I.T., Cambridge, Massachusetts, January 1968.
79. Marschak, J., Rational behavior, uncertain prospects, and measurable utility, Econometrica 18, 111-141 (1950).
80. Matheson, J. E., and W. T. Roths, Decision analysis of space projects, in Proceedings of the National Symposium, Saturn/Apollo and Beyond, American Astronautical Society, 1967.
81. Miller, J., Assessing Alternative Transportation Systems, RM-5865-DOT, Rand Corporation, Santa Monica, California, April 1969.
82. Nair, K., and A. Sicherman, Environmental Assessment Methodology: Solar Power Plant Applications, Woodward-Clyde Consultants Study conducted for Electric Power Research Institute under grant number RP551, 1977.
83. Oppenheimer, K. R., A Proxy Approach to Multi-Attribute Decision Making, Ph.D. Dissertation, Department of Engineering-Economic Systems, Stanford University, March 1977.
84. Philip, J., Algorithms for the vector maximization problem, Math. Programming 2, 207-229 (1972).
85. Raiffa, H., Preferences for Multi-Attributed Alternatives, Memorandum RM-5868-DOT, Rand Corporation, Santa Monica, California, 1969.

86. Roy, B., Problems and methods with multiple objective functions, Math. Programming 1, 239-266 (1971).
87. Roy, B., How outranking relation helps multiple criteria decision making, in Multiple Criteria Decision Making (J. L. Cochrane and M. Zeleny, eds.), University of South Carolina Press, Columbia, South Carolina, 1973, pp. 179-201.
88. Sarin, R. K., Interactive Procedures for Evaluation of Multi-Attributed Alternatives, Working Paper 232, Western Management Science Institute, University of California, Los Angeles, June 1975.
89. Sarin, R. K., Interactive evaluation and bound procedure for selecting multiattribute alternatives, TIMS Stud. Manage. Sci. 6, 211-224 (1977).
90. Sarin, R. K., Elicitation of subjective probabilities in the context of decision making, Decision Sci. 9, 37-48 (1978).
91. Sarin, R. K., A. Sicherman, and K. Nair, Evaluating proposals using decision analysis, IEEE Trans. Syst., Man Cybern. SMC-8, 128-131 (1978).
92. Saska, J., Linear programming, Econ. Mat. Obzor 4, 359-373 (1968).
93. Savage, L. J., Elicitation of personal probabilities and expectations, J. Am. Stat. Assoc. 66, 783-801 (1971).
94. Schlaifer, R. O., Computer Programs for Elementary Decision Analysis, Harvard Business School, Boston, Massachusetts, 1971.
95. Schneider, J. B., D. G. Miller, and T. W. Friedman, Locating and sizing park-ride lots with interactive computer graphics, Transportation 5, 389-406 (1976).
96. Sen, A. K., Collective Choice and Social Welfare, Holden-Day, San Francisco, 1970.
97. Slovic, P., and S. Lichtenstein, Comparison of Bayesian and regression approaches to the study of information processing in judgement, Organ. Behav. Human Performance 6, 649-744 (1971).
98. Spetzler, C. S., and C.-A. S. Stael von Holstein, Probability encoding in decision analysis, Manage. Sci. 22, 340-358 (1975).
99. Srinivasan, V., and A. D. Shocker, Linear programming techniques for multidimensional analysis of preference, Psychometrika 38, 337-369 (1973).
100. Stanford Research Institute, Decision Analysis of Nuclear Plants in Electrical Systems Expansion, Report on Project 6496, Menlo Park, California, 1968.
101. Stanford Research Institute, Readings in Decision Analysis, 2nd ed., Menlo Park, California, 1976.
102. Steuer, R. E., ADEX: An Adjacent Efficient Extreme Point Algorithm for Solving Vector-Maximum and Interval Weighted-Sums Linear Programming Problems, SHARE Program Library Agency, Distribution Code 360D-15.2.014, 1974.
103. Steuer, R. E., ADBASE: An adjacent efficient basis algorithm for solving vector-maximum and interval weighted-sums linear programming problems, J. Marketing Res. 12, 454-455 (1975) (abstract).
104. Steuer, R. E., A five phase procedure for implementing a vector maximum algorithm for multiple objective linear programming problems, in Multiple Criteria Decision Making (H. Tniriez and S. Zionts, eds.), Jouy-en-Josas; France, Springer, Heidelberg, 1975, pp. 159-168.
105. Steuer, R. E., Multiple objective linear programming with interval criterion weights, Manage. Sci. 23, 305-316 (1976).

106. Steuer, R. E., and A. T. Schuler, An Interactive Multiple Objective Linear Programming Approach to a Problem in Forest Management, Working Paper No. BA2, College of Business and Economics, University of Kentucky, Lexington, Kentucky, 1977.
107. Stevens, S., Measurement, psychophysics, and utility, in Measurement: Definitions and Theories (C. Churchman and P. Ratoash, eds.), Wiley, New York, 1959.
108. Suppes, P., and M. Winet, An axiomatization of utility based on the notion of utility differences, Manage. Sci. 1, 259-270 (1955).
109. Suppes, P., and J. L. Zinnes, Basic measurement theory, in Handbook of Mathematical Psychology, Vol. 1 (R. D. Luce, R. R. Bush, and E. Galanter, eds.), Wiley, New York, 1963, pp. 1-76.
110. Tversky, A., Elimination by aspects: A theory of choice, Psychol. Rev. 79, 281-299 (1972).
111. Tversky, A., Choice by elimination, J. Math. Psychol. 9, 341-367 (1972).
112. Von Neumann, J., and O. Morgenstern, Theory of Games and Economic Behavior, Princeton University Press, Princeton, New Jersey, 1947.
113. Von Winterfeldt, D., An Overview, Integration, and Evaluation of Utility Theory for Decision Analysis, Research Report 75-9, Social Science Research Institute, University of Southern California, Los Angeles, California, 1975.
114. Von Winterfeldt, D., and G. W. Fischer, Multi-Attribute Utility Theory: Models and Assessment Procedures, Technical Report 011313-7-T, Engineering Psychology Laboratory, University of Michigan, Ann Arbor, Michigan, 1973.
115. Wagner, H. M., Principles of Operations Research, Prentice-Hall, Englewood Cliffs, New Jersey, 1969.
116. Wainer, H., Estimating coefficients in linear models: It don't make no nevermind, Psychol. Bull. 83, 213-217 (1976).
117. Wald, A., Statistical Decision Functions, Wiley, New York, 1950.
118. Wallenius, J., Comparative evaluation of some interactive approaches to multicriterion optimization, Manage. Sci. 11(4), 1387-1396 (December 1976).
119. Wallenius, J., H. Wallenius, and P. Vartia, An Approach to Solving Multiple Criteria Macroeconomic Policy Problems and an Application, presented at the Conference on Multiple Criteria Problem Solving—Theory, Methodology, and Practice, The State University of New York at Buffalo, August 22-26, 1977.
120. Wehrung, D. A., Mathematical Programming Procedures for the Interactive Identification and Optimization of Preferences in a Multi-Attributed Decision Problem, Ph.D. Dissertation, Graduate School of Business, Stanford University, May 1975.
121. Weinzapfel, G., and S. Handel, IMAGE: Computer assistant for architectural design, in Spatial Analysis in Computer-Aided Building Design (C. Eastman, ed.), Wiley, New York, 1975.
122. Winkler, R. L., The assessment of prior distributions in Bayesian analysis, J. Am. Stat. Assoc. 62, 776-800 (1967).
123. Winkler, R. L., The quantification of judgment: Some methodological suggestions, J. Am. Stat. Assoc. 62, 1105-1120 (1967).
124. Wytock, D. M., and B. F. Grall, Computer Assistance for Planning School Attendance Boundaries, presented at Urban and Regional Information Systems Association, Seattle, Washington, September 1975.

125. Yu, P. L., Introduction to domination structures in multicriteria decision problems, in Multiple Criteria Decision Making (J. L. Cochrane and M. Zeleny, eds.), University of South Carolina Press, Columbia, South Carolina, 1973, pp. 249-261.
126. Yu, P. L., and M. Zeleny, The set of all nondominated solutions in linear cases and the multicriteria simplex method, J. Math. Anal. Appl. 49, 430-468 (1974).
127. Zeleny, M., Compromise programming, in Multiple Criteria Decision Making (J. L. Cochrane and M. Zeleny, eds.), University of South Carolina Press, Columbia, South Carolina, 1973, pp. 262-301.
128. Zeleny, M., Linear Multiobjective Programming, Springer, New York, 1974.
129. Zionts, S., Multiple Criteria Decision Making for Discrete Alternatives with Multiple Objectives, Working Paper No. 299, School of Management, State University of New York at Buffalo, February 1977.
130. Zionts, S., and D. Deshpande, A Time Sharing Computer Programming Application of a Multiple Criteria Decision Method to Energy Planning, presented at the Conference on Multiple Criteria Problem Solving—Theory, Methodology, and Practice, The State University of New York at Buffalo, August 22-26, 1977.
131. Zionts, S., and J. Wallenius, An interactive programming method for solving the multiple criteria problem, Manage. Sci. 22, 652-663 (1976).
132. Zoutendijk, G., Methods of Feasible Directions, Elsevier, Amsterdam, 1960.

6
QUADRATIC PROGRAMMING

VINCE SPOSITO

Department of Statistics
Iowa State University
Ames, Iowa

INTRODUCTION

Determining the minimum (or maximum) numerical value of some mathematical function in some restricted space has been a frequent problem attempted by scientists for centuries.

Most of these problems are often classified or denoted as mathematical programming problems. The general form of such models can be expressed as:

Minimize (or maximize)

$$F(y) \tag{1}$$

subject to

$$f_i(y) \geq c_i; \quad i = 1, 2, \ldots, m$$

Here F(y) and $\{f_i(y)\}$ are real-valued functions of y, and the vector c is known. A special case of the above formulation is a class of nonlinear programming problems commonly called "quadratic programming problems." In matrix notation, these problems take the form:

Minimize

$$b'y + y'Dy \tag{2}$$

subject to

$$A'y \geq c$$
$$y \geq 0$$

where D is positive semidefinite or definite, b and c are known vectors in E^n and E^m, respectively, A' is a known $m \times n$ matrix, and y is an unknown vector in E^n.

The set of constraints, $\{y \mid A'y \geq c,\ y \geq 0\}$, is called the "feasible region" associated with the model. If D is a $n \times n$ symmetric matrix and $y'Dy \geq 0$ for all y in E^n, then D is said to be positive semidefinite. Under this formulation, the function to be minimized, or the objective function, is quadratic, and moreover convex. Also, the constraints associated with (2) are linear.

A special case of this model is when D is null. In this situation, (2) reduces to a classical linear programming problem.

Some of the features associated with the quadratic programming problem are outlined below from an excellent exposition given by Dantzig [3]:

1. A quadratic programming problem may have no solution. When this happens, we say the problem is infeasible or the feasible region is empty.
2. The feasible region is generated by the intersection of a finite number of half planes.
3. The feasible region is a convex polyhedral set; it contains a finite number of sides, and any convex combination of any two vectors in the feasible region also lies in this region.
4. An optimal solution is a feasible solution which minimizes the objective function. This solution does not necessarily have to lie at a corner or extreme point of the feasible region.
5. The set of all optimal solutions forms a convex set.

The general area of research in linear programming was pioneered during World War II by George Dantzig when large-scale military operations required careful planning of logistic support. Dantzig's algorithm is known as the simplex procedure. Following these major initial steps developed by Dantzig, interest grew tremendously with applications developing in fields of economics, engineering, statistics, mathematics, and business.

In 1951, Kuhn and Tucker [10] addressed themselves to nonlinear programming models. In particular, their classical research underscored necessary and sufficient optimality conditions associated with various mathematical programming problems. Computational algorithms in quadratic programming have been primarily based on the "Kuhn-Tucker Conditions." These conditions have been incorporated into Dantzig's simplex procedure to necessarily identify an optimal solution.

Moreover, the Kuhn-Tucker theory has led to the area of quadratic duality through the saddle value problem or Lagrangian approach. In particular, if an optimal solution can be identified to any of the above problems, then the associated problem also has an optimal solution.

The following section will outline the general results of Kuhn-Tucker duality which are applicable to quadratic programming.

The section entitled Quadratic Programming Algorithms will give an exposition regarding how "the Kuhn-Tucker Conditions" are used in Dantzig's simplex procedure, and how quadratic programming computational algorithms are modified by incorporating these optimality conditions.

The section entitled Statistical Application will present a statistical application of quadratic programming which often can yield closed form solutions. These solutions are derived appealing only to the Kuhn-Tucker conditions, and are extremely useful in solving these problems, or related problems, without using an iterative or simplex approach.

KUHN-TUCKER DUALITY

In this section optimality conditions of the saddle value type are presented for quadratic programming problems. In this discussion this class of problems has the following structure:

Problem I

Minimize

$$F(y) = b'y + y'Dy$$

subject to

$$c - A'y \leq 0$$

$$y \geq 0$$

where F(y) is a differentiable convex function. A sample problem is

Minimize

$$(Y - X\beta)'(Y - X\beta)$$

subject to

$$A'\beta \geq c$$

$$\beta \text{ unrestricted}$$

In this form the problem represents the classical problem of determining least squares regression coefficients subject to certain side conditions. Also, y_h is the h-th observation on a dependent variable, x_h the h-th observation on an independent variable, and β_j are the unknown regression coefficients.

The presentation given in this section will consider the equivalence relationship between Problem I, its associated saddle value problem, and the appropriate Kuhn-

Tucker optimality conditions. It should be noted that a more general development is given by Kunzi and Krelle [11], Mangasarian [13], and Sposito [18].

Consider the following problem:

Problem S (Saddle Value Problem)

Find vectors $y^0 \geq 0$ and $x^0 \geq 0$ such that

$$\emptyset(y^0, x) \leq \emptyset(y^0, x^0) \leq \emptyset(y, x^0) \quad \forall x \geq 0 \ \forall y \geq 0$$

where $\emptyset(y, x) = F(y) + x'(c - A'y)$.

Problem I has an optimal solution y^0 if and only if there exists $x^0 \geq 0$ such that (y^0, x^0) is a saddle point solution of Problem S. Furthermore, as shown by Sposito and David [19], without assuming differentiability, if (y^0, x^0) is a solution of Problem S, then the following four conditions hold:

(1) $c - A'y^0 \leq 0$

(2) $y^0 \geq 0$

(3) $x^{0\prime}(c - A'y^0) = 0$

(4) $x^0 \geq 0$

Hence, if y^0 is an optimal solution of Problem I, then there exists $x^0 \geq 0$ such that (y^0, x^0) is a saddle point solution of Problem S and (y^0, x^0) satisfies the above four conditions. These four conditions are indeed necessary conditions when y^0 solves Problem I, but are not sufficient conditions as presented by Kuhn and Tucker. In their development, assuming differentiability, then the following conditions are necessary and sufficient to characterize an optimal solution of Problem I. In particular, let $\emptyset(y, x) = F(y) + x'(c - A'y)$ be a function from $Q_n^+ \times Q_m^+$ into the reals, and suppose that $\emptyset(y, x)$ has continuous first derivatives $\emptyset_x$ and $\emptyset_y$ at (y^0, x^0). Then the classical Kuhn-Tucker conditions are:

(a) $\emptyset_x(y^0, x^0) \leq 0$

(b) $y^0 \geq 0$

(c) $\emptyset_x'(y^0, x^0)(x^0) = 0$

(d) $x^0 \geq 0$

(e) $\emptyset_y(y^0, x^0) \geq 0$

(f) $\emptyset_y'(y^0, x^0)(y^0) = 0$

The first four conditions are equivalent to conditions (1), (2), (3), and (4). However, the above Kuhn-Tucker conditions are necessary and sufficient to insure that (y^0, x^0) is a saddle point solution of Problem S and, moreover, that y^0 solves Problem I.

Example 1. Consider the problem

minimize

y^2

subject to

$y \geq 1$

$y \geq 0$

Then the optimal solution is clearly seen to be $y^0 = 1$. Moreover, in view of the Kuhn-Tucker conditions, since $(y^0, x^0) = (1, 2)$ satisfies the following system of constraints:

(a) $(1 - y^0) \leq 0$

(b) $y^0 \geq 0$

(c) $(1 - y^0)x^0 = 0$

(d) $x^0 \geq 0$

(e) $2y^0 - x^0 \geq 0$

(f) $(2y^0 - x^0)y^0 = 0$

where $\emptyset(y, x) = y^2 + x'(1 - y)$, then necessarily $y^0 = 1$ solves the original problem.

Let us consider the Lagrangian of Problem I with $b'y = F(y)$; i.e. $\emptyset(y, x) = b'y + x'(c - A'y)$. Since $\emptyset(y, x) = c'x + y'(b - Ax) = \psi(x, y)$, it follows that

1. if (y^0, x^0) is a saddle point solution of its associated saddle value problem, then (x^0, y^0) must necessarily solve the following saddle value problem:

   $$\psi(x, y^0) \leq \psi(x^0, y^0) \leq \psi(x^0, y) \quad \forall x \geq 0 \quad \forall y \geq 0$$

2. (x^0, y^0) must also satisfy the same Kuhn-Tucker conditions associated with $\emptyset(y^0, x^0)$

This equivalence relationship suggests that there is a symmetric dual problem which is closely associated with Problem I. In particular, the classical dual linear programming problem:

Problem II

Maximize

$c'x$

subject to

$Ax \leq b$

$x \geq 0$

Hence the Kuhn-Tucker conditions are necessary and sufficient optimality conditions that insure that y^0 solves Problem I and x^0 solves the corresponding dual problem, Problem II.

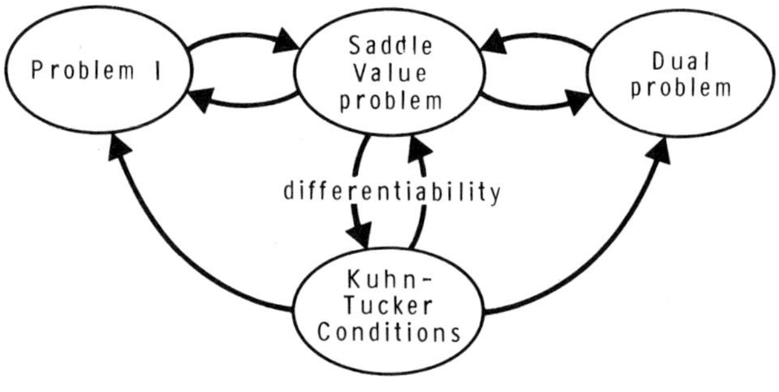

FIGURE 1.

Figure 1 illustrates the equivalence relationship between Problem I, Problem S, the associated dual problem, and the Kuhn-Tucker conditions. If D is not null, then the dual problem associated with Problem I would be:

Problem III

Maximize

$c'x - \lambda' D\lambda$

subject to

$2D\lambda - Ax + b \geq 0$

$x \geq 0$

λ unrestricted.

In this formulation, the Kuhn-Tucker conditions would be composed of the vectors y, x, and λ (where y is the Lagrangian vector).

Kunzi and Krelle [11] and Sposito [18] underscore several fundamental theorems associated with Problem I and Problem III. In particular:

Lemma 1. (Weak Duality Theorem). If there exists $\bar{y} \in \Omega_I$, the feasible region of Problem I, and $(\hat{x}, \hat{\lambda}) \in \Omega_{III}$, then

$$c'\hat{x} - \hat{\lambda}' D\hat{\lambda} \leq b'\bar{y} + \bar{y}' D\bar{y} \tag{3}$$

Lemma 2. (Existence Theorem). If Problem I and III (or II) both have feasible solutions, then both problems have optimal solutions.

Lemma 3. If $\bar{y} \in \Omega_I$, and $(\hat{x}, \hat{\lambda}) \in \Omega_{III}$ and the values of the corresponding objective functions are equal (3), then $\bar{y}$, $(\hat{x}, \hat{\lambda})$ are optimal solutions of Problem I and Problem III, respectively.

Lemma 4. If $(\hat{x}, \hat{\lambda})$ solves Problem III, then there exists $v^0 \geq 0$ such that

1. $Dv^0 = Dy^0$
2. $b'v^0 + v^{0\prime} Dv^0 = c'\hat{x} - \hat{\lambda}' D\hat{\lambda}$
3. v^0 solves Problem I

Clearly, if D is nonsingular, v^0 is equivalent to y^0; also, if D is null, the above lemmas are familiar fundamental results from linear programming.

QUADRATIC PROGRAMMING ALGORITHMS

Two of the most frequently used algorithms will be outlined below, in particular algorithms developed by Hildreth [7] and Wolfe [20].

Hildreth's algorithm for solving quadratic programming problems is an asymptotic method which makes direct use of the duality concepts introduced in the previous section. The steps at each iteration are extremely simple in form and thus make this algorithm highly suitable for automatic computation.

For simplicity we shall assume that if positivity is assumed on any of the activities, then these restrictions are incorporated into the linear system $A'y \geq c$. Thus we have the following model:

minimize

$$b'y + y'Dy$$

subject to

$$A'y \geq c$$

Now let us assume y^0 is the optimal solution, then from the Kuhn-Tucker theory we know that there exists a vector x^0 such that (y^0, x^0) is a saddle point solution of $\phi(y, x) = b'y + y'Dy + x'(c - A'y)$. Since $\phi(y, x)$ is a convex function in y for a fixed x, then (y^0, x^0) must also satisfy the Kuhn-Tucker conditions. Moreover, since y is unrestricted in sign, then the Kuhn-Tucker conditions can be reduced to four conditions, i.e.,

$$\phi_y = 0$$

$$\phi_x \leq 0, \quad \phi_x' x^0 = 0, \quad x^0 \geq 0$$

Therefore (y^0, x^0) must satisfy the following conditions:

$$b + 2Dy^0 - Ax^0 = 0 \tag{4}$$

$$c - A'y^0 \leq 0 \tag{5}$$

$$x^{0\prime}(c - A'y^0) = 0 \tag{6}$$

$$x^0 \geq 0$$

If D is positive definite, then the determinant of D is nonzero and D^{-1} exists. We then have from (4) that

$$y^0 = -\tfrac{1}{2}D^{-1}(b - Ax^0) \tag{7}$$

Substituting (7) into Eqs. (5) and (6) gives us the following equivalent necessary and sufficient conditions:

$$c + \tfrac{1}{2}A'D^{-1}(b - Ax^0) \leq 0$$

$$x^{0\prime}(c + \tfrac{1}{2}A'D^{-1}(b - Ax^0)) = 0$$

$$x^0 \geq 0$$

Let

$$\varphi_x = -c - \tfrac{1}{2}A'D^{-1}(b - Ax^0),$$

then there exists some "dual" problem such that

$$-\varphi_x \leq 0, \ \varphi_x' x^0 = 0, \ x^0 \geq 0$$

are necessary and sufficient conditions which ensure that x^0 is an optimal solution. A ready-made Lagrangian function of this dual problem would be:

$$\varphi(x) = -x'c - \tfrac{1}{2}x'(A'D^{-1}(b - \tfrac{1}{2}Ax))$$

and its associated dual problem is

minimize

$$(-x'c - \tfrac{1}{2}x'A'D^{-1}b + \tfrac{1}{4}x'A'D^{-1}Ax) \tag{8}$$

subject to

$$x \geq 0$$

The above formulation gives us the following result:

Lemma 4. y^0 is an optimal solution of (I) if and only if $y^0 = -\tfrac{1}{2}D^{-1}(b - Ax^0)$ where x^0 is an optimal solution of (8).

The objective function in the above dual problem is convex, since the quadratic term $A'D^{-1}A$ is positive semidefinite.

Solution of the Dual Problem

Let $h = -\tfrac{1}{2}A'D^{-1}b - c$ and $G = \tfrac{1}{4}A'D^{-1}A$, then (8) can be expressed equivalently as:

minimize

$$h'x + x'Gx \tag{9}$$

subject to

$$x \geq 0$$

Now

$$\frac{\partial(h'x + x'Gx)}{\partial x_i} = h_i + 2\sum_{j=1}^{n} g_{ij}x_j \geq 0; \quad i = 1, \ldots, \bar{n}$$

Thus (9) can be solved iteratively by considering $\bar{n}$ linear programming problems, i.e.,

minimize	minimize
$\sum_{j=1}^{\bar{n}} g_{1j}x_j$	$\sum_{j=1}^{\bar{n}} g_{\bar{n}j}x_j$
subject to	subject to

$$\sum_{j=1}^{\bar{n}} g_{1j}x_j \geq -h_1/2 \;\; \ldots \;\; \sum_{j=1}^{n} g_{\bar{n}j}x_j \geq -h_{\bar{n}}/2$$

$$x_j \geq 0 \qquad\qquad x_j \geq 0$$

$$j = 1, \ldots, \bar{n} \qquad\qquad j = 1, \ldots, \bar{n}$$

where the restrictions in each programming model restrict x_j to equal zero or φx_i to equal zero.

Hildreth has proven that one can obtain a solution if one uses Gauss-Seidel's iterative procedure. Namely, choose an arbitrary feasible solution, possibly $\bar{x} = 0$. Then in each linear programming problem hold all but one variable fixed and determine a new value for this variable. This involves nothing more than determining the new value of $x_i^{\ell+1}$ where

$$x_i^{\ell+1} = \max\left\{0, \frac{-1}{g_{ii}}\left(\frac{h_i}{2} + \sum_{j<i} g_{ij}x_j^{(\ell+1)} + \sum_{j>1} g_{ij}x_j^{(\ell)}\right)\right\}$$

$$i = 1, \ldots, \bar{n}$$

$$\ell = 0, 1, 2, \ldots$$

where $(\ell + 1)$ denotes the $(\ell + 1)$ iteration.

The procedure is continued until

$$\left|x_i^{(\ell+1)} - x_i^{\ell}\right| \leq \text{some predetermined value for all } i$$

Example 2. Consider the problem

minimize

$$-y_1 + \tfrac{1}{2}y_1^2 + \tfrac{1}{2}y_2^2$$

subject to

$$y_1 - y_2 \leq 4$$

$$y_1, y_2 \geq 0$$

then $A' = \begin{pmatrix} -1 & 1 \\ 1 & 0 \\ 0 & 1 \end{pmatrix}$, $D = \begin{pmatrix} \frac{1}{2} & 0 \\ 0 & \frac{1}{2} \end{pmatrix}$

$$h = -\tfrac{1}{2}A'D^{-1}b - c = \begin{pmatrix} 3 \\ 1 \\ 0 \end{pmatrix}$$

and $G = \tfrac{1}{4}A'D^{-1}A = \begin{pmatrix} 1 & -\frac{1}{2} & \frac{1}{2} \\ -\frac{1}{2} & \frac{1}{2} & 0 \\ \frac{1}{2} & 0 & \frac{1}{2} \end{pmatrix}$

Hence the solution of the dual problem can be obtained by considering the following three iterative equations

$$x_1^{\ell+1} = \max\{0, -1(3/2 - 1/2x_2^{\ell} + 1/2x_3^{\ell})\}$$

$$x_2^{\ell+1} = \max\{0, -2(-\tfrac{1}{2}x_1^{\ell+1} + \tfrac{1}{2})\}$$

$$x_3^{\ell+1} = \max\{0, -2(\tfrac{1}{2}x_1^{\ell+1})\}$$

Letting $\bar{x} = (0, 0, 0)$ be our initial feasible solution, then

Iteration	x_1	x_2	x_3
1	0	0	0
2	0	0	0

and

$$y^0 = -\tfrac{1}{2}D^{-1}(b - Ax^0) = \binom{1}{0}$$

solves the original quadratic programming problem.

Another well-used quadratic programming algorithm, developed by Wolfe, uses the linear simplex procedure of Dantzig. In particular, to characterize an optimal solution of Problem I, let us consider the Lagrangian function, $\phi(y, x) = b'y + y'Dy + x'(c - A'y)$, and the Kuhn-Tucker conditions. If we let

$$\partial\phi/\partial y = v$$

and

$$\partial\phi/\partial x = -x_s$$

then

$$\partial\phi/\partial y = v = b + 2Dy - Ax$$

and

$$\partial\phi/\partial x = -x_s = c - A'y$$

With these, the Kuhn-Tucker conditions can be expressed as:

(a) $2Dy - Ax - v = -b$

(b) $A'y - x_s = c$

(c) $y \geq 0, \; x \geq 0, \; v \geq 0, \; x_s \geq 0$

(d) $y'v = 0 = x'x_s$

Conditions (a) through (c) form a linear system. This suggests that one could use the simplex method to find an optimal solution of Problem I. This is indeed the case even though Condition (d) is nonlinear by making certain that at each iteration a vector is allowed to enter the basis only if Condition (d) is satisfied.

A basic solution to the linear system will have no more than (m + n) positive variables, and the remaining variables will be zero. Here we are defining a feasible solution of (a)-(c) to be a basic feasible solution if (d) is also satisfied. Furthermore, if a vector (y^0, x_s^0, x^0, v^0) is a feasible solution of the following linear constraints:

$$\begin{pmatrix} 2D & 0_{(n\times m)} & -A & -I_n \\ A' & -I_m & 0_{(m\times m)} & 0_{(m\times n)} \end{pmatrix} \begin{pmatrix} y \\ x_s \\ x \\ v \end{pmatrix} = \begin{pmatrix} -b \\ \\ \\ c \end{pmatrix} \qquad (10)$$

with y, x_s, x, and $v \geq 0$ and also satisfying Condition (d), then from the Kuhn-Tucker theory y^0 solves the original quadratic programming problem.

The system of linear equations expressed above is the form:

minimize

$$0'z$$

subject to

$$\bar{A}z = \bar{b}$$

$$z \geq 0$$

Hence, using a modification of the simplex algorithm, one can solve (10) by incorporating the following two additional rules to insure that $y'v = 0 = x'x_s$.

1. If a variable y_j (or v_j) is currently in the basis at a positive level, do not consider y_j (or x_j) as a candidate for entry into the basis. If y_j (or v_j) is currently in the basis at a zero level, then v_j (or y_j) may enter the basis only if y_j (or v_j) remains at a zero level.
2. If a variable x_i (or x_{s_i}) is currently in the basis at a positive level, do not consider x_{s_i} (or x_i) as a candidate for entry into the basis. If x_i (or x_{s_i}) is in the basis at a zero level, x_{s_i} (or x_i) may enter the basis if x_i (or x_{s_i}) remains at a zero level.

The above two procedures, which appeal to duality results and the simplex procedure, have been in wide use since 1959. Dantzig and Van Slyke [5] developed a simplex procedure to solve bounded linear programming problems with a set of p orthogonal constraints. This procedure has been extensively used by major computer companies; in particular, IBM and Management Science Corporation. Basically, the algorithm uses a "working basis" of size $m \times m$ rather than of $(m + p) \times (m + p)$. Hence, when p is large, it becomes considerably less expensive to solve the condensed problem. These types of models are often called problems with simple upper bounds (SUB) or generalized upper bounds (GUB).

In 1969 a large-scale quadratic computer package was developed by Soults [17]. This package has solved models with up to 3550 variables. However, GUB features have not, at the present time, been incorporated into the program. Klemm [9] has derived the necessary theoretical aspects of incorporating the generalized upper bound features into any quadratic computer package. At the present time, modification of Soults' program is taking place at Iowa State University and Management Science Corporation.

STATISTICAL APPLICATION

The classical regression problem involves estimating β where $Y = X\beta + \epsilon$, Y is a known $n \times \ell$ vector, X is a known $n \times k$ matrix, β is a $k \times \ell$ vector, ϵ is an $n \times \ell$ random vector. Employing least squares as the basis for determining the best regression coefficients, the problem to solve becomes minimize $(Y - X\beta)'(Y - X\beta)$. In this situation, the best linear unbiased estimator for β is $\hat{\beta} = (X'X)^{-1}X'Y$, provided $(X'X)^{-1}$ exists.

In the classical problem, β is completely unrestricted. Often, however, this is not a reasonable assumption, and we have reason to derive estimates of the least squares regression problem subject to constraints of the form $A'\beta \geq P$ where A' is a known $m \times k$ matrix and P is a known $m \times \ell$ vector. With these additional constraints the traditional regression problem becomes a constrained quadratic programming problem which could be solved iteratively employing algorithms such as those developed by Hildreth [7] and Wolfe [20] (or see Judge and Takayama [8] or Mantel [14]).

Zellner [21] offers a closed form for the simple linear regression case when the slope is restricted by some lower bound. The case when $A' = [1, 0, \ldots, 0]$ and $P \geq 0$ was examined by Lowell and Prescott [12], resulting in a closed form for this special case. A similar investigation by O'Hagan [15] estimates the regression coefficients by a Bayesian method when $A' = [0, 0, 1]$ and $P \geq 0$.

By utilizing the Kuhn-Tucker conditions [10] of nonlinear programming, closed form solutions for problems with constraints of the type $A'\beta \geq P$, where A' is either a $1 \times k$ or $2 \times k$ matrix, are covered in detail in this section. A concluding section addresses the general case.

Constraints of the Form $A'\beta \geq P$ Where A' Is a $2 \times k$ Matrix

The least squares problem with side conditions can be written as

minimize

$$(Y - X\beta)'(Y - X\beta) = Y'Y - 2\beta'X'Y + \beta'X'X\beta \tag{11}$$

subject to

$$A'\beta \geq P$$

Appealing to the theory of Kuhn-Tucker [10], a set of necessary and sufficient conditions for a vector β^* to solve (11) is that there exists $\lambda^0 \geq 0$ such that

$$\frac{\partial \phi}{\partial \lambda} \leq 0 \quad \text{or } P - A'\beta^* \leq 0 \tag{12}$$

$$\frac{\partial \phi}{\partial \beta} = 0 \quad \text{or } X'X\beta^* - X'Y - A\lambda^0 = 0 \tag{13}$$

and

$$\lambda^{0\prime} \frac{\partial \phi}{\partial \lambda} = 0 \quad \text{or } \lambda^{0\prime}(P - A'\beta^*) = 0 \tag{14}$$

where $\phi(\beta, \lambda) = \frac{1}{2}\beta'X'X\beta - \beta'X'Y + \lambda'(P - A'\beta)$.

Suppose $\beta^* = \tilde{\beta}$, the restricted least squares estimator. Then solving for the restricted estimator in view of (13) we have

$$\tilde{\beta} = \hat{\beta} + (X'X)^{-1}A\lambda^0 \tag{15}$$

If $\hat{\beta}$ is such that $A'\hat{\beta} \geq P$, then a sufficient λ^0 would be zero, and necessarily $\tilde{\beta} = \hat{\beta}$.

If, however, one or both constraints are violated when evaluated at $\hat{\beta}$, we must then derive a vector λ^0 which will satisfy the above Kuhn-Tucker conditions.

Case I: Both Constraints Are Violated. Since $\hat{\beta}$ is such that $A'\hat{\beta} < P$, we must therefore determine an adjustment to $\hat{\beta}$ so that $\tilde{\beta}$ is now either on the boundary or in the interior of the constrained feasible region. Therefore, $\tilde{\beta}$ must be such that

$$A'\tilde{\beta} = A'\hat{\beta} + A'(X'X)^{-1}A\lambda^0 \geq P \tag{16}$$

Let $D = A'(X'X)^{-1}A$ and assume that the inverse of D exists. This situation will prevail when, for example, A' is of full row rank. In particular, given any vector $y \neq 0$, let $\bar{y} = Ay$. Since A' and A are of rank m, then $Ay = \bar{y} \neq 0$. Therefore $y'(A'(X'X)^{-1}A)y = \bar{y}'(X'X)^{-1}\bar{y} > 0$. Consequently, D^{-1} exists. D is commonly denoted as the variance-covariance matrix of $A_1'\hat{\beta}$ and $A_2'\hat{\beta}$.

Let us now substitute D into (16). We then have that λ^0 must be such that

$$D\lambda^0 \geq P - A'\hat{\beta} \tag{17}$$

In order to satisfy the Kuhn-Tucker conditions we must also require that

$$\text{(i)}\quad \lambda_1^0 \geq 0, \quad \lambda_2^0 \geq 0$$

$$\text{(ii)}\quad \lambda_1^0(P_1 - A_1'\tilde{\beta}) = 0 = \lambda_2^0(P_2 - A_2'\tilde{\beta}) \tag{18}$$

This implies that either $\lambda_i^0 = 0$ or $P_i - A_i'\tilde{\beta} = 0$ for each i.

Let us consider different combinations of λ_1^0 and λ_2^0.

Subcase 1: $\lambda_1^0 = 0 = \lambda_2^0$. In this situation, inequality (17) can never hold since

$$P - A'\hat{\beta} > 0$$

Subcase 2: $\lambda_1^0 = 0$, $\lambda_2^0 > 0$. In this situation, (17) and (18) in view of (16) imply that

$$\begin{pmatrix} d_{11} & d_{12} \\ d_{12} & d_{22} \end{pmatrix} \begin{pmatrix} 0 \\ \lambda_2^0 \end{pmatrix} \begin{matrix} \geq \\ = \end{matrix} \begin{pmatrix} P_1 - A_1'\hat{\beta} \\ P_2 - A_2'\hat{\beta} \end{pmatrix} = \begin{pmatrix} r_1 \\ r_2 \end{pmatrix} \tag{19}$$

(i) If $d_{12} < 0$, then from (19)

$$d_{12}\lambda_2^0 < r_1, \qquad \text{for any } \lambda_1^0 > 0$$

i.e., this inequality can never hold for this subcase.

(ii) If $d_{12} > 0$, then from (19)

$$\lambda_2^0 \geq r_1/d_{12} \quad \text{and} \quad \lambda_2^0 = r_2/d_{22} > 0$$

This implies that whenever $d_{12} > 0$ and

$$r_1/r_2 \leq d_{12}/d_{22}$$

then

$$\tilde{\beta} = \hat{\beta} + (X'X)^{-1}A\begin{pmatrix} 0 \\ \lambda_2^0 \end{pmatrix}$$

$$= \hat{\beta} + (X'X)^{-1}A_2\left(\frac{1}{d_{22}}\right)(P_2 - A_2'\hat{\beta})$$

Subcase 3. $\lambda_1^0 > 0$, $\lambda_2^0 > 0$. From (17) and (18) we have that λ^0 must satisfy the following system:

$$\begin{pmatrix} d_{11} & d_{12} \\ d_{12} & d_{22} \end{pmatrix}\begin{pmatrix} \lambda_1^0 \\ \lambda_2^0 \end{pmatrix} = \begin{pmatrix} r_1 \\ r_2 \end{pmatrix}$$

Hence

$$\lambda^0 = \frac{1}{\text{Det } D}\begin{pmatrix} d_{22} & -d_{12} \\ -d_{12} & d_{11} \end{pmatrix}\begin{pmatrix} r_1 \\ r_2 \end{pmatrix}$$

Since $\lambda_1^0 > 0$ and $\lambda_2^0 > 0$, then necessarily we must have that

$$d_{22}r_1 - d_{12}r_2 > 0 \tag{20}$$

and

$$d_{11}r_2 - d_{12}r_1 > 0 \tag{21}$$

(i) If $d_{12} \leq 0$, then (20) and (21) always hold.

(ii) If $d_{12} > 0$, then (20) and (21) hold only if

$$\frac{d_{12}}{d_{22}} < \frac{r_1}{r_2} < \frac{d_{11}}{d_{12}}$$

Consequently, if

(i) $d_{12} \leq 0$

or

(ii) $d_{12} > 0$ and $d_{12}/d_{22} < r_1/r_2 < d_{11}/d_{12}$

then

$$\tilde{\beta} = \hat{\beta} + (X'X)^{-1}AD^{-1}(P - A'\hat{\beta})$$

Subcase 4. $\lambda_1^0 > 0$, $\lambda_2^0 = 0$. λ_1^0 must be such that

$$\begin{pmatrix} d_{11} & d_{12} \\ d_{12} & d_{22} \end{pmatrix}\begin{pmatrix} \lambda_1^0 \\ 0 \end{pmatrix} \begin{matrix} = \\ \geq \end{matrix} \begin{pmatrix} r_1 \\ r_2 \end{pmatrix}$$

Therefore, λ_1^0 must satisfy the following conditions:

$$d_{11}\lambda_1^0 = r_1 \quad \text{and} \quad d_{12}\lambda_1^0 \geq r_2 \tag{22}$$

(i) If $d_{12} > 0$, and $r_1/r_2 \geq d_{11}/d_{12}$, then $\lambda_1^0 = r_1/d_{11}$

(ii) If $d_{12} \leq 0$, then the relationship $d_{12}\lambda_1^0 \ngeq r_2$ for any $\lambda_1^0 > 0$.

Therefore, if $d_{12} > 0$ and $r_1/r_2 \geq d_{11}/d_{12}$, then

TABLE 1
Both Constraints Violated

Condition	$d_{12} \leq 0$	$d_{12} > 0$
(None)	$\hat{\beta} + (X'X)^{-1}A[A'(X'X)^{-1}A]^{-1}(P - A'\hat{\beta})$	
$\frac{d_{12}}{d_{22}} < \frac{r_1}{r_2} < \frac{d_{11}}{d_{12}}$		$\hat{\beta} + (X'X)^{-1}A[A'(X'X)^{-1}A]^{-1}(P - A'\hat{\beta})$
$\frac{d_{12}}{d_{22}} \geq \frac{r_1}{r_2}$		$\hat{\beta} + (X'X)^{-1}A_2\left(\frac{1}{d_{22}}\right)(P_2 - A_2'\hat{\beta})$
$\frac{d_{11}}{d_{12}} \leq \frac{r_1}{r_2}$		$\hat{\beta} + (X'X)^{-1}A_1\left(\frac{1}{d_{11}}\right)(P_1 - A_1'\hat{\beta})$

$$\tilde{\beta} = \hat{\beta} + (X'X)^{-1}A_1\left(\frac{1}{d_{11}}\right)(P_1 - A_1'\hat{\beta})$$

Table 1 summarizes the above discussion.

Case II: Second Constraint is Violated. In this situation we must adjust $\hat{\beta}$ so that the second constraint is satisfied while the first constraint continues to be satisfied after this adjustment.

To derive the necessary closed form of restricted expressions, it suffices to consider again different combinations of λ_1^0 and λ_2^0.

Subcase 1: $\lambda_1^0 = 0 = \lambda_2^0$. Again, since this subcase reflects no adjustment to $\hat{\beta}$, this situation can never arise.

Subcase 2: $\lambda_1^0 = 0$, $\lambda_2^0 > 0$. From (18) and (19), λ_2^0 must satisfy the following expressions:

$$d_{12}\lambda_2^0 \geq r_1 \quad \text{and} \quad \lambda_2^0 d_{22} = r_2$$

or

$$d_{12}/d_{22} \geq r_1/r_2 \tag{23}$$

(i) If $d_{12} \geq 0$, then inequality (23) will always hold since $r_1 \leq 0$.

(ii) If $d_{12} < 0$, and if $d_{12}/d_{22} \geq r_1/r_2$, then λ_2^0 will necessarily be positive.

Hence, whenever $d_{12} \geq 0$ or $d_{12} < 0$ and $d_{12}/d_{22} \geq r_1/r_2$, then the closed form solution is

$$\tilde{\beta} = \hat{\beta} + (X'X)^{-1}A_2\left(\frac{1}{d_{22}}\right)(P_2 - A_2'\hat{\beta})$$

Subcase 3: $\lambda_1^0 > 0$, $\lambda_2^0 > 0$. In this situation, λ^0 must satisfy the following system:

$$D\lambda^0 = P - A'\hat{\beta} \quad \text{or} \quad \lambda^0 = D^{-1}(P - A'\hat{\beta})$$

Therefore, it must be that

TABLE 2
Second Constraint Violated

Condition	$d_{12} < 0$	$d_{12} \geq 0$
$\frac{d_{12}}{d_{22}} \geq \frac{r_1}{r_2}$	$\hat{\beta} + (X'X)^{-1}A_2\left(\frac{1}{d_{22}}\right)(P_2 - A_2'\hat{\beta})$	
$\frac{d_{12}}{d_{22}} < \frac{r_1}{r_2}$	$\hat{\beta} + (X'X)^{-1}A[A'(X'X)^{-1}A]^{-1}(P - A'\hat{\beta})$	
(None)		$\hat{\beta} + (X'X)^{-1}A_2\left(\frac{1}{d_{22}}\right)(P_2 - A_2'\hat{\beta})$

$$d_{22}r_1 - d_{12}r_2 > 0 \quad \text{and} \quad d_{11}r_2 - d_{12}r_1 > 0 \tag{24}$$

(i) If $d_{12} < 0$, then we must also require that

$$r_1/r_2 > d_{12}/d_{22} \quad \text{and} \quad r_1/r_2 > d_{11}/d_{12}$$

Since Det D > 0 and $d_{12} < 0$, then d_{11}/d_{12} is always less than d_{12}/d_{22}.

(ii) When $d_{12} \geq 0$, the relationship $d_{22}r_1 - d_{12}r_2$ can never be positive. Consequently, if $d_{12} < 0$ and $r_1/r_2 > d_{12}/d_{22}$, then

$$\tilde{\beta} = \hat{\beta} + (X'X)^{-1}AD^{-1}(P - A'\hat{\beta})$$

Subcase 4: $\lambda_1^0 > 0$, $\lambda_2^0 = 0$. (17) and the cross-products imply that

$$\begin{pmatrix} d_{11} & d_{12} \\ d_{12} & d_{22} \end{pmatrix} \begin{pmatrix} \lambda_1^0 \\ 0 \end{pmatrix} \begin{matrix} = \\ \geq \end{matrix} \begin{pmatrix} r_1 \\ r_2 \end{pmatrix}$$

or

$$d_{11}\lambda_1^0 = r_1 \quad \text{and} \quad d_{12}\lambda_1^0 \geq r_2 \tag{25}$$

However, since r_1 is negative, this situation can never hold.

TABLE 3
First Constraint Violated

Condition	$d_{12} < 0$	$d_{12} \geq 0$
$\frac{d_{12}}{d_{11}} \geq \frac{r_2}{r_1}$	$\hat{\beta} + (X'X)^{-1}A_1\left(\frac{1}{d_{11}}\right)(P_1 - A_1'\hat{\beta})$	
$\frac{d_{12}}{d_{11}} < \frac{r_2}{r_1}$	$\hat{\beta} + (X'X)^{-1}A[A'(X'X)^{-1}A]^{-1}(P - A'\hat{\beta})$	
(None)		$\hat{\beta} + (X'X)^{-1}A_1\left(\frac{1}{d_{11}}\right)(P_1 - A_1'\hat{\beta})$

Table 2 summarizes the above subcases when only the second restriction is violated. By a similar argument as given above, the closed form solution for β when only the first restriction is violated is given in Table 3.

The General Case

The above presentation yields closed form expressions which are rather easy to establish when A' is a $2 \times k$ matrix, and these solutions are quite useful when deriving least square estimators for simple regression models or linear models with one or two inequality constraints.

If the set of constraints consists of one restriction, then we have the following closed form solution:

$$\tilde{\beta} = \begin{cases} \hat{\beta}, & \text{if } A'\hat{\beta} \geq P \\ \hat{\beta} + (X'X)^{-1}A[A'(X'X)^{-1}A]^{-1}(P - A'\hat{\beta}), & \text{when } A'\hat{\beta} < P \end{cases} \tag{26}$$

It should be noted that simple interval bounds imposed on any regression coefficient are a special case of our presentation. The feasible region over an interval constraint can be written as $\bar{A}'\beta \geq \bar{P}$ where $\bar{A} = (A, -A)$ and $\bar{P}' = (P_0, -P_1)$.

Hence, appealing to the above discussion, we have the following useful closed form solutions:

$$\tilde{\beta} = \begin{cases} \hat{\beta} + (X'X)^{-1}A(P_0 - A'\hat{\beta})/a, & \text{if } P_0 > A'\hat{\beta} \\ \hat{\beta}, & \text{if } P_0 \leq A'\hat{\beta} \leq P_1 \\ \hat{\beta} - (X'X)^{-1}A(-P_1 + A'\hat{\beta})/a, & \text{if } P_1 < A'\hat{\beta} \end{cases}$$

where $a = A'(X'X)^{-1}A$.

We conclude this section by establishing some sufficient conditions on λ^0. In particular, conditions to insure that λ^0 will satisfy the Kuhn-Tucker conditions for the general case.

Let A' be an $m \times k$ matrix with full row rank. Then, whenever all the constraints are violated by $\hat{\beta}$ and $\lambda^0 = D^{-1}(P - A'\hat{\beta}) > 0$,

$$\tilde{\beta} = \hat{\beta} + (X'X)^{-1}AD^{-1}(P - A'\hat{\beta})$$

This follows from the fact that $(\tilde{\beta}, \lambda^0)$ satisfies the Kuhn-Tucker conditions.

Using Cramer's rule, the solution of the system $D\lambda^0 = (P - A'\hat{\beta})$ is

$$\lambda_i^0 = \frac{|M_i|}{|D|}, \quad \text{for all i}$$

where

$$|M_i| = (-1)^{i+1}Q_1D^*(1, i) + (-1)^{i+2}Q_2D^*(2, i) + \cdots + (-1)^{i+m}Q_mD^*(m, i)$$

i.e., the Laplace expansion of Det D along with the first column where $(-1)^{i+j}D^*(j, i)$ is the classical adjoint of D, and Q_j is the j-th component of the vector $(P - A'\hat{\beta})$.

When all constraints are violated, in view of the above we have that a sufficient condition for each λ_i^0 to be positive is that the following condition holds:

Condition I: $(-1)^{i+j}D^*(j, i)$ is nonnegative for all j and i and strictly positive for some j and i.

Suppose the constraint equations are partitioned into two sets:

$$A_1'\beta \geq P_1$$

$$A_2'\beta \geq P_2$$

and suppose $\hat{\beta}$ satisfies the first set, but not the second. Let

$$D = \begin{pmatrix} D_{11} & D_{12} \\ D_{12}' & D_{22} \end{pmatrix}$$

Appealing to the section entitled Constraints of the Form $A'\beta \geq P$ Where A' is a $2 \times k$ Matrix, it is impossible for $\lambda_1^0 = 0 = \lambda_2^0$, or $\lambda_1^0 > 0$ and $\lambda_2^0 = 0$. Consequently, there are only two situations to consider.

Subcase 1: $\lambda_1^0 = 0$, $\lambda_2^0 > 0$. In this situation, we must have that

$$\begin{pmatrix} D_{11} & D_{12} \\ D_{12}' & D_{22} \end{pmatrix} \begin{pmatrix} 0 \\ \lambda_2^0 \end{pmatrix} \begin{matrix} \geq \\ = \end{matrix} \begin{pmatrix} P_1 - A_1'\hat{\beta} \\ P_2 - A_2'\hat{\beta} \end{pmatrix}$$

or

$$D_{12}\lambda_2^0 \geq P_1 - A_1'\hat{\beta} \tag{27}$$

$$D_{22}\lambda_2^0 = P_2 - A_2'\hat{\beta} \tag{28}$$

Condition II: All the coefficients of D_{12} are nonnegative.

If D_{22} satisfies Condition I, then from (28), $\lambda_2^0 = D_{22}^{-1}(P_2 - A_2'\hat{\beta})$. Substituting λ_2^0 into (27) we have

$$D_{12}D_{22}^{-1}(P_2 - A_2'\hat{\beta}) \geq P_1 - A_1'\hat{\beta}$$

Hence this expression holds if all the entries of D_{12} are nonnegative. Consequently, in this situation, we have the following closed form:

$$\tilde{\beta} = \hat{\beta} + (X'X)^{-1}A_2D_{22}^{-1}(P_2 - A_2'\hat{\beta})$$

Subcase 2: $\lambda_1^0 > 0$, $\lambda_2^0 > 0$. In this case we have

$$D_{11}\lambda_1^0 + D_{12}\lambda_2^0 = P_1 - A_1'\hat{\beta} \tag{29}$$

and

$$D_{12}'\lambda_1^0 + D_{22}\lambda_2^0 = P_2 - A_2'\hat{\beta} \tag{30}$$

If all the coefficients of D_{11} are positive, then $D_{11}\lambda_1^0 > 0$.

However, since λ_2^0 is assumed to be positive and $P_1 - A_1'\hat{\beta}$ is nonnegative, then (29) will never hold if D_{12} satisfies Condition II. In view of Subcase 1, we have:

Lemma 5. If D_{12}, the covariance matrix of $(A_1'\hat{\beta}, A_2'\hat{\beta})$, satisfies Condition II and if D_{22} satisfies Condition I, then

$$\tilde{\beta} = \hat{\beta} + (X'X)^{-1}A_2D_{22}^{-1}(P_2 - A_2'\hat{\beta})$$

Note that a special case of Lemma 5 is when m = 2 and only the second constraint is violated.

For m > 2 the extension to higher dimensions is restrictive, but for simple linear regression the closed form expressions presented in this section are extremely useful and alleviate the researcher from seeking the present set of available iterative procedures which require high-speed computers or tedious computations.

Example 3. Let us determine the optimal restricted estimator for β, where $Y = X\beta + \epsilon$ and $\beta_0 \leq 2$, $\beta_1 \leq 1/4$ when

x	1	2	3
y	1	3	2

Here

$$\hat{\beta} = (X'X)^{-1}X'Y = \begin{pmatrix} 1 \\ \frac{1}{2} \end{pmatrix}$$

and only the second restriction is violated. Also,

$$D = \begin{pmatrix} 7/3 & -1 \\ -1 & \frac{1}{2} \end{pmatrix}$$

and

$$\frac{d_{12}}{d_{22}} = -2, \quad \frac{P_1 - A_1'\hat{\beta}}{P_2 - A_2'\hat{\beta}} = -4$$

Therefore, from Table 2

$$\tilde{\beta} = \hat{\beta} + (X'X)^{-1}A_2\left(\frac{1}{d_{22}}\right)(P_2 - A_2'\hat{\beta}) = \begin{pmatrix} 3/2 \\ 1/4 \end{pmatrix}$$

CONCLUSION

Research topics in quadratic programming have taken various directions since the pioneer work of Dantzig and Kuhn and Tucker. One area of quadratic duality has considered quadratic programming problems which are self-dual. These formulations have been proposed by Dorn [6], Cottle [2], Dantzig, Eisenberg, and Cottle [4], and others. Another area of quadratic duality has considered quadratic problems over cone domains. In particular, if L_1 and L_2 are arbitrary closed, convex cones in E^n and E^m, with polar cones L_1^* and L_2^*, respectively, then Problems I and III can be formulated as:

Problem I*

Minimize

$$b'y + y'Dy$$

subject to

$$c - A'y \in L_1^*$$
$$y \in L_2$$

Problem III*

Maximize

$$c'x - \lambda' D\lambda$$

subject to

$$-2D\lambda + Ax - b \in L_2^*$$
$$x \in L_1$$

In the above formulations, L_1^* (or L_2^*) is defined as $L_1^* = \{x^* \in E^n \mid x^{*\prime}x \leq 0 \text{ for any } x \in L_1\}$. Consequently, some special cases include:

1. Problems I and III where, for example, L_1 (or L_2) denotes the positive orthant, then the polar cone of L_1 (or L_2) is the negative orthant.
2. Polyhedral cone domains with or without interiors. Under this structure, the constraints in Problems I and III can be formulated with equality constraints and hence, unrestricted dual variables.

Abrams and Ben-Israel [1] showed that the above programming problems are mutually dual when L_1 and L_2 are convex, polyhedral cone domains; in particular, Lemmas 1-4 remain valid under this formulation. However, in the general case, when L_1 and L_2 are arbitrary convex cones, these fundamental lemmas hold under the assumption that

1. L_1^* has an interior $\tilde{L}_1^*$ relative to E^m, and there exists at least one $\bar{y} \in L_2$ such that $c - A'\bar{y} \in \tilde{L}_1^*$.
2. L_2^* has an interior $\tilde{L}_2^*$ relative to E^n, and there exists at least one $\bar{x} \in L_1$, $\bar{\lambda} \in E^m$ such that $-2D\bar{\lambda} + A\bar{x} - b \in \tilde{L}_2^*$.

These conditions are based on a natural modification of Slater's condition [16]. Under these assumptions, one is able to display quadratic programming problems which are constrained by nonlinear constraints.

Example 4. Let

$$L_2 = \{t \in E^3 \mid t'Bt \leq 0,\ t_3 \geq 0\}$$

then

$$L_2^* = \{w \in E^3 \mid w'B^{-1}w \leq 0,\ w_3 \geq 0\}$$

where

$$B = \begin{pmatrix} 1 & 0 & 0 \\ 0 & 1 & 0 \\ 0 & 0 & -d^2 \end{pmatrix}, \quad d \neq 0$$

and let L_1 denote the negative orthant in E^n. Then Problems I* and III* can be written equivalently as

minimize

$$b'y + y'Dy$$

subject to

$$c - A'y \leq 0$$
$$y'By \leq 0$$
$$y'e \geq 0$$

and

maximize

$$c'x - \lambda'D\lambda$$

subject to

$$(-2D\lambda + Ax - b)B^{-1}(-2D\lambda + Ax - b) \leq 0$$
$$(-2D\lambda + Ax - b)'e \leq 0$$
$$x \geq 0$$

where $e' = (0, 0, 1)$.

Other avenues of research have extended the "Kuhn-Tucker optimality conditions" to problems with nonlinear objective functions. For example, under differentiability, a numerical function is pseudoconvex if

$$[y \in E^n,\ F'_y(y^0)(y - y^0) \geq 0] \Rightarrow [y \in E^n,\ F(y) - F(y^0) \geq 0]$$

Some functions which are pseudoconvex at y^0 are

1. Convex functions
2. If $F = G/H$, where G is convex, H is linear, and $H(y) > 0$ for all $y \in E^n$
3. If $F = G/H$, where $G(y) \leq 0$ for all $y \in E^n$, G is convex, H is convex, and $H(y) > 0$ for all $y \in E^n$.
4. If $F = G \cdot H$, where G is convex, H is concave, and $H(y) > 0$ for all $y \in E^n$
5. If ln F is convex on E^n, then F is pseudoconvex

When the objective function of Problem I is pseudoconvex, and the feasible region is convex, then the "Kuhn-Tucker Conditions" are sufficient conditions to insure that y^0 solves Problem I. An excellent exposition on this topic is given by Mangasarian [13].

Avenues for further research still remain open. For example, the development of large-scale quadratic computer programs with special generalized upper bounding features, and the derivation of closed form solutions for quadratic programming problems.

In the section entitled Statistical Application, closed form solutions were developed appealing only to the "Kuhn-Tucker Conditions." Hence this procedure is independent of many computer computations. Also, this section suggests that related problems in other areas of research can be solved in the same manner.

However, many questions are still left open in this area. In particular,

1. What is the bias and mean square error associated with these estimators?
2. What is the bias of the individual parameters?

Some research is presently being pursued to answer these questions, but many new questions will develop.

ACKNOWLEDGMENT

This research was partially supported by AFOSR grant No. 78-3518.

REFERENCES

1. Abrams, R., and A. Ben-Israel, A duality theorem for complex quadratic programming, J. Optim. Theory Appl. 4, 244-252 (1968).
2. Cottle, R. W., Symmetric dual quadratic programs, Q. Appl. Math. 21, 237-243 (1963).
3. Dantzig, G. B., Linear Programming and Extensions, Princeton University Press, Princeton, New Jersey, 1963.
4. Dantzig, G. B., E. Eisenberg, and R. W. Cottle, Symmetric dual nonlinear programs, Pac. J. Math. 15, 809-812 (1965).
5. Dantzig, G. B., and R. M. Van Slyke, Generalized upper bounding techniques, J. Comput. Syst. Sci. 1, 213-226 (1967).
6. Dorn, W. S., Self-dual quadratic programs, J. Soc. Ind. Appl. Math. 9, 51-54 (1961).
7. Hildreth, C., A quadratic programming procedure, Nav. Res. Logistics Q. 4, 79-85 (1957).
8. Judge, G., and T. Takayama, Inequality restrictions in regression analysis, J. Am. Stat. Assoc. 61, 161-181 (1966).
9. Klemm, R. J., Aspects of Quadratic Programming with Statistical Applications, Ph.D. Thesis, Iowa State University of Science and Technology, 1976.
10. Kuhn, H. W., and A. W. Tucker, Nonlinear programming, in Proceedings of the Second Berkeley Symposium on Mathematical Statistics and Probability, University of California Press, Berkeley, California, 1951.
11. Kunzi, H., and W. Krelle, Nonlinear Programming, Blaisdell, Waltham, Massachusetts, 1966.
12. Lowell, M., and E. Prescott, Multiple regression with inequality constraints: Pretesting bias, hypothesis testing and efficiency, J. Am. Stat. Assoc. 65, 913-925 (1970).
13. Mangasarian, O. L., Nonlinear Programming, McGraw-Hill, New York, 1969.
14. Mantel, N., Restricted least squares regression and convex quadratic programming, Technometrics 11, 763-773 (1969).
15. O'Hagan, A., Bayes estimation of a convex quadratic, Biometrika 60, 565-571 (1973).
16. Slater, M., Lagrange Multipliers Revisited: A Contribution to Nonlinear Programming, RAND Corporation Report RM-676, Santa Monica, California, 1951.
17. Soults, D. J., and V. A. Sposito, An Algorithm for the Optimization of a Quadratic Form Subject to Linear Restrictions: Zorilla, Numerical Analysis Program 9, Statistical Laboratory, Iowa State University, Ames, Iowa, 1969.

18. Sposito, V. A., Linear and Nonlinear Programming, Iowa State Press, Ames, Iowa, 1975.
19. Sposito, V. A., and H. T. David, Saddlepoint optimality criteria of nonlinear programming problems over cone without differentiability, SIAM J. Appl. Math. 20, 698-708 (1971).
20. Wolfe, P., The simplex algorithm for quadratic programming, Econometrica 27, 382-398 (1959).
21. Zellner, A., Linear Regression with Inequality Constraints on the Coefficients: An Application of Quadratic Programming and Linear Decision Rules, International Center for Management Science Report 6109 (MSN09), Rotterdam, England, 1961.

BIBLIOGRAPHY

Arrow, K. J., and A. C. Enthoven, Quasi-concave programming, Econometrica 29, 779-800 (1961).

Ben-Israel, A., A. Charnes, and K. Kortanek, Duality and asymptotic solvability over cones, Bull. Am. Math. Soc. 75, 318-324 (1969).

Cooper, L., and D. Steinberg, Introduction to Methods of Optimization, Saunders, Philadelphia, 1970.

Cottle, R. W., and J. A. Ferland, Matrix-theoretic criteria for the quasi-convexity and pseudoconvexity of quadratic functions, J. Linear Algebra Appl. 5, 123-136 (1972).

Dantzig, G. B., and R. M. Van Slyke, A generalized upper-bounded technique for linear programming, in Proceedings of the IBM Scientific Computing Symposium on Combinatorial Problems, IBM, White Plains, New York, 1966.

Dorn, W. S., Duality in quadratic programming, Q. Appl. Math. 18, 155-160 (1960).

Frank, M., and P. Wolfe, An algorithm for quadratic programming, Nav. Res. Logistics Q. 3, 95-110 (1956).

Gould, F., and J. Tolle, Geometry of optimality conditions and constraint qualifications, J. Math. Programming 2, 1-18 (1972).

Krafft, O., Programming in Statistics and Probability. Nonlinear Programming, Academic, New York, 1970.

Lasdon, L. A., Optimization Theory for Large Systems, Macmillan, New York, 1970.

Orchard-Hays, W., Advanced Linear Programming Computing Techniques, McGraw-Hill, New York, 1968.

Sposito, V. A., Quadratic duality over arbitrary cone domains, J. Math. Programming 10, 277-283 (1976).

Theil, H., and C. Van de Panne, Quadratic programming as an extension of classical quadratic programming, Manage. Sci. 7, 1-20 (1960).

Tomlin, J. A., Survey of Computational Methods for Solving Large Scale Systems, Technical Report 72-75, Operations Research House, Stanford University, 1972.

Wolfe, P., The Simplex Method for Quadratic Programming, RAND Report P-1205, Santa Monica, California, 1957.

7
COMPLEMENTARITY PROBLEMS

KATTA G. MURTY

Department of Industrial
and Operations Engineering
The University of Michigan
Ann Arbor, Michigan

INTRODUCTION

Let R^n denote the n-dimensional Euclidean real vector space. Let M be a given square matrix of order n and q a column vector in R^n. Consider the problem: find $w_1, \ldots, w_n, z_1, \ldots, z_n$ satisfying

$$w - Mz = q, \qquad w \geqq 0,\ z \geqq 0, \quad \text{and} \quad w_i z_i = 0 \text{ for all } i$$

As a specific example, let

$$n = 2, \qquad M = \begin{pmatrix} 2 & 1 \\ 1 & 2 \end{pmatrix}, \qquad q = \begin{pmatrix} -5 \\ -6 \end{pmatrix}$$

For this case, the above problem is to solve:

$$\begin{aligned} w_1 \qquad & -2z_1 - z_2 = -5 \\ w_2 \; & -z_1 - 2z_2 = -6 \end{aligned} \tag{1}$$

all variables $w_1, w_2, z_1, z_2 \geqq 0$ and $w_1 z_1 = w_2 z_2 = 0$.

Problems of this type, known as *linear complementarity problems* (L.C.P.) arise in linear programming, in quadratic programming, in game theory, and in various other areas. Problem (1) can be expressed in the form of a vector equation as

$$w_1 \begin{pmatrix} 1 \\ 0 \end{pmatrix} + w_2 \begin{pmatrix} 0 \\ 1 \end{pmatrix} + z_1 \begin{pmatrix} -2 \\ -1 \end{pmatrix} + z_2 \begin{pmatrix} -1 \\ -2 \end{pmatrix} = \begin{pmatrix} -5 \\ -6 \end{pmatrix} \tag{2}$$

$$w_1, w_2, z_1, z_2 \geqq 0 \qquad \text{and} \qquad w_1 z_1 = w_2 z_2 = 0 \tag{3}$$

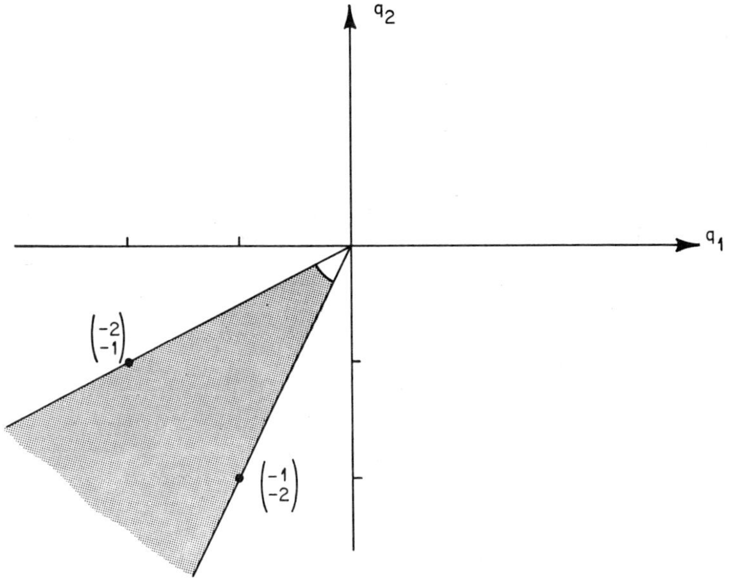

Fig. 1. $\text{Pos}\left\{\begin{pmatrix}-2\\-1\end{pmatrix}, \begin{pmatrix}-1\\-2\end{pmatrix}\right\}$.

In any solution satisfying (3), at least one of the variables in the pair (w_1, z_1) has to be equal to zero, since $w_1 z_1 = 0$. A similar statement holds for the pair (w_2, z_2). One approach for solving this problem is to pick one variable from each of the pairs (w_1, z_1), (w_2, z_2) and to fix these variables at zero value in the system of equations (2). The remaining variables in the system may be called *active variables*. After eliminating the zero variables from (2), if the remaining system has a solution in which the active variables are nonnegative, that would provide a solution to (2), (3).

Denote the right-hand side constant vector in (2), $(-5, -6)$, as (q_1, q_2). Pick w_1, w_2 as the zero valued variables. Hence the active variables are z_1, z_2. After setting $w_1, w_2 = 0$ in (2), the remaining system is

$$z_1 \begin{pmatrix}-2\\-1\end{pmatrix} + z_2 \begin{pmatrix}-1\\-2\end{pmatrix} = \begin{pmatrix}-5\\-6\end{pmatrix} = \begin{pmatrix}q_1\\q_2\end{pmatrix} = q \qquad (4)$$
$$z_1 \geqq 0, \qquad z_2 \geqq 0$$

This restricted system (4) has a solution if and only if the vector q can be expressed as a nonnegative linear combination of the vectors $\begin{pmatrix}-2\\-1\end{pmatrix}$ and $\begin{pmatrix}-1\\-2\end{pmatrix}$.

The set of all nonnegative linear combinations of $\begin{pmatrix}-2\\-1\end{pmatrix}$ and $\begin{pmatrix}-1\\-2\end{pmatrix}$ is a cone in the q_1, q_2-space as in Fig. 1. Only if the given vector $q = \begin{pmatrix}-5\\-6\end{pmatrix}$ lies in this cone does the L.C.P. (1) have a solution in which the active variables are z_1, z_2 and $w_1 = w_2 = 0$.

We verify that the point $(-5, -6)$ does lie in the cone, that the solution to (4) is $(z_1, z_2) = (\frac{4}{3}, \frac{7}{3})$, and hence a solution for (1) is $(w_1, w_2, z_1, z_2) = (0, 0, \frac{4}{3}, \frac{7}{3})$.

The cone in Fig. 1 is known as a *complementary cone* associated with the L.C.P. (1). Complementary cones are generalizations of the well known class of quadrants or orthants (see next section).

The L.C.P. (1) is of order 2. In a L.C.P. of order n, there will be $2n$ variables. Problems of order 2 can be solved by drawing all the complementary cones and checking whether the vector q lies in them. To solve problems of higher order, we need efficient algorithms and computers which can obtain solutions using these algorithms.

IMPORTANCE OF COMPLEMENTARITY PROBLEMS AND THE ROLE OF COMPUTERS IN SOLVING THEM

An Example

Let K denote the shaded convex polyhedral region in Fig. 2. Let P be the point $(-2, -1)$. It is required to find the point in K which is closest to P (in terms of the usual Euclidean distance). Problems of this type appear very often in engineering and in operations research applications.

Every point in K can be expressed as a convex combination of its *extreme points* (or *corner points*) A, B, C, D, i.e., the coordinates of a general point in K are $(\lambda_1 + 4\lambda_2 + 5\lambda_3 + 5\lambda_4, 3\lambda_1 + 0\lambda_2 + 2\lambda_3 + 4\lambda_4)$ where the λ_i satisfy $\lambda_1 + \lambda_2 + \lambda_3 + \lambda_4 = 1$ and $\lambda_i \geqq 0$ for all i. Hence the problem of finding the point in K closest to P is equivalent to solving

$$\begin{array}{ll} \text{minimize} & (\lambda_1 + 4\lambda_2 + 5\lambda_3 + 5\lambda_4 - (-2))^2 + (3\lambda_1 + 2\lambda_3 + 4\lambda_4 - (-1))^2 \\ \text{subject to} & \lambda_1 + \lambda_2 + \lambda_3 + \lambda_4 = 1 \\ & \qquad\qquad \lambda_i \geqq 0 \text{ for all } i \end{array}$$

λ_4 can be eliminated from this problem by substituting the expression $\lambda_4 = 1 - \lambda_1 - \lambda_2 - \lambda_3$ for it. Doing this and simplifying leads to the quadratic program

$$\begin{array}{ll} \text{minimize} & (-66, -54, -20)\lambda + (\tfrac{1}{2})\lambda^T \begin{pmatrix} 34 & 16 & 4 \\ 16 & 34 & 16 \\ 4 & 16 & 8 \end{pmatrix} \lambda \\ \text{subject to} & -\lambda_1 - \lambda_2 - \lambda_3 \geqq -1 \\ & \qquad\qquad \lambda \geqq 0 \end{array}$$

where $\lambda = \begin{pmatrix}\lambda_1\\ \lambda_2\\ \lambda_3\end{pmatrix}$. It will be shown in the section entitled Quadratic Programming

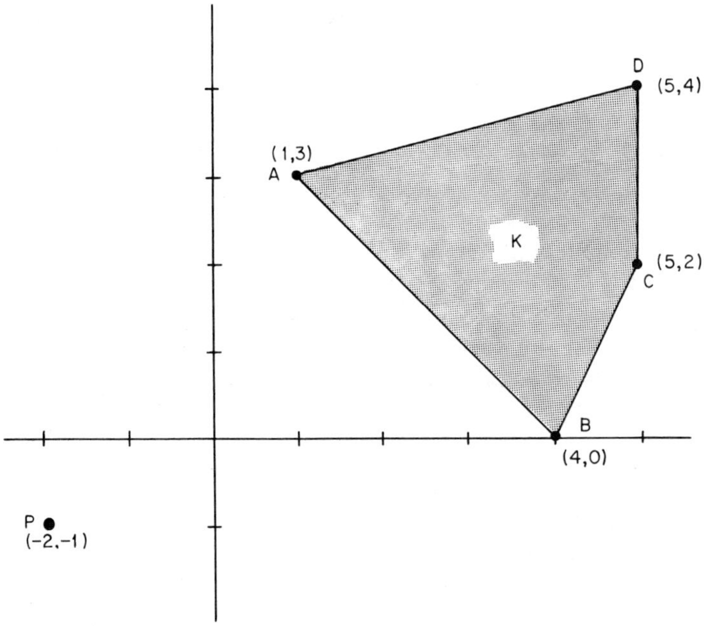

Fig. 2. Minimum distance problem.

that solving this quadratic program is equivalent to solving the L.C.P.

$$\begin{pmatrix} U_1 \\ U_2 \\ U_3 \\ V_1 \end{pmatrix} - \begin{pmatrix} 34 & 16 & 4 & 1 \\ 16 & 34 & 16 & 1 \\ 4 & 16 & 8 & 1 \\ -1 & -1 & -1 & 0 \end{pmatrix} \begin{pmatrix} \lambda_1 \\ \lambda_2 \\ \lambda_3 \\ Y_1 \end{pmatrix} = \begin{pmatrix} -66 \\ -54 \\ -20 \\ 1 \end{pmatrix}$$

all variables

$$U_1,\ U_2,\ U_3,\ V_1,\ \lambda_1,\ \lambda_2,\ \lambda_3,\ Y_1 \geqq 0$$

and

$$U_1\lambda_1 = U_2\lambda_2 = U_3\lambda_3 = V_1 Y_1 = 0$$

If $(\tilde{U}_1, \tilde{U}_2, \tilde{U}_3, \tilde{V}_1, \tilde{\lambda}_1, \tilde{\lambda}_2, \tilde{\lambda}_3, \tilde{Y}_1)$ is a solution to this L.C.P., let $\tilde{\lambda}_4 = 1 - \tilde{\lambda}_1 - \tilde{\lambda}_2 - \tilde{\lambda}_3$, and $\tilde{x} = (\tilde{\lambda}_1 + 4\tilde{\lambda}_2 + 5\tilde{\lambda}_3 + 5\tilde{\lambda}_4, 3\tilde{\lambda}_1 + 2\tilde{\lambda}_3 + 4\tilde{\lambda}_4)$ is the point in K which is closest to P.

In the section entitled Applications it is shown that all linear programming problems and all convex quadratic programming problems can be transformed into L.C.P.'s. Such L.C.P.'s can be solved by the complementary pivot algorithm discussed in the section entitled Algorithms for Solving L.C.P.

However, most practical applications of the L.C.P. lead to problems of large orders. It is virtually impossible to solve such large problems manually, and it becomes essential to use a computer. Also it is important to develop algorithms for solving the L.C.P. which lead to efficient computer programs.

Linear programming finds a vast number of applications. At present, variants of the simplex method are popularly used to solve linear programs encountered in real life applications. An experimental study (see Ref. 11) has shown that the complementary pivot algorithm for solving L.C.P.'s (see the section entitled Algorithms for Solving L.C.P.) is far superior over the simplex method for solving linear programs. It is possible that in the near future, computer programs for solving the L.C.P. will be used routinely to solve linear programming problems.

In the sections entitled Applications, Algorithms for Solving L.C.P., and Conditions, it is shown that the complementary pivot algorithm for solving the L.C.P. can be used to solve linear programs, convex quadratic programs, and bimatrix game problems. So the L.C.P. provides a unified theory for studying all these different problems.

It has also been shown that the arguments used in the complementary pivot algorithm for solving the L.C.P. can be generalized, and these generalizations have led to algorithms which can compute approximate Brouwer and Kakutani fixed points! Until now, the greatest single contribution of the complementarity problem is probably the insight that it has provided for the development of fixed point computing algorithms (see the section entitled Fixed Point Computing Methods). In mathematics, fixed point theory is very highly developed, but the absence of efficient algorithms for computing these fixed points has so far frustrated all attempts to apply this rich theory to real life problems. With the development of these new algorithms, fixed point theory is finding numerous applications in mathematical programming, in mathematical economics, and in various other areas.

NOTATION

If D is a matrix, its jth column vector will be denoted by $D_{\cdot j}$. D^T will denote the transpose of D. I denotes the *identity matrix*.

Example 1. If $n = 3$, $I = \begin{pmatrix} 1 & 0 & 0 \\ 0 & 1 & 0 \\ 0 & 0 & 1 \end{pmatrix}$ and $I_{\cdot 3} = \begin{pmatrix} 0 \\ 0 \\ 1 \end{pmatrix}$.

If $\{E_{\cdot 1}, \ldots, E_{\cdot k}\}$ is a set of column vectors in R^n, $\text{Pos}\{E_{\cdot 1}, \ldots, E_{\cdot k}\}$ will denote the set of all vectors in R^n which can be expressed as a nonnegative linear combination of $E_{\cdot 1}, \ldots, E_{\cdot k}$, i.e.,

$$\text{Pos}\{E_{\cdot 1}, \ldots, E_{\cdot k}\} = \left\{ \begin{matrix} y: \quad y = \alpha_1 E_{\cdot 1} + \cdots + \alpha_k E_{\cdot k} \\ \text{where} \quad \alpha_1 \geqq 0, \ldots, \alpha_k \geqq 0 \end{matrix} \right\}$$

Example 2. $\text{Pos}\left\{\begin{pmatrix} 1 \\ 0 \end{pmatrix}, \begin{pmatrix} 0 \\ 1 \end{pmatrix}\right\}$ is the nonnegative orthant in R^2. See Fig. 3.

Example 3. $\text{Pos}\left\{\begin{pmatrix} 2 \\ 1 \end{pmatrix}, \begin{pmatrix} -2 \\ -2 \end{pmatrix}\right\}$ is the cone which is shaded in Fig. 4.

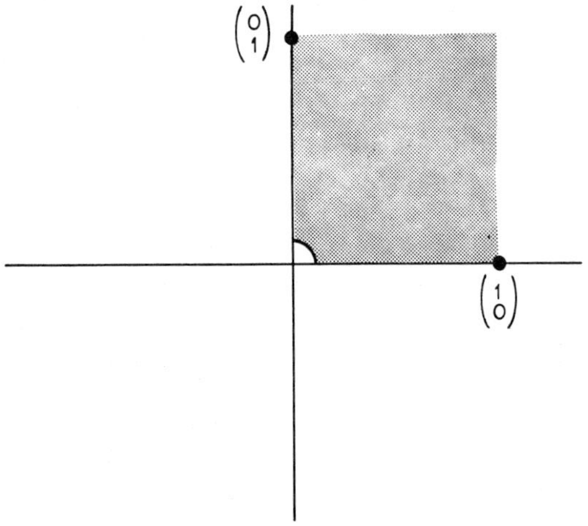

Fig. 3. Pos$\left\{\begin{pmatrix}1\\0\end{pmatrix}, \begin{pmatrix}0\\1\end{pmatrix}\right\}$.

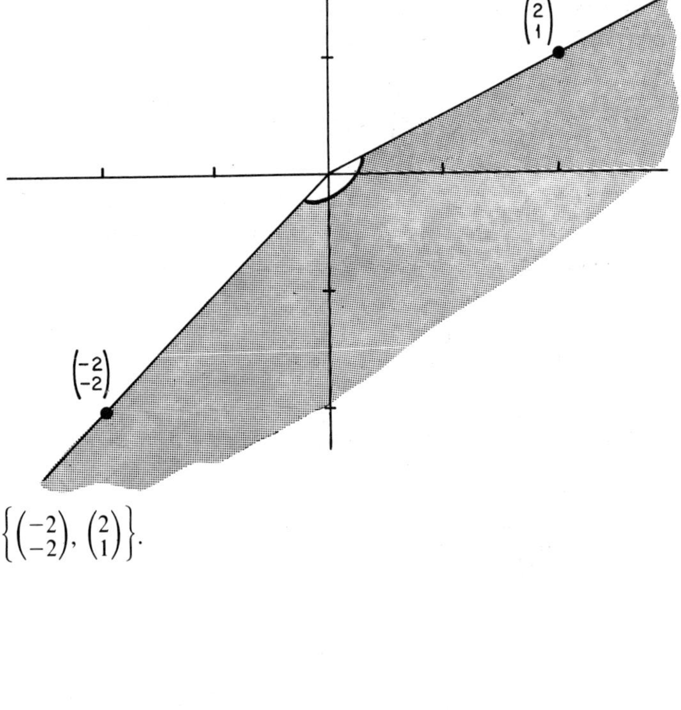

Fig. 4. Pos$\left\{\begin{pmatrix}-2\\-2\end{pmatrix}, \begin{pmatrix}2\\1\end{pmatrix}\right\}$.

For any r, e_r is the column vector in R^r all of whose components are equal to 1.

Example 4. $e_3 = (1, 1, 1)^T$.

COMPLEMENTARY CONES

Let M be a given square matrix of order n. For obtaining $\mathscr{C}(M)$, the class of complementary cones corresponding to M, the pair of column vectors $(I_{\cdot j}, -M_{\cdot j})$ is known as the jth *complementary pair of vectors*, $1 \leqq j \leqq n$. Pick a vector from the pair $(I_{\cdot j}, -M_{\cdot j})$ and denote it by $A_{\cdot j}$. The ordered set of vectors $(A_{\cdot 1}, \ldots, A_{\cdot n})$ is known as a *complementary set of vectors*. The cone $\text{Pos}(A_{\cdot 1}, \ldots, A_{\cdot n})$ is known as a *complementary cone* in the class $\mathscr{C}(M)$. Clearly there are 2^n complementary cones.

Example 5. Let $n = 2$ and $M = I$. In this case, the class $\mathscr{C}(I)$ is just the class of orthants in R^2. See Fig. 5.

In general for any n, $\mathscr{C}(I)$ = the class of orthants in R^n. Thus the class of complementary cones is a generalization of the class of orthants.

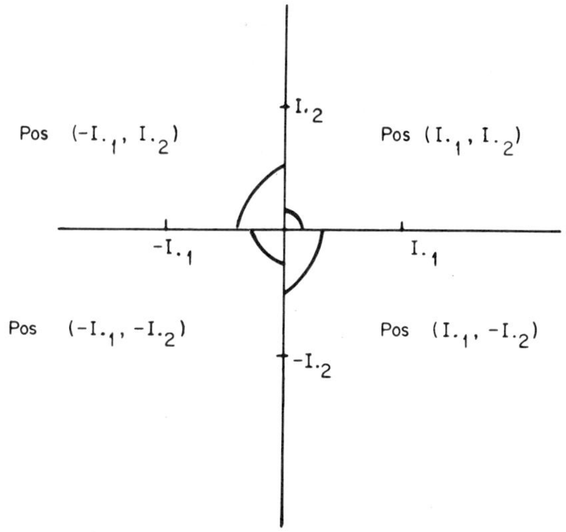

Fig. 5. Complementary cones when $M = \begin{pmatrix} 1 & 0 \\ 0 & 1 \end{pmatrix}$.

Example 6. $n = 2$ and $M = \begin{pmatrix} 2 & -1 \\ 1 & 3 \end{pmatrix}$. See Fig. 6.

Example 7. $n = 2$ and $M = \begin{pmatrix} -2 & 1 \\ 1 & -2 \end{pmatrix}$. See Fig. 7.

THE LINEAR COMPLEMENTARITY PROBLEM

Given the square matrix M of order n and the column vector $q \in R^n$, the L.C.P. (q, M), is to find a complementary cone in $\mathscr{C}(M)$ which contains the point q, i.e., to find a complementary set of column vectors $(A_{\cdot 1}, \ldots, A_{\cdot n})$ such that

(i) $A_{\cdot j} \in \{I_{\cdot j}, -M_{\cdot j}\}$, for $1 \leqq j \leqq n$, and
(ii) q can be expressed as a nonnegative linear combination of $(A_{\cdot 1}, \ldots, A_{\cdot n})$.

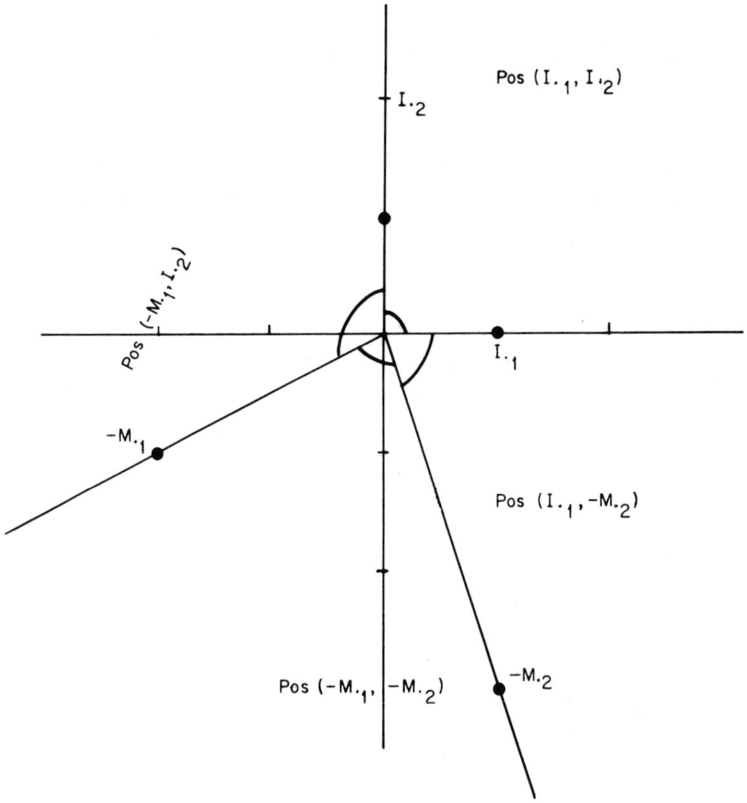

Fig. 6. Complementary cones when $M = \begin{pmatrix} 2 & -1 \\ 1 & 3 \end{pmatrix}$.

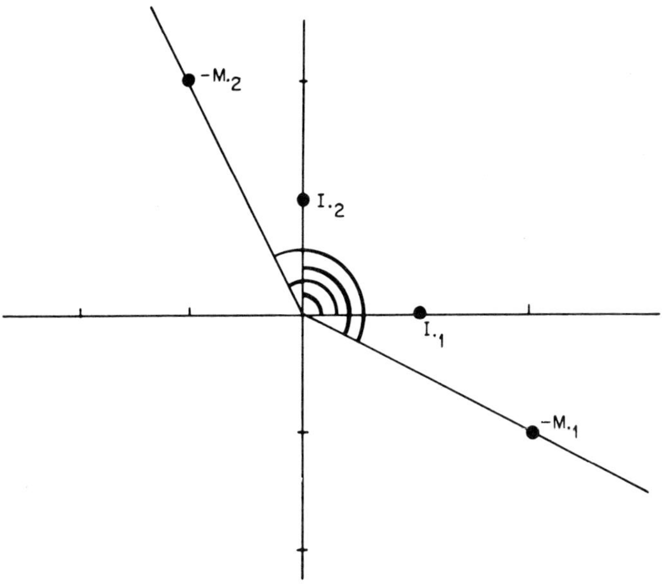

Fig. 7. Complementary cones when $M = \begin{pmatrix} -2 & 1 \\ 1 & -2 \end{pmatrix}$.

This is equivalent to finding $w \in R^n$, $z \in R^n$ satisfying

$$\sum_{j=1}^{n} I_{.j} w_j - \sum_{j=1}^{n} M_{.j} z_j = q$$

$$w_j \geqq 0, \quad z_j \geqq 0 \quad \text{(for all } 1 \leqq j \leqq n)$$

and either

$$w_j = 0 \text{ or } z_j = 0 \quad \text{(for all } 1 \leqq j \leqq n)$$

In matrix notation this is

$$w - Mz = q \tag{5}$$

$$w \geqq 0, \quad z \geqq 0 \tag{6}$$

$$w_j z_j = 0 \quad \text{(for all } j) \tag{7}$$

Because of (6), the condition (7) is equivalent to

$$\sum_{j=1}^{n} w_j z_j = w^T z = 0$$

and this condition is known as the *complementarity constraint*. Thus the L.C.P. (q, M) is to find $w \in R^n$, $z \in R^n$ satisfying (5), (6), (7).

In any solution to the L.C.P. (q, M), if one of the variables in the pair (w_j, z_j) is positive, the other should be zero. Hence the pair (w_j, z_j) is known as a *complementary pair of variables* and each variable in this pair is the *complement* of the other.

APPLICATIONS

Linear Programming

It is well known that every linear programming problem can be expressed in the *symmetric form*:

$$\begin{array}{ll} \text{minimize} & cx \\ \text{subject to} & Ax \geqq b \\ & x \geqq 0 \end{array} \tag{8}$$

Suppose A is a matrix of order $m \times N$. If $\bar{x}$ is an optimum solution of problem (8), there exists a dual vector $\bar{y} \in R^m$ and slack vectors $\bar{v} \in R^m$, $\bar{u} \in R^N$ such that $\begin{pmatrix} \bar{u} \\ \cdots \\ \bar{v} \end{pmatrix}$ and $\begin{pmatrix} \bar{x} \\ \cdots \\ \bar{y} \end{pmatrix}$ together satisfy

$$\begin{pmatrix} u \\ \cdots \\ v \end{pmatrix} - \begin{pmatrix} 0 & \vdots & -A^T \\ \cdots & \cdots & \cdots \\ A & \vdots & 0 \end{pmatrix} \begin{pmatrix} x \\ \cdots \\ y \end{pmatrix} = \begin{pmatrix} c^T \\ \cdots \\ -b \end{pmatrix}$$
$$\begin{pmatrix} u \\ \cdots \\ v \end{pmatrix} \geqq 0, \quad \begin{pmatrix} x \\ \cdots \\ y \end{pmatrix} \geqq 0, \quad \text{and} \quad \begin{pmatrix} u \\ \cdots \\ v \end{pmatrix}^T \begin{pmatrix} x \\ \cdots \\ y \end{pmatrix} = 0 \tag{9}$$

and conversely if $\begin{pmatrix} \bar{u} \\ \cdots \\ \bar{v} \end{pmatrix}, \begin{pmatrix} \bar{x} \\ \cdots \\ \bar{y} \end{pmatrix}$ together satisfy all the conditions in (9), then $\bar{x}$ is an optimum solution to the linear program (8).

Note: In (9) all the vectors and matrices are written in *partitioned form*. For example, $\begin{pmatrix} u \\ \cdots \\ v \end{pmatrix}$ is the vector $(u_1, \ldots, u_n, v_1, \ldots, v_m)^T$. If $n = m + N$,

$$w = \begin{pmatrix} u \\ \cdots \\ v \end{pmatrix}, \quad z = \begin{pmatrix} x \\ \cdots \\ y \end{pmatrix}, \quad M = \begin{pmatrix} 0 & \vdots & -A^T \\ \cdots & \cdots & \cdots \\ A & \vdots & 0 \end{pmatrix}, \quad q = \begin{pmatrix} c^T \\ \cdots \\ -b \end{pmatrix}$$

(9) is seen to be an L.C.P. Solving the linear program (8) can be achieved by solving the L.C.P. (9).

Quadratic Programming

Consider the quadratic programming problem

$$\begin{array}{ll} \text{minimize} & cx + (\tfrac{1}{2})x^T Dx \\ \text{subject to} & Ax \geqq b \\ & x \geqq 0 \end{array} \tag{10}$$

where A is a matrix of order $m \times N$, and D is a square symmetric matrix of order N. A point $\tilde{x} \in R^n$ is said to be a *Karush-Kuhn-Tucker point* for (10), if there exists a vector $\tilde{y} \in R^m$, and slack vectors $\tilde{u} \in R^N$, $\tilde{v} \in R^m$ such that $\tilde{x}$, $\tilde{y}$, $\tilde{u}$, $\tilde{v}$ together satisfy

$$\begin{pmatrix} u \\ \cdots \\ v \end{pmatrix} - \begin{pmatrix} D & -A^T \\ A & 0 \end{pmatrix} \begin{pmatrix} x \\ \cdots \\ y \end{pmatrix} = \begin{pmatrix} c^T \\ \cdots \\ -b \end{pmatrix}$$

$$\begin{pmatrix} u \\ \cdots \\ v \end{pmatrix} \geqq 0, \quad \begin{pmatrix} x \\ \cdots \\ y \end{pmatrix} \geqq 0, \quad \text{and} \quad \begin{pmatrix} u \\ \cdots \\ v \end{pmatrix}^T \begin{pmatrix} x \\ \cdots \\ y \end{pmatrix} = 0 \qquad (11)$$

(11) is a L.C.P. By the *Karush-Kuhn-Tucker necessity theorem*, if x is a local minimum for (10), it must be a Karush-Kuhn-Tucker point. If D is a *positive semidefinite matrix* (see the section entitled Classes of Matrices for a definition of this concept), then (10) is known as a *convex quadratic program* and, in this case, the *Karush-Kuhn-Tucker sufficiency theorem* guarantees that every Karush-Kuhn-Tucker point is an optimum solution of (10). Solving the L.C.P. (11) provides a Karush-Kuhn-Tucker point for (10), and if D is a positive semidefinite matrix, this is an optimum solution for (10).

Two Person Games

In each play of the game, player I picks one out of a possible set of his m choices and independently player II picks one out of a possible set of his N choices. In a play, if it happens that Player I picks his choice i, and player II picks his choice j, then player I loses an amount a'_{ij}\$ and player II loses an amount b'_{ij}\$, where $A' = (a'_{ij})$ and $B' = (b'_{ij})$ are given *loss matrices*.

If $a'_{ij} + b'_{ij} = 0$ for all i and j, the game is known as a *zero sum game*, and in this case it is possible to develop the concept of an *optimum strategy* for playing the game using *Von Neumann's minimax theorem*. Games which are not zero sum games are called *nonzero sum games* or *bimatrix games*. In bimatrix games it is difficult to define an optimum strategy. However, in this case an *equilibrium pair of strategies* can be defined (see next paragraph) and the problem of computing an equilibrium pair of strategies can be transformed into a L.C.P.

Suppose player I picks his choice i with a probability of x_i. The column vector $x = (x_i) \in R^m$ completely defines player I's strategy. Similarly let the probability vector $y = (y_j) \in R^N$ be player II's strategy. If player I adopts strategy x and player II adopts strategy y, the expected loss of player I is obviously $x^TA'y$ and that of player II is $x^TB'y$.

The strategy pair $(\bar{x}, \bar{y})$ is said to be an *equilibrium pair* if no player benefits by unilaterally changing his own strategy while the other player keeps his strategy in the pair $(\bar{x}, \bar{y})$ unchanged, i.e., if

$$\bar{x}A'\bar{y} \leqq xA'\bar{y} \qquad \text{(for all probability vectors } x \in R^m)$$

and

$$\bar{x}B'\bar{y} \leqq \bar{x}B'y \qquad \text{(for all probability vectors } y \in R^N)$$

Let α, β be arbitrary positive numbers such that $a_{ij} = a'_{ij} + \alpha > 0$ and $b_{ij} = b'_{ij} + \beta > 0$ for all i, j. Let $A = (a_{ij})$, $B = (b_{ij})$. Consider the L.C.P.

$$\begin{pmatrix} u \\ \cdots \\ v \end{pmatrix} - \begin{pmatrix} 0 & \vdots & A \\ \cdots & \cdots & \cdots \\ B^T & \vdots & 0 \end{pmatrix} \begin{pmatrix} \xi \\ \cdots \\ \eta \end{pmatrix} = \begin{pmatrix} -e_m \\ \cdots \\ -e_N \end{pmatrix}$$
$$\begin{pmatrix} u \\ \cdots \\ v \end{pmatrix} \geqq 0, \quad \begin{pmatrix} \xi \\ \cdots \\ \eta \end{pmatrix} \geqq 0, \quad \begin{pmatrix} u \\ \cdots \\ v \end{pmatrix}^T \begin{pmatrix} \xi \\ \cdots \\ \eta \end{pmatrix} = 0 \tag{12}$$

See the section entitled Notation for the definitions of e_m, e_N. It has been shown that if $(\tilde{u}, \tilde{v}, \tilde{\xi}, \tilde{\eta})$ solves the L.C.P. (12), then $\tilde{x} = \tilde{\xi}/(\Sigma\tilde{\xi}_i)$ and $\tilde{y} = \tilde{\eta}/(\Sigma\tilde{\eta}_i)$, are an equilibrium pair of strategies for the original game.

Other Applications

L.C.P. has been used to study problems in the plastic analysis of structures, in the inelastic flexural behavior of reinforced concrete beams, in the free boundary problems for journal bearings, in the study of finance models, and in several other areas.

CLASSES OF MATRICES

We define several classes of matrices which are useful in the study of the L.C.P. Let M be a square matrix of order n. It is said to be:

1. A *positive definite* matrix if

$$y^T M y = \sum_{i=1}^{n} \sum_{j=1}^{n} y_i m_{ij} y_j > 0 \text{ for all } 0 \neq y \in R^n$$

2. A *positive semi-definite matrix* if

$$y^T M y \geqq 0 \text{ for all } y \in R^n$$

3. A *copositive matrix* if

$$y^T M y \geqq 0 \text{ for all } y \geqq 0$$

4. A *copositive plus matrix* if it is a copositive matrix and if

$$y^T(M + M^T) = 0, \text{ whenever } y \geqq 0 \text{ satisfies } y^T M y = 0$$

5. A *P-matrix* if all principal subdeterminants of M are positive.
6. A *Q-matrix* if the L.C.P. (q, M) has a solution for every $q \in R^n$.

ALGORITHMS FOR SOLVING L.C.P.

L.C.P.'s of order 2 can be solved by drawing all the complementary cones in the q_1, q_2 plane as discussed in the Introduction.

Example 8. Let $q = \begin{pmatrix} 4 \\ -1 \end{pmatrix}$, $M = \begin{pmatrix} -2 & 1 \\ 1 & -2 \end{pmatrix}$ and consider the L.C.P. (q, M). The class of complementary cones corresponding to this problem is in Fig. 7.

w_1	w_2	z_1	z_2	q
1	0	2	−1	4
0	1	−1	2	−1

(13)

$$w_1, w_2, z_1, z_2 \geqq 0, \; w_1 z_1 = w_2 z_2 = 0$$

q lies in two complementary cones $\text{Pos}(-M_{\cdot 1}, I_{\cdot 2})$ and $\text{Pos}(-M_{\cdot 1}, -M_{\cdot 2})$. This implies that the sets of active variables (z_1, w_2) and (z_1, z_2) lead to solutions of the L.C.P.

Putting $w_1 = z_2 = 0$ in (13) and solving the remaining system for the values of the active variables (z_1, w_2) leads to the solution $(z_1, w_2) = (2, 1)$. Hence $(w_1, w_2, z_1, z_2) = (0, 1, 2, 0)$ is a solution to this L.C.P. Similarly putting $w_1 = w_2 = 0$ in (13) and solving it for the values of the active variables (z_1, z_2) leads to the second solution $(w_1, w_2, z_1, z_2) = (0, 0, \frac{7}{3}, \frac{2}{3})$ to this L.C.P.

Example 9. Let $q = \begin{pmatrix} -1 \\ -1 \end{pmatrix}$, $M = \begin{pmatrix} -2 & 1 \\ 1 & -2 \end{pmatrix}$ and consider the L.C.P. (q, M). The class of complementary cones corresponding to this problem is in Fig. 7. We verify that q is not contained in any complementary cone. Hence this L.C.P. has no solution.

As discussed in the Introduction, the method described above can be conveniently used only for L.C.P.'s of order 2. For solving L.C.P.'s of higher order, we will describe the *complementary pivot algorithm* which uses *row operations* on the system of equations in (5).

In the L.C.P. (5), (6), (7), if $q \geqq 0$, then $w = q$, $z = 0$ is a solution and we are done. So we assume $q \not\geqq 0$.

Bases

An artificial variable z_0, associated with the column vector $-e_n$ (see the section entitled Notation), is introduced into the system (5) to get a feasible basis for starting the algorithm. In detached coefficient tableau form, (5) then becomes:

w	z	z_0	$=$
I	$-M$	$-e_n$	q

(14)

$$w \geqq 0 \qquad z \geqq 0 \qquad z_0 \geqq 0$$

In each stage the algorithm deals with a *basis*, which is a square nonsingular submatrix of order n of the coefficient matrix in (14). The column vectors in the basis are called the *basic column vectors*, and the variables in (14) associated with them are called the *basic variables*. The remaining column vectors in the tableau are called the *nonbasic column vectors* and the variables associated with them are the *nonbasic variables*.

The solution of (14) corresponding to a given basis is obtained by setting all the nonbasic variables equal to zero, and then solving the remaining system of linear equations in (14) for the values of the basic variables. The basis is a *feasible basis* if the values of all the basic variables in the solution turn out to be nonnegative. The algorithm deals only with *feasible bases*.

When the basis is transformed into the identity matrix [either by performing the necessary row operations or by multiplying the entire tableau in (14) on the left by the inverse of the basis], (14) is transformed into the *canonical form* with respect to that basis. The basic variable associated with $I_{.r}$ in the canonical form is known as the *rth basic variable*, or the basic variable in the rth row of the canonical tableau. All the column vectors in the canonical tableau are called the *updated column vectors*. Let $\bar{q}$ denote the updated right-hand side constant vector. Then $\bar{q}$ is the vector of values of the basic variables in the solution corresponding to this basis, $\bar{q} \geqq 0$ if the basis is a feasible basis.

Pivot Operations

The primary computational step used in the algorithm is the *pivot step* which is also the main step in the *simplex algorithm* for solving *linear programs*.

In each stage of the algorithm, the basis is changed by bringing into the basis exactly one nonbasic column vector. The variable in (14) associated with this column vector is called the *entering variable*. Its updated column vector is called the *pivot column* for this basis change.

In the solution of (14), with respect to the new basis, all the nonbasic variables other than the entering variable will be equal to zero.

Since a basis for (14) always consists of n column vectors, when a new column is introduced into the basis, one of the present basic column vectors has to be dropped from the basis. The dropping vector has to be determined according to a special rule known as the *minimum ratio test* in linear programming terminology. This test guarantees that the new basis obtained after the pivot step will also be a feasible basis.

We will review this minimum ratio test here briefly. The only variables whose values in the solution change in this step are the entering variable and the present basic variables. Suppose the value of the entering variable is changed from its present value of 0 to some nonnegative value, λ. The values of the basic variables are to be determined as functions of λ, so that the new solution still satisfies the equality constraints in (14).

As an example, let us assume that the present basic variables are $y_1, \ldots, y_n$ with y_r as the rth basic variable, and let the entering variable be x_s. [The variables in (14) are $w_1, \ldots, w_n; z_1, \ldots, z_n, z_0$. Exactly n of these variables are the present basic variables. For convenience in referring, we are assuming that these basic variables are called $y_1, \ldots, y_n$.] After rearranging the variables in (14), if necessary, the canonical form of (14) with respect to the present basic is of the form

Basic variable	$y_1 \cdots y_i \cdots y_n$	x_s	Other variables	Right-hand constant vector
y_1	1 0	$\bar{a}_{1s}$		$\bar{q}_1$
.	0 .	.		.
.	. .			.
.	. .			.
.	. .			.
y_n	0 1	$\bar{a}_{ns}$		$\bar{q}_n$

Since all the nonbasic variables other than x_s will be equal to zero, the new solution is clearly

$$
\begin{aligned}
x_s &= \lambda \\
y_1 &= \bar{q}_1 - \bar{a}_{1s}\lambda \\
&\vdots \\
y_i &= \bar{q}_i - \bar{a}_{is}\lambda \\
&\vdots \\
y_n &= \bar{q}_n - \bar{a}_{ns}\lambda \\
&\text{all other variables} = 0
\end{aligned}
\tag{15}
$$

There are two possibilities here.

1. The pivot column may be nonpositive; i.e., $\bar{a}_{is} \leqq 0$ for all $1 \leqq i \leqq n$. In this case the solution in (15) remains nonnegative for all $\lambda \geqq 0$. As λ varies from 0 to ∞, this solution traces an *extreme ray* (or an *unbounded edge*) of the set of feasible solutions of (14).

If this happens, the algorithm terminates. This type of termination is called *ray termination*.

2. There is at least one positive entry in the pivot column. In this case, if the solution in (15) should remain nonnegative, the maximum value that λ can take is θ = minimum ($\bar{q}_i/\bar{a}_{is}$: i such that $\bar{a}_{is} > 0$).

This θ is called the *minimum ratio* in this pivot step. If $\theta = \bar{q}_r/\bar{a}_{rs}$, then y_r becomes equal to zero in the solution in (15) when $x_s = \theta$. Hence y_r drops from the basic vector and x_s becomes the rth basic variable in its place. Then the rth row is known as the *pivot row* for this pivot step.

Pivoting consists of transforming the pivot column into $I_{.r}$ by row operations. This leads to the canonical tableau with respect to the new basis and this completes this stage.

As λ varies from 0 to θ in (15), we move from the solution in this stage to the solution of (14) with respect to the new basis along the line segment joining them. These two solutions are adjacent extreme points of the set of feasible solutions of (14), and the line segment joining them is an *edge*.

Initial Basis

The artificial variable z_0 has been introduced into (14) for the sole purpose of obtaining a feasible basis to start the algorithm.

Identify row t such that q_t = minimum (q_i: $1 \leqq i \leqq n$). Since we assumed $q \ngeq 0$, $q_t < 0$. When a pivot is made in (14) with the column vector of z_0 as the pivot column and the tth row as the pivot row, the right-hand side constant vector becomes a nonnegative vector. The resulting tableau is the canonical tableau with respect to the basis in which the basic variables are $w_1, \ldots, w_{t-1}, z_0, w_{t+1}, \ldots, w_n$. This is the initial basis for starting the algorithm.

Properties of Bases

The initial basis satisfies the following properties:

1. There is at most one basic variable from each complementary pair of variables (w_j, z_j).
2. z_0 is a basic variable in it. It contains exactly one basic variable from each of ($n - 1$) complementary pairs of variables, and both the variables in the remaining complementary pair are nonbasic.

A feasible basis for (14) in which there is exactly one basic variable from each complementary pair (w_j, z_j) is known as a *complementary feasible basis*. The feasible solution of (14) corresponding to a complementary feasible basis is a solution to the L.C.P. (5), (6), (7).

A feasible basis for (14) satisfying the two properties given above is known as an *almost complementary feasible basis*. All the bases obtained in the algorithm with the possible exception of the final basis are almost complementary feasible bases. If at some stage of the algorithm a complementary feasible basis is obtained, it is a final basis and the algorithm terminates.

Adjacent Almost Complementary Feasible Bases

Let $(y_1, \ldots, y_{j-1}, z_0, y_{j+1}, \ldots, y_n)$ be the vector of basic variables in an almost complementary feasible basis for (14), where $y_i \in \{w_i, z_i\}$ for each $i \neq j$. In this basis, both the variables in the complementary pair $\{w_j, z_j\}$ are nonbasic variables. Adjacent almost complementary feasible bases can only be obtained by picking as the entering variable either w_j or z_j. Thus from each almost complementary feasible basis there are exactly two possible ways of generating adjacent almost complementary feasible bases.

In the initial almost complementary feasible basis, both w_t and z_t are nonbasic variables. In the canonical tableau with respect to the initial basis, the updated column vector of w_t can be verified to be $-e_n$, which is <0. Hence, if w_t is picked as the entering variable into the initial basis by possibility 1 of the section entitled Pivot Operations, a ray is generated. Hence the initial almost complementary feasible basis is at the end of an almost complementary ray.

So there is a unique way of obtaining an adjacent almost complementary feasible basis from the initial basis, and that is to pick z_t as the entering variable.

Complementary Pivot Rule

In all subsequent stages of the algorithm, there is a unique way to continue the algorithm, which is to pick as the entering variable the complement of the variable that just dropped from the basis. This is known as the *complementary pivot rule*.

The main property of the path generated by the algorithm is the following. Each basic feasible solution obtained in the algorithm has two almost complementary edges containing it. We arrive at this solution along one of these edges. And we leave it by the other edge. So the algorithm continues in a unique manner.

Termination

There are exactly two possible ways in which the algorithm can terminate. They are:

1. At some stage of the algorithm, z_0 may leave the basic vector and a complementary feasible basis is then obtained. The solution of (14) corresponding to this final basis is a solution of the L.C.P. (5), (6), (7).
2. At some stage of the algorithm, the pivot column may turn out to be nonpositive (discussed in possibility 1 of the section entitled Pivot Operations), and in this case the algorithm terminates with a ray. This is called *ray termination*. When this happens, the algorithm is unable to solve the L.C.P.

NUMERICAL EXAMPLES

Example 10

Consider the following L.C.P.

w_1	w_2	w_3	w_4	z_1	z_2	z_3	z_4	q
1	0	0	0	−1	1	1	1	3
0	1	0	0	1	−1	1	1	5
0	0	1	0	−1	−1	−2	0	−9
0	0	0	1	−1	−1	0	−2	−5

$$w_i \geqq 0, \quad z_i \geqq 0, \quad w_i z_i = 0 \quad \text{for all} \quad i$$

Introducing the artificial variable z_0 the tableau becomes

w_1	w_2	w_3	w_4	z_1	z_2	z_3	z_4	z_0	q
1	0	0	0	−1	1	1	1	−1	3
0	1	0	0	1	−1	1	1	−1	5
0	0	1	0	−1	−1	−2	0	(−1)	−9
0	0	0	1	−1	−1	0	−2	−1	−5

Since the most negative q_i is $q_3 = -9$, we have to pivot in the column vector of z_0 with the third row as the pivot row. The pivot element is circled in the tableau. Performing the pivot leads to the canonical tableau with respect to the initial basis. The basic variables are marked in their order on the left-hand side column.

Basic variables	w_1	w_2	w_3	w_4	z_1	z_2	z_3	z_4	z_0	q	Ratios
w_1	1	0	−1	0	0	2	3	1	0	12	12/3
w_2	0	1	−1	0	2	0	3	1	0	14	14/3
z_0	0	0	−1	0	1	1	2	0	1	9	9/2
w_4	0	0	−1	1	0	0	(2)	−2	0	4	4/2

As in the section entitled Adjacent Almost Complementary Feasible Bases, we now pick z_3 as the entering variable. The column vector of z_3 is the pivot column. The ratios discussed in the section entitled Pivot Operations are entered on the rightmost column. The minimum of these ratios is in the fourth row. So the pivot row is the fourth row. The pivot element is circled. w_4 drops from the basic vector and is replaced by z_3.

Basic variables	w_1	w_2	w_3	w_4	z_1	z_2	z_3	z_4	z_0	q	Ratios
w_1	1	0	1/2	−3/2	0	2	0	(4)	0	6	6/4
w_2	0	1	1/2	−3/2	2	0	0	4	0	8	8/4
z_0	0	0	0	−1	1	1	0	2	1	5	5/2
z_3	0	0	−1/2	1/2	0	0	1	−1	0	2	

Since w_4 dropped from the basic vector, its complement, z_4, is the entering variable for the next step. The column of z_4 is the pivot column. The pivot element is circled. w_1 drops from the basic vector.

Basic variables	w_1	w_2	w_3	w_4	z_1	z_2	z_3	z_4	z_0	q	Ratios
z_4	1/4	0	1/8	−3/8	0	1/2	0	1	0	6/4	
w_2	−1	1	0	0	(2)	−2	0	0	0	2	2/2
z_0	−1/2	0	−1/4	−1/4	1	0	0	0	1	2	2/1
z_3	1/4	0	−3/8	1/8	0	1/2	1	0	0	14/4	

Since w_1 dropped from the basic vector, its complement, z_1, is the new entering variable. Now w_2 drops from the basic vector.

Basic variables	w_1	w_2	w_3	w_4	z_1	z_2	z_3	z_4	z_0	q	Ratios
z_4	1/4	0	1/8	−3/8	0	1/2	0	1	0	6/4	3
z_1	−1/2	1/2	0	0	1	−1	0	0	0	1	
z_0	0	−1/2	−1/4	−1/4	0	(1)	0	0	1	1	1
z_3	1/4	0	−3/8	1/8	0	1/2	1	0	0	14/4	7

Since w_2 dropped from the basic vector, its complement, z_2, is the entering variable. Now z_0 drops from the basic vector.

Basic variable	w_1	w_2	w_3	w_4	z_1	z_2	z_3	z_4	z_0	q
z_4	1/4	1/4	1/4	1/4	0	0	0	1	−1/2	1
z_1	−1/2	0	−1/4	−1/4	1	0	0	0	1	2
z_2	0	−1/2	−1/4	−1/4	0	1	0	0	1	1
z_3	1/4	1/4	−1/4	1/4	0	0	1	0	−1/2	3

Since the present basis is a complementary feasible basis, the algorithm terminates. The corresponding solution is:

nonbasic variables $w_1, w_2, w_3, w_4 = 0$
basic vector $z_4 = 1,\ z_1 = 2,\ z_2 = 1,\ z_3 = 3$

and this is a solution to the L.C.P.

Example 11

Consider the following L.C.P.

w_1	w_2	w_3	z_1	z_2	z_3	q
1	0	0	1	0	3	−3
0	1	0	−1	2	5	−2
0	0	1	2	1	2	−1

$$w_i \geqq 0,\ z_i \geqq 0,\ w_i z_i = 0 \text{ for all } i$$

The tableau with the artificial variable z_0 is

w_1	w_2	w_3	z_1	z_2	z_3	z_0	q
1	0	0	1	0	3	(−1)	−3
0	1	0	−1	2	5	−1	−2
0	0	1	2	1	2	−1	−1

The initial canonical tableau is

Basic variables	w_1	w_2	w_3	z_1	z_2	z_3	z_0	q	Ratios
z_0	−1	0	0	−1	0	−3	1	3	
w_2	−1	1	0	−2	2	2	0	1	
w_3	−1	0	1	(1)	1	−1	0	2	2/1

The next tableau is

Basic variables	w_1	w_2	w_3	z_1	z_2	z_3	z_0	q
z_0	−2	0	1	0	1	−4	1	5
w_2	−3	1	2	0	4	0	0	5
z_1	−1	0	1	1	1	−1	0	2

The entering variable here is z_3. The pivot column is nonpositive. Hence the algorithm stops here with a ray termination. The algorithm has been unable to solve this L.C.P.

CONDITIONS

It has been proved that the complementary pivot algorithm for solving the L.C.P. provides a solution to the L.C.P., if such a solution exists, whenever M is either a nonnegative matrix with positive diagonal entries, or a P-matrix, or a positive semidefinite matrix, or a copositive plus matrix, or when M belongs to some other special classes of matrices not discussed here.

Linear programs and convex quadratic programs can be solved by this algorithm using the L.C.P. format for them discussed in the section entitled Applications. The complementary pivot algorithm may provide a solution to the L.C.P. (5), (6), (7) even when M is not in any of these classes of matrices, but in this case it is possible that it will stop with ray termination even if a solution to the L.C.P. exists.

COMPUTER PROGRAMS

A computer program (in FORTRAN) for the complementary pivot algorithm for solving the L.C.P. has been published in Ref. 10.

OTHER ALGORITHMS

A special starting routine has been developed which, when combined with the complementary pivot routine of the section entitled Algorithms for Solving L.C.P., yields a solution to the L.C.P. associated with a bimatrix game. Algorithms based on principal pivoting are available, and they seem to be particularly useful when M is positive semidefinite. Algorithms which use only complementary (but infeasible) bases in intermediate stages are being studied. Algorithms using systematic overrelaxation methods have also been developed.

The only methods which will solve an L.C.P. when M is a general matrix are enumerative algorithms. Even when M belongs to a special class of matrices, the only algorithms which can generate the set of all solutions to an L.C.P. are again enumerative-type algorithms.

SUMMARY OF THEORETICAL RESULTS

In the linear complementarity problem (q, M) if M is fixed, the problem has a unique solution for each $q \in R^n$ if M is a P-matrix, i.e., the class of complementary cones $\mathscr{C}(M)$ partitions R^n (like the class of orthants does), if M is a P-matrix. Under certain conditions on M it has been shown that the problem either has an even number of solutions for all $q \in R^n$ or an odd number of solutions for all $q \in R^n$. Several such constant parity results are known. The properties of the linear complementarity problem when M belongs to special classes of matrices have been extensively studied.

UNSOLVED PROBLEMS

An algorithm for a problem is said to be good algorithm if the number of computations required by it is bounded above by a polynomial in n, the order (or size) of the problem.

Good algorithms are known for testing whether a given square matrix is a positive definite or a positive semidefinite matrix. These algorithms are based on row operations on the matrix and are discussed in textbooks on linear algebra. So far there are no good algorithms to test whether a general square matrix is copositive, copositive plus, a P-matrix, or a Q-matrix. Also, for a general matrix, there are no simple necessary and sufficient conditions known for it to be a Q-matrix.

Given a solution to an L.C.P., one would like to know whether this solution is unique or whether alternate solutions exist. If alternate solutions exist, one would like to compute them. The enumerative algorithms available at present to answer these questions, and to solve the L.C.P. when M is a general matrix, are not very efficient when n is large.

Besides these, there are a large number of unsolved theoretical problems on the spanning properties of complementary cones and on the matrix theoretic properties associated with the L.C.P.

THE NONLINEAR COMPLEMENTARITY PROBLEM

Let $f(z) = (f_1(z), \ldots, f_n(z))^T \in R^n$ be a vector valued function of $z \in R^n$. The problem of solving

$$\begin{aligned} z &\geqq 0 \\ f(z) &\geqq 0 \\ f(z)^T z &= \sum_{i=1}^{n} z_i f_i(z) = 0 \end{aligned}$$

is known as a nonlinear complementarity problem. If $f(z) = Mz + q$, this becomes a L.C.P. Various results on necessary and sufficient conditions for the existence of a solution to the nonlinear complementarity problem have been established. It has been shown that the nonlinear complementarity problem can be transformed into the problem of computing a *Kakutani fixed point* (see next section) of a *point to set map* (see next section) defined on R^n. Algorithms for computing these fixed points using ideas analogous to those discussed in the complementary pivot algorithm for solving the L.C.P. have been developed, and these are discussed very briefly in the next section.

FIXED POINT COMPUTING METHODS

Consider a map which associates a subset $F(x) \in R^n$ for each point $x \in R^n$. A map like this is known as a *point to set map*. A *Kakutani fixed point* of this map is a point $y \in R^n$ which satisfies $y \in F(y)$.

Algorithms for computing these fixed points efficiently have been developed recently. These algorithms use *triangulations* of R^n. A triangulation is a partition of R^n into *simplices*. Each simplex in R^n is the convex hull of $n + 1$ points, called its *vertices*. Each vertex is given a *label* from the set $\{1, \ldots, n + 1\}$. The labeling is done in such a way that if a simplex, all of whose vertices have distinct labels is found, any point in that simplex is an approximation of the fixed point. So the problem reduces to that of finding a simplex in the triangulation all of whose vertices have distinct labels. Since each simplex has exactly $n + 1$ vertices, the set of labels on the vertices of a distinctly labeled simplex must be $\{1, \ldots, n + 1\}$. Hence, such simplices are called *completely labeled simplices*.

The algorithm begins with a specially constructed simplex which is *almost completely labeled,* i.e., the vertices of this simplex have distinct labels with the exception of exactly one pair of vertices which have the same label. As an example, the set of labels on the vertices of the starting simplex might be $\{1, 1, 2, \ldots, n\}$. The labels on the vertices of an almost completely labeled simplex satisfy the following properties.

1. Exactly one label appears twice.
2. Exactly one label from the set $\{1, \ldots, n + 1\}$ is missing.

The properties of the triangulation used guarantee that each almost completely labeled simplex has at most two adjacent simplices which are either completely labeled or almost completely labeled. The starting simplex has exactly one adjacent almost completely labeled simplex and the algorithm moves to that. All the simplices obtained by the algorithm along the path will be almost completely labeled excepting the terminal one which will be completely labeled. We arrive at a simplex on the path from an adjacent almost completely labeled simplex. This simplex has exactly one other adjacent simplex which is either almost completely labeled or completely labeled, and the algorithm moves to that. Thus the path continues in a unique, unambiguous manner until it terminates with a completely labeled simplex.

The resemblance of the almost completely labeled simplices path obtained in the algorithm here to the almost complementary bases path obtained in the complementary pivot algorithm for solving the L.C.P. is quite clear.

It has been shown that nonlinear programming problems (either constrained or unconstrained) can be transformed into problems of computing fixed points. The algorithms for computing fixed points hold a great promise for wide applicability in mathematical programming and its applications. For more details on these algorithms, see the article on *Fixed Point Computing Methods*.

BACKGROUND AND REFERENCES

The complementary pivot algorithm for computing equilibrium strategies in bimatrix games appeared in Ref. 7. The algorithm discussed in the section entitled Algorithms for Solving L.C.P. is due to Lemke [5]. This algorithm, various applications of the L.C.P., and a discussion of principal pivoting methods are covered in Ref. 1 very clearly. For theoretical results on the L.C.P., see Refs. 3, 8, and 12. For results on the nonlinear complementarity problem, see Ref. 4. Reference 6 is a survey paper. Numerous references on the complementarity problem appear in the bibliographies at the end of these papers. The pioneering paper on fixed point computing methods is Ref. 13. See the article on *Fixed Point Computing Methods* for complete references on that topic. For applications of L.C.P. in structural engineering, see Ref. 2. For efficient algorithms to check positive semidefiniteness of a matrix, and proofs of the results discussed in this article, see Ref. 9, Chap. 16.

REFERENCES

1. Cottle, R. W., and G. B. Dantzig, Complementary pivot theory of mathematical programming, *Linear Algebra and Its Applications* **1,** 103–125 (1968).
2. Donato, O. D., and G. Maier, Mathematical programming methods for the inelastic analysis of reinforced concrete frames allowing for limited rotation capacity, *Int. J. Numer. Methods Eng.* **4,** 307–329 (1972).
3. Eaves, B. C., On the basic theorem of complementarity, *Math. Programming* **1,** 68–75 (1971).
4. Karamardian, S., The complementarity problem, *Math. Programming* **2,** 107–129 (1972).
5. Lemke, C. E., Bimatrix equilibrium points and mathematical programming, *Manage. Sci.* **11,** 681–689 (1965).
6. Lemke, C. E., Recent results on complementarity problems, in *Nonlinear Programming* (J. B. Rosen, O. L. Mangasarian, and K. Ritter, eds.), Academic, New York, 1970.
7. Lemke, C. E., and J. T. Howson, Jr., Equilibrium points of bimatrix games, *SIAM J. Appl. Math.* **12,** 413–423 (1964).
8. Murty, K. G., On the number of solutions to the complementarity problem and spanning properties of complementary cones, *Linear Algebra and Its Applications* **5,** 65–108 (1972).
9. Murty, K. G., *Linear and Combinatorial Programming,* Wiley, New York, 1976.
10. Ravindran, A. A computer routine for quadratic and linear programming problems [H], *Commun. ACM* **15**(9), 818–820 (September 1972).
11. Ravindran, A., A comparison of the primal-simplex and complementary pivot methods for linear programming, *Naval Res. Logistics Q.* **20**(1), 95–100 (March 1973).
12. Saigal, R., On the class of complementary cones and Lemke's algorithm, *SIAM J. Appl. Math.* **23,** 46–60 (1972).
13. Scarf, H., The approximation of fixed points of a continuous map, *SIAM J. Appl. Math.* **15,** 328–343 (1967).

8
GEOMETRIC PROGRAMMING

DON T. PHILLIPS

Department of Industrial Engineering
Texas A & M University
College Station, Texas

Geometric programming is a relatively new technique only recently developed to solve nonlinear programming problems with or without inequality constraints. In its most general form, elegant theoretical properties have been developed which lead to solution procedures for signed algebraic formulations characterized by linear and/or nonlinear inequality constraints. Maximization or minimization problems can theoretically be solved via a primal formulation or an equivalent (constrained) dual formulation involving only linear equality constraints. This article will attempt to perform several major goals in a relatively short space. The fundamental structure and representation of geometric programming problems will be presented, a review of recent research activities will be summarized, and suggestions for future research will be given. Two numerical examples will serve to illustrate the simplest solution procedures.

Geometric programming is an elegant solution procedure developed by Duffin et al. [18] to deal with highly nonlinear programming problems subject to linear or nonlinear inequality constraints. The name geometric programming stems from the utilization of Cauchy's arithmetic-geometric inequality in establishing fundamental primal-dual relationships. In engineering practice, most design problems can be effectively optimized via geometric programming. Restricted forms of geometric programming design problems have been successfully formulated in industrial engineering tool design [37], mechanical journal bearing design [6], nuclear engineering reactor design [7], electrical engineering transformer design [18, 48], water pollution control [19], chemical engineering process design [8, 10], civil engineering structural design [45, 48], and reliability systems design [20]. Several other applications are reported in Refs. 5 and 43. Numerical solution techniques advocated are numerous. Yet, to date, no single approach can efficiently solve large-scale problems.

MATHEMATICAL FORMULATION OF THE PROBLEM

The general statement of the problem is adapted from Passy and Wilde [32], and the notation used is of Beightler and Phillips [5]. First the primal program is considered; thereafter, the dual form is presented.

Primal Program (PGP)

Minimize

$$Y_0(\tilde{x}) = \sum_{t=1}^{T_0} \sigma_{0t} c_{0t} \prod_{n=1}^{N} x_n^{a_{0tn}} \tag{1}$$

subject to

$$Y_m(\tilde{x}) = \sum_{t=1}^{T_m} \sigma_{mt} c_{mt} \prod_{n=1}^{N} x_n^{a_{mtn}} \leq \sigma_m \tag{2}$$

where

$$\sigma_{mt} = \pm 1; \quad m = 0, 1, 2, \ldots, M, \quad t = 1, 2, \ldots, T_m \tag{3}$$

$$\sigma_m = \pm 1; \quad m = 1, 2, \ldots, M \tag{4}$$

$$c_{mt} > 0; \quad m = 0, 1, 2, \ldots, M, \quad t = 1, 2, \ldots, T_m \tag{5}$$

$$\tilde{x} = (x_1, x_2, \ldots, x_N); \quad x_n > 0, \quad n = 1, 2, \ldots, N \tag{6}$$

$$-\infty < a_{mtn} < \infty; \quad \text{for all } m, t, n \tag{7}$$

$$T = \sum_{m=0}^{M} T_m \equiv \text{total number of terms} \tag{8}$$

Finally, we introduce the notion of "degrees of difficulty," expressed as $D = T - (N + 1)$

This number actually measures the degrees of freedom of the dual constraints provided the exponent matrix has full rank.

An optimal solution to PGP exists provided that the functions $Y_m(\tilde{x})$, $m = 0, 1, \ldots, M$, are assumed to satisfy the Kuhn-Tucker constraint qualification [27], and that the program is canonical. PGP is called a canonical program if $Y_0(\tilde{x})$ has a minimum at a finite point in the positive orthant, otherwise it is called degenerate. However, certain techniques when applied can reduce degenerate programs into canonical [18]. It is important to note that the finding of an optimal solution to PGP is in general a nonconvex programming problem with nonlinear inequality constraints. On the other hand, the problem structure could be simplified considerably when viewed as a dual program.

Dual Program (DGP)

Maximize

$$d(\tilde{w}) = \sigma \left[\prod_{m=0}^{M} \prod_{t=1}^{T_m} \left(\frac{c_{mt} W_{m0}}{W_{mt}} \right)^{\sigma_{mt} W_{mt}} \right]^{\sigma} \tag{10}$$

subject to

$$\sum_{t=1}^{T_0} \sigma_{0t} w_{0t} = \sigma \quad \text{(normality condition)} \tag{11}$$

$$\sum_{m=0}^{M} \sum_{t=1}^{T_m} \sigma_{mt} a_{mtn} w_{mt} = 0; \quad n = 1, 2, \ldots, N \text{ (orthogonality conditions)} \tag{12}$$

$$w_{m0} \equiv \sigma_m \sum_{t=1}^{T_m} \sigma_{mt} w_{mt} \geq 0; \quad m = 1, 2, \ldots, M \tag{13}$$

$$w_{mt} \geq 0; \quad m = 0, 1, 2, \ldots, M, \quad t = 1, 2, \ldots, T_m \tag{14}$$

where it is understood that

$$w_{00} \equiv \sigma \sum_{t=1}^{T_0} \sigma_{0t} w_{0t} = 1 \tag{15}$$

$$\sigma = \pm 1 \tag{16}$$

$$\lim_{w_{mt} \to 0} \left(\frac{c_{mt} w_{m0}}{w_{mt}} \right)^{\sigma_{mt} w_{mt}} = 1 \tag{17}$$

and that $w_{mt} = 0$ if and only if $w_{m0} = 0$, $m = 1, \ldots, M$.

Relation (17), which can be verified L'Hôpital's rule, is established to assure the continuity of the product function $d(\tilde{w})$ on the boundary of the domain of definition of $d(\tilde{w})$.

It can be shown that if PGP is canonical and has a minimum at $(\tilde{x}^*)$, then there exist dual variables $(\tilde{w}^*)$ such that

$$Y(\tilde{x}^*) = d(\tilde{w}^*) \tag{18}$$

$$c_{0t} \prod_{n=1}^{N} x_n^{a_{0tn}} = w_{0t} \sigma \, d(\tilde{w}^*); \quad t = 1, 2, \ldots, T_0 \tag{19}$$

and

$$c_{mt} \prod_{n=1}^{N} x_n^{a_{mtn}} = \frac{w_{mt}}{w_{m0}}; \quad t = 1, 2, \ldots, T_m, \quad m = 1, 2, \ldots, M, \quad w_{m0} \neq 0 \tag{20}$$

Equations (19) and (20) are called an "objective functional relationship" and a "constraint functional relationship," respectively. Although these equations are nonlinear, they are nonlinear in only one term. Thus the primal values can easily be recovered by taking logarithms and solving any (N) independent equations in the (N) desired solution variables.

In general, the dual problem is to maximize a nonconvex function subject to a convex set of constraints. Accordingly, DGP seems amenable to conventional solution techniques. Nevertheless, the difficulty of obtaining the solution proportionally increases with the increase of the number of terms in PGP (number of dual variables = number of primal terms = T). This situation is exemplified by the notion of degrees of difficulty mentioned above. In addition, if a primal constraint is loose (inactive, nonbinding) at optimality, all dual variables associated with that constraint will be zero. Subsequently, the terms of the dual objective function corresponding to such dual variables will be mathematically undefined. The impact of this property on solution procedures is discussed later in this article.

Finally, the nonconvexity of the primal and dual objective functions leads to the conclusion that the vectors $(\tilde{x}^*)$ and $(\tilde{w}^*)$ may not be the global optima. Such vectors when they exist are called pseudo-optima [32], quasi-optima [30], or equilibrium solutions [17]. In that sense, a global minima to PGP is obtained by taking the minimum of all the local maxima of DGP.

LITERATURE REVIEW

Generally, the geometric programming problem is classified in accordance to the values of the signum functions [1, 2, 13]. When all these functions are positive, the problem is called either a posynomial geometric program or a prototype geometric program. On the other hand, when these functions are mixed in sign, the problem is termed either a signomial program, a generalized polynomial program, or a generalized geometric program. According to such distinctions, the state-of-the-art of the numerical solution techniques to the problem is presented.

Posynomial Geometric Program

The dual formulation has received almost all the attention when numerical solutions are considered. This is due to the fact that when a posynomial structure exists, the problem reduces to maximizing a concave function subject to a convex constraint set. The concavity of the dual objective function is established via a logarithmic transformation with the restriction that the dual vector lies in the positive orthant [17]. The properties of logarithmic concavity (convexity) are developed in Ref. 24. When the problem formulation gives rise to a signomial geometric program, the above characterization is no longer valid.

One of the first attempts to solve this problem was through the use of the Linear Logarithmic Programming Code (LLPC) developed by Clasen to solve the chemical equilibrium problem [9]. Under a logarithmic transformation of the dual objective function, LLPC can be applied. Hence the procedure involves the iterative solution of T nonlinear equations in T unknowns using the Newton-Raphson technique. Though still widely used, the code has proven inefficient for large degrees of difficulty and it also fails when primal loose constraints are present at optimality. Beck and Ecker [4], using the same transformation, utilized a modified version of Zangwill's convex simplex algorithm [47]. Taking advantage of this simplex-like procedure, the problem of primal inactive constraints is avoided by implementing simultaneous changes in certain dual variables. Though reported to behave well on moderately sized problems [3], it often requires the solution of a subsidiary maximization problem when loose constraints are encountered [3, 4]. The procedure fails entirely when applied to signomial problems. Kochenberger et al. [26] separated the dual function through a logarithmic transformation. The problem is then piecewise linearized and solved as a separable programming problem. The increases in program dimensionality and round-off errors are some of the many drawbacks of this procedure [23, 44]. Rijckaert and Martens [44] presented another linear approximation method. After the logarithmic transformation, the dual function is linearized using first-order Taylor's series. Experimentation with this approach showed unsatisfactory results. These procedures also fail on signomial formulations.

Duffin et al. [17] have presented another formulation to the dual program. In this version, each dual variable is expressed in terms of (D) basic vectors (r_j) as follows:

$$W_i \equiv b_i^{(0)} + \sum_{j=1}^{D} r_j \cdot b_i^{(j)} \geq 0; \quad i = 1, 2, \ldots, T \tag{21}$$

The vector $b^{(0)}$, called the normality vector, is a feasible solution to Eqs. (11) and (12), while $b^{(j)}$, $j = 1, 2, \ldots, D$, called nullity vectors, are linearly independent solutions to the homogeneous counterpart of Eqs. (11) and (12). Hence the problem is reduced in complexity and it involves maximizing the substituted dual objective function over D unrestricted variables, constrained by T nonnegativity conditions. Frank [21] and Dinkel et al. [13] have employed such a program. Computational experience showed applicability to a wide class of problems, but was characterized by extremely slow convergence [38].

The problem of primal inactive constraints motivated the consideration of augmented programs. An augmented program is a program whose constraints are forced to be tight. Kochenberger [25] developed an augmented primal program by adding slack variables to each constraint, and terms involving the inverse of these variables to the objective function. The inclusion of slack variables increases problem dimensionality and subsequently hinders its solution. However, Duffin and Peterson [16] used this approach to give a simple proof of the "refined duality theory" of geometric programming. A similar approach was proposed by McNamara [28]. His treatment involves the multiplication of each term in the system of primal constraints by a slack variable, then adding a monomial constraint composed of the product of the slack variables. The limitations of this procedure are quite obvious. Another approach to combat the problem of inactive primal constraints was developed utilizing Glover's surrogate constraints concept [22]. Phillips and Beightler [38] presented such a procedure via their surrogated geometric program. This program has the advantage of dealing only with the constraints that are binding at optimality. Though the approach looks promising, limited computational experience has been reported. A completely different treatment to the posynomial problem is presented in the Reklaitis and Wilde differential algorithm [40, 41]. In this algorithm the primal program is reduced using exponential transformations, then the constrained derivative concept [46] is utilized. The major difficulty with this approach is that it requires the calculation of the generalized inverse of an $N \times T$ matrix. Computational experience with this technique is limited.

Signomial Geometric Program

Solution procedures developed for this case could be classified in accordance to the particular mathematical concept being used. The algorithms presented here are primarily based on either the Lagrangian multipliers, constrained derivatives, or condensation.

Blau and Wilde [8] developed a Lagrangian algorithm applicable to PGP. It involves the iterative solution of (N + M - 1) nonlinear simultaneous equations using the Newton-Raphson procedure. The algorithm encounters serious computational difficulties when loose constraints are present. Phillips [35], using the Lagrangian function formed of DGP, reduced the problem size considerably. He established direct correspondence between the Lagrangian multipliers and the dual variables. The problem reduces to a search for those multipliers that yield dual variables satisfying the dual constraints as equalities, as well as the Kuhn-Tucker necessary conditions.

The constrained derivative concept was applied by Phillips on the dual problem [34, 36, 39]. His treatment reduces the DGP problem into solving simultaneously (N + 1) linear equations and T - (N + 1) nonlinear equations in T unknowns. When the total number of terms in PGP is large, serious computational difficulties are encountered.

Condensation as developed by Duffin [14] is a technique that approximates any posynomial function by means of a monomial. Such an approximation is normally based on the arithmetic-geometric inequality or the arithmetic-harmonic inequality. On the other hand, signomial programs could be algebraically transformed into posynomial programs [33]. The resulting posynomial programs might not be in the standard PGP form, but in a similar form called reversed geometric programs. A reversed geometric program is essentially PGP with some of the constraint inequality directions reversed. These particular programs could be condensed using the harmonic inequality [1, 15]. It is evident that transformation coupled with condensation gives rise to several solution procedures. In fact, Avriel and Williams [2], Passy [30, 31], Pascual and Ben-Israel [29], Reklaitis and Wilde [42], Dembo [10, 11], and Duffin and Peterson [17] have all independently applied this approach to signomial programs. From these results, Dembo recently presented the only known computationally efficient results using condensation [5, 12]. In his algorithm a signomial program is condensed into a posynomial one, then partially solved using cutting plane procedures. Thereafter, a new approximation is computed and the process repeated until it converges to a Kuhn-Tucker solution. With such an approach he managed to completely solve a modular design problem with 119 degrees of difficulty. This problem is the largest signomial program successfully solved to date. However, when applied to a larger problem (149 degrees of difficulty), convergence was too slow to yield a successful completion [5]. Such poor convergence is partially attributed to the known peculiarities of the cutting plane approach. These results, though limited, suggest that condensation could prove very effective.

The above review suggests that the general geometric programming problem with large degrees of difficulty has not been satisfactorily solved to date. Such a situation warrants more research to find new avenues through which a successful procedure could be developed.

POSYNOMIAL GEOMETRIC PROGRAMS WITH ZERO DEGREES OF DIFFICULTY

An interesting and useful formulation of geometric programming problems occurs when the signum functions of Eqs. (1) through (8) are all positive and $T - (N + 1) \equiv 0$. In such a case the geometric programming problem is one of "zero degrees of difficulty," and the dual formulation possesses unique solution properties. Note that under these conditions the dual program given by Eqs. (10) through (17) generates exactly (N + 1) equality constraints in T unknowns. (One constraint for each primal variable plus Eq. 11 and one dual variable for each primal term.) In this case, since $T \equiv N + 1$, a unique solution to the dual problem can be obtained through the solution of (N + 1) equations in (N + 1) unknowns. Using these dual variables and the primal coefficients, the optimal value of the dual objective function (Eq. 10) can immediately be calculated. It has been shown [5, 18] that under these conditions the optimal value of the primal objective function is exactly equal to the value of the dual objective function. Using this knowledge and Eqs. (19) and (20), the complete primal solution is easily obtained. An important characteristic of this solution procedure is that it is globally optimal to the primary problem. Further, even if $T - (N + 1) > 0$, the dual objective function is concave in the dual solution space, and any extreme point solution to the dual maximization problem will provide a global minimum solution to the primal problem. Consider the following simple example: Minimize

$$y(\tilde{X}) = 5X_1^{-1}X_2^{-3} + 2X_1^2X_2X_3^{-2} + 10X_1X_2^4X_3^{-1} + 20X_3^2$$

Note that the problem is unconstrained, all signum functions are positive, and the formulation exhibits zero degrees of difficulty ($T - N - 1 \equiv 0$). Using Eqs. (11) through (17), we obtain

$$W_{01} + W_{02} + W_{03} + W_{04} = 1$$

$$-W_{01} + 2W_{02} + W_{03} = 0$$

$$-3W_{01} + W_{02} + 4W_{03} = 0$$

$$-2W_{02} - W_{03} + 2W_{04} = 0$$

Solving these equations, we obtain

$$W_{01}^* = 14/33, \ W_{02}^* = 2/33, \ W_{03}^* = 10/33, \ W_{04}^* = 7/33$$

By using Eq. (11) the dual objective function is calculated as

$$d^*(\tilde{W}) = \left[\frac{5(33)}{14}\right]^{14/33} \left[\frac{2(33)}{2}\right]^{2/33} \left[\frac{10(33)}{10}\right]^{10/33} \left[\frac{20(33)}{7}\right]^{7/33}$$

Hence $d^*(\tilde{W}) = 26.60$.

From geometric programming duality theory, this value of $d^*(\tilde{W})$ is equal to $Y^*(\tilde{X})$. Using this theory and Eqs. (18) through (20), the primal solution vector can be determined. Using the first, second, and fourth terms in the objective function, one obtains

$$Y^*(\tilde{X})W_{04}^* = 20X_3^2 = 7/33(26.60)$$

$$Y^*(\tilde{X})W_{01}^* = 5X_1^{-1}X_2^{-3} = 14/33(26.60)$$

$$Y^*(\tilde{X})W_{02}^* = 2X_1^2X_2X_3^{-2} = 2/33(26.60)$$

Although these three equations in three unknowns are nonlinear, they are linear in the logarithms of solution variables. Solving these equations:

$$X_1^* = 0.4765$$

$$X_2^* = 0.9750$$

$$X_3^* = 0.5310$$

It is interesting to ask the following question. Suppose that the technological coefficient on the third term of the objective function (C_{03}) has been underestimated; more accurate data shows this coefficient to be 15.0 instead of 10.0. How does my optimal solution change? Note that even if C_{03} changes, Eqs. (11) through (17) are unaffected! Hence the optimal cost distribution in the objective function remains the same. However, the optimal solution in the primal variables will change due to the presence of C_{03} in Eq. (10). In other words, the primal solution variables will adjust in such a way that the optimal cost distribution across the objective function remains unchanged. This sort of invariance can only be discovered via geometric programming, and only in the zero degree of difficulty problem. Nevertheless, this sort of information can be very useful.

For a second example, consider the following problem first formulated by Passy [32] and later presented by Wilde and Beightler [46].

Maximize

$$Y(\tilde{X}) = 5X_1^2 - X_2^2X_3$$

subject to the usual nonnegativity conditions and the inequality

$$-5X_1X_2^{-1} + 3X_2^{-1}X_3^2 \geq -2$$

To get this into the proper form we must minimize the negative of $y(\tilde{X})$ and multiply all terms of the inequality by $-\frac{1}{2}$ in order to reverse its sense. The equivalent problem is to minimize $y(\equiv -y(\tilde{X}))$:

$$y = -5X_1^2 + X_2^2X_3$$

subject to

$$\tfrac{5}{2}X_1X_2^{-1} - \tfrac{3}{2}X_2^{-1}X_3^2 \leq 1$$

Note that:

$$\sigma_{01} = -1$$

$$\sigma_{02} = 1$$

$$\sigma_{11} = 1$$

$$\sigma_{12} = -1$$

The dual variables must satisfy

$$\begin{aligned}
-W_{01} + W_{02} \qquad\qquad\qquad &= \sigma \\
-2W_{01} \qquad + W_{11} \qquad\qquad &= 0 \\
2W_{02} - W_{11} + W_{12} &= 0 \\
W_{02} \qquad - 2W_{12} &= 0
\end{aligned}$$

The sign of σ is unknown until the final solution is examined. Since we were maximizing the original objective, we guess $\sigma = -1.0$. The solution is $\sigma = -1$, $W_{01} = 5$, $W_{02} = 4$, $W_{11} = 10$, and $W_{12} = 2$. This solution confirms our choice of $\sigma = -1$. If we had guessed wrong, all dual variables would be negative. By definition, $W_{00} \equiv 1$ and

$$W_{10} = \sigma_1(W_{11} - W_{12}) = 8 > 0$$

The value of $Y(\tilde{X})$ at the stationary point X^0 is

$$y^0 = -\left[\left(\frac{5\cdot 1}{5}\right)^{-5}\left(\frac{1\cdot 1}{4}\right)^{4}\left(\frac{5}{2}\cdot\frac{8}{10}\right)^{10}\left(\frac{3}{2}\cdot\frac{8}{2}\right)^{-2}\right]^{-1} = -9$$

To find the optimal policy, start by considering the value of the first term of the objective function

$$5X_1^2 = W_{01}\sigma y^0 = 5(-1)(-9)$$

whence

$$X_1 = 3$$

The weight on the first term of the constraint is

$$\frac{W_{11}}{W_{10}} = \frac{10}{8} = \frac{5}{2}\cdot X_1(X_2)^{-1}$$

which gives

$$X_2 = 6$$

The second term of the objective gives X_3.

$$(X_2)^2X_3 = (6)^2X_3 = 4(9)$$

$$X_3 = 1$$

Since there are zero degrees of difficulty, this is the only stationary point.

Since the geometric program is signomial in nature, this stationary point should be checked to insure that the proper solution vector has been determined. Using standard procedures, one can verify that this solution is not a global minimum (maximum) but is indeed a stationary point.

This example clearly demonstrates that even when the primal formulation yields zero degrees of difficulty, if negative signs are present in the primal problem, solution of the dual constraint equations does not necessarily yield a minimizing primal solution. This is characteristic of the signed (signomial) geometric programming problem since it is generally nonconvex.

SUMMARY AND CONCLUSIONS

Geometric programming provides a unique and powerful solution procedure for constrained nonlinear programming problems which can be formulated in algebraic terms. Recent research has focused on solution procedures for posynomial programs, or procedures which iteratively solve signomial programs through a series of posynomial approximations.

When zero degrees of difficulty are present, geometric programming represents the most powerful and efficient solution procedure known. When degrees of difficulty are present, serious solution problems still exist. Future research efforts will undoubtedly focus on the more difficult signomial (nonconvex) programming problem. In spite of spectacular advances in current solution procedures, the full potential of geometric programming is yet to be realized.

REFERENCES

1. Abrams, R., and M. Bunting, Reducing reversed posynomial programs, SIAM J. Appl. Math. 27(4), 629-640 (December 1974).
2. Avriel, M., and A. C. Williams, Complementary geometric programming, SIAM J. Appl. Math. 19(1), 125-141 (July 1970).
3. Beck, P. A., and J. G. Ecker, Some Computational Experience with a Modified Concave Simplex Algorithm for Geometric Programming, Rensselaer Polytechnic Institute, Operations Research and Statistics Research, Paper No. 73-P1, 1973.
4. Beck, P. A., and J. G. Ecker, A modified concave simplex algorithm for geometric programming, J. Optim. Theory Appl. 15(2), 189-202 (1975).
5. Beightler, C. S., and D. T. Phillips, Applied Geometric Programming, Wiley, New York, 1976.
6. Beightler, C. S., T.-C. Lo, and H. G. Rylander, Optimal Design by Geometric Programming, ASME Paper 69-WA/DE-7, 1969.
7. Bouchey, G. D., C. S. Beightler, and B. V. Koen, Optimization of nuclear systems by geometric programming, Nucl. Sci. Eng. 44, 267-272 (1971).
8. Blau, G. E., and D. J. Wilde, A Lagrangian algorithm for equality constrained generalized polynomial optimization, AIChE J. 17, 235-240 (1971).
9. Clasen, R. J., The Linear-Logarithmic Programming Problem, The RAND Corp., Memo. RM-3707-PR, 1963.
10. Dembo, R. S., Solution of Complementary Geometric Programming Problems, unpublished MS Thesis, Israel Institute of Technology, Haifa, June 1972.
11. Dembo, R. S., GGP-A Computer Program for the Solution of Generalized Geometric Programming Problems, Users Manual, Report 72/55, Department of Chemical Engineering, Technion, Haifa, 1972.
12. Dembo, R. S., Modular Design by Decomposition, Working Paper No. 81, Management Sciences Department, University of Waterloo, Waterloo, Ontario, December 1973.
13. Dinkel, J. G., G. A. Kochenberger, and B. McCarl, An approach to the numerical solutions of geometric programs, Math. Prog. 7, 181-190 (1974).
14. Duffin, R. J., Linearizing geometric programs, SIAM Rev. 12(2), 211-227 (April 1970).
15. Duffin, R. J., and E. L. Peterson, Reversed geometric programs treated by harmonic means, Indiana Univ. Math. J., 22(6), 531-550 (1972).
16. Duffin, R. J., and E. L. Peterson, Geometric programs treated with slack variables, Appl. Anal. 2, 225-267 (1972).
17. Duffin, R. J., and E. L. Peterson, Geometric programming with signomials, J. Optim. Theory Appl. 11(1), 3-35 (1973).
18. Duffin, R. J., E. L. Peterson, and C. Zener, Geometric Programming, Wiley, New York, 1967.
19. Ecker, J. G., and J. R. McNamara, Geometric programming and the preliminary design of industrial waste treatment plants, Water Resour. Res. 7, 18-22 (1971).
20. Federowicz, A. J., and M. Mazumdar, The use of geometric programming to maximize reliability achieved by redundancy, Oper. Res. 16, 948-954 (1968).
21. Frank, C. J., An algorithm for geometric programming, in Recent Advances in Optimization Techniques (A. Lavi and T. P. Vogel, eds.), Wiley, New York, 1966, pp. 145-162.

22. Glover, F., Surrogate constraints, Oper. Res. 16(4), 741-749 (1968).
23. Humber, J. B., Application of Separable Programming to Optimization by Geometric Programming, unpublished MS Thesis, Mathematics Department, Texas A&M University, December 1971.
24. Klinger, A., and O. L. Mangasarian, Logarithmic convexity and geometric programming, J. Math. Anal. Appl. 24, 388-408 (1968).
25. Kochenberger, G. A., Geometric Programming, Extension to Deal with Degrees of Difficulty and Loose Constraints, unpublished Ph.D. Dissertation, University of Colorado, 1969.
26. Kochenberger, G. A., R. E. D. Woolsey, and B. A. McCarl, On the solution of geometric programs via separable programming, Oper. Res. Q. 24(2), 285-293 (1973).
27. Kuhn, H. W., and A. W. Tucker, Nonlinear programming, in Proceedings of the 2nd Berkeley Symposium on Mathematical Statistics and Probability, 1950, pp. 481-492.
28. McNamara, J. R., A solution procedure for geometric programming, Oper. Res. 24(1), 15-25 (1976).
29. Pascual, L. D., and A. Ben-Israel, Constrained maximization of posynomials by geometric programs, J. Optim. Theory Appl. 5(2), 73-80 (1970).
30. Passy, U., Generalized weighted mean programming, SIAM J. Appl. Math. 20(4), 763-778 (1971).
31. Passy, U., Condensing generalized polynomials, J. Optim. Theory Appl. 9(4), 221-237 (1972).
32. Passy, U., and D. J. Wilde, Generalized polynomial optimization, SIAM J. Appl. Math. 15(5), 1344-1356 (1967).
33. Peterson, E. L., Geometric programming and some of its extensions, in Optimization and Design (M. Avriel, M. J. Rijckaert, and D. J. Wilde, eds.), Prentice-Hall, Englewood Cliffs, New Jersey, 1973, pp. 228-289.
34. Phillips, D. T., Geometric programming with slack constraints and degrees of difficulty, AIIE Trans. 5(1), 7-13 (1973).
35. Phillips, D. T., Solution to generalized geometric programming problems via a dual auxiliary program, AIIE Trans. 8(1), 70-75 (1976).
36. Phillips, D. T., Sensitivity analysis in generalized polynomial programming, AIIE Trans. 6(2), 114-119 (1974).
37. Phillips, D. T., and C. S. Beightler, Optimization in tool engineering using geometric programming, AIIE Trans. 2(4), 355-360 (1970).
38. Phillips, D. T., and C. S. Beightler, Geometric programming: A technical state-of-the-art survey, AIIE Trans. 5(2), 97-112 (1973).
39. Phillips, D. T., and G. V. Reklaitis, Constrained derivatives and equilibrium conditions in generalized geometric programming, J. Eng. Math. 8(4), 311-314 (1974).
40. Reklaitis, G. V., Singularity in Differential Optimization Theory: Differential Algorithm for Posynomial Programs, unpublished Ph.D. Dissertation, Stanford University, 1969.
41. Reklaitis, G. V., and D. J. Wilde, A Differential Algorithm for Posynomial Programs, paper presented at DECHMA Congress, Frankfurt, Germany, June 1970.
42. Reklaitis, G. V., and D. J. Wilde, Geometric programming via a primal auxiliary problem, AIIE Trans. 6(4), 308-317 (1974).

43. Rijckaert, M. J., Engineering applications of geometric programming, in Optimization and Design (M. Avriel, M. J. Rijckaert, and D. J. Wilde, eds.), Prentice-Hall, Englewood Cliffs, New Jersey, 1973, pp. 196-220.
44. Rijckaert, M. J., and X. M. Martens, A Comparison of Generalized Geometric Programming Algorithms, Katholieke Universiteit Leuven, Belgium, Report CE-RM-7503, 1975.
45. Templeman, A. B., Structural design for minimum cost using the method of geometric programming, Proc. Inst. Civil Eng. 459-472 (August 1970).
46. Wilde, D. J., and C. S. Beightler, Foundations of Optimization, Prentice-Hall, Englewood Cliffs, New Jersey, 1967.
47. Zangwill, W. I., The convex simplex method, Manage. Sci., A 14(3), 221-238 (1967).
48. Zener, C., Engineering Design by Geometric Programming, Wiley, New York, 1971.

BIBLIOGRAPHY

Abadie, J., and J. Carpentier, Generalization of the Wolfe reduced gradient method to the case of nonlinear constraints, in Optimization (R. Fletcher, ed.), Academic, New York, 1969, pp. 37-47.

Abadie, J., and J. Guigo, Numerical experiments with the GRG method, in Integer and Nonlinear Programming (J. Abadie, ed.), North Holland, Amsterdam, 1970, Appendix III.

Colville, A. R., A comparative study of nonlinear programming codes, in Proceedings of the Princeton Symposium on Mathematical Programming (H. W. Kuhn, ed.), Princeton University Press, Princeton, New Jersey, 1970, pp. 487-501.

Faure, P., and P. Haurd, Résolution de programmes mathématiques à fonction nonlinéare par la méthode du gradient réduit, Rev. Fr. R. O. 36, 167-206 (1965).

Gabriele, G. A., and K. M. Ragsdell, The Generalized Reduced Gradient Method: A Reliable Tool for Optimal Design, paper presented at ASME Conference, Houston, Texas, 1975.

Himmelblau, D. M., Applied Nonlinear Programming, McGraw-Hill, New York, 1972.

Huard, P., Convergence of the reduced gradient method, in Nonlinear Programming, Part 2 (O. L. Mangasarian, R. R. Meyer, and S. M. Robinson, eds.), Academic, New York, 1975, pp. 29-54.

Lasdon, L. S., R. L. Fox, and M. W. Ratner, Nonlinear optimization using the generalized reduced gradient method, R.A.I.R.O. 3, 73-104 (1974).

Wolfe, P., Methods of nonlinear programming, in Recent Advances in Mathematical Programming (R. L. Grave and P. Wolfe, eds.), McGraw-Hill, New York, 1963, pp. 76-77.

9
FIXED POINT COMPUTING METHODS

R. SAIGAL

Technical Staff, Operations Research Center
Bell Telephone Labs
Holmdel, New Jersey

INTRODUCTION

Let R^n be the space of all n-dimensional vectors $x = (x_1, \ldots, x_n)$ (the n-dimensional Euclidean space) and let

$$f(x_1, \ldots, x_n) = [f_1(x_1, \ldots, x_n), \ldots, f_n(x_1, \ldots, x_n)]$$

be a function from R^n into R^n, i.e., the i-th component of this function is a real valued function $f_i(x_1, \ldots, x_n)$ of n-variables $x_1, \ldots, x_n$. Then the problem we consider is to compute a vector x in R^n such that

$$f(x) = x \tag{1}$$

that is, x is a fixed point of f. This problem is closely related to solving a system of n equations in n variables; that is, computing a vector x in R^n such that

$$f(x) = y$$

where y is some given vector in R^n. The traditional methods for solving (1) generally require the differentiability of f or some rather strong property like contractability. In this article we will consider methods based on complementary pivoting, a concept introduced by Lemke and Howson [11] and Lemke [10], and refer the reader to the excellent book by Ortega and Rheinboldt [13] for a complete study of the traditional methods. As we will subsequently see, the methods of complementary pivoting seem to require the minimal assumptions on f, and in many cases the conditions on f under which the existence of a solution has been established are sufficient for these methods to successfully compute a solution. We now introduce two such theorems.

Current affiliation: Department of Industrial Engineering, Northwestern University, Evanston, Illinois

Probably, the most celebrated fixed point theorem is that of Brouwer [2]. A subset S of R^n is called an n-dimensional simplex if it is a convex hull of n + 1 affinely independent points $v^1, v^2, \ldots, v^{n+1}$. The points v^i are called vertices of the simplex. (A set of n + 1 points $v^1, \ldots, v^{n+1}$ in R^n are affinely independent if the $(n + 1) \times (n + 1)$ matrix $\begin{bmatrix} v^1 & \cdots & v^{n+1} \\ 1 & & 1 \end{bmatrix}$ has full rank. In two dimensions, three points are affinely independent if they are not collinear.) A 2-dimensional simplex is a triangle and a 3-dimensional one a tetrahedron. We now state the theorem in its simplest form.

Brouwer's Theorem. Let S be an n-dimensional simplex, and let f be a continuous mapping from S into S. Then there is an x in S such that $f(x) = x$; i.e., f has a fixed point x in S.

The above theorem holds under the relaxed assumption that S is a compact, convex subset of R^n. To see the need for the assumption of convexity in 2 dimensions, it is sufficient to note that the rotation of a ring (see Fig. 1) is a continuous mapping of the ring onto itself, which has no fixed points. The need for boundedness follows from the fact that the mapping $x + a$ $(a \neq 0)$ of R^n into R^n has no fixed points, and for closedness is illustrated by the mapping $f(x_1, x_2) = \frac{1}{2}[(x_1, x_2) + (1, 0)]$ of the open set $\{(x_1, x_2): |x_1| + |x_2| < 1\}$ onto itself whose only fixed point is (1, 0).

The algorithm based on complementary pivoting, developed by Scarf [17] in 1967, was primarily motivated by a desire to compute a fixed point guaranteed by the Brouwer's theorem. He showed that for continuous f and any $\epsilon > 0$, sufficiently small, the algorithm could be used to compute an approximate fixed point x, in the sense that $\|f(x) - x\| \leq \epsilon$. Hansen [6] and Kuhn [8] related the procedure of Scarf to pivoting on a subdivision of the simplex, and thus considerably simplified the implementation and discussion. A major drawback of this algorithm is that a substantial amount of computational effort is involved before a reasonable approximation is reached. In addition, the computation has to be initiated at a point on the boundary of the simplex. In subsequent works, Eaves [3], Merrill [12], and Eaves and Saigal [5] proposed variations where one can initiate the computation at any point, and where the grid of search, δ, can be continuously refined to enable one to obtain a sequence of approximate fixed points which converges to a fixed point of f.

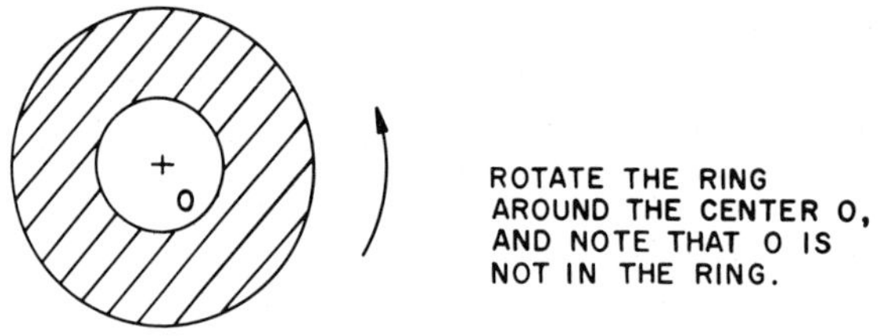
O
ROTATE THE RING
AROUND THE CENTER O,
AND NOTE THAT O IS
NOT IN THE RING.

FIGURE 1.

The algorithms of Refs. 5 and 12 can, in addition, be used to compute fixed points in problems where the region is not bounded. This feature results in the ability to compute fixed points of (1) under other hypotheses. One such example follows.

Leray-Schauder Theorem [13, 6.3.3]. Let C be a bounded, open subset of R^n containing 0 and let f be a continuous mapping from the closure of C into R^n such that $f(x) \neq \lambda x$ whenever $\lambda > 1$ and x is in the boundary of C. Then f has a fixed point in the closure of C.

These later improvements in the algorithm are also based on the methods of complementary pivoting and have a very appealing geometric interpretation. These algorithms can be viewed as starting with a simply constructed mapping f^0 and its fixed point x_0 in C, deforming f^t to $f^{\infty} = f$ as t goes from 0 to ∞, and following the path x_t of fixed points of f^t. Under rather general conditions on f (like the conditions of the Leray-Schauder theorem), the path x_t can be shown to converge to a fixed point of f. An extensive study of the path x_t thus generated has been made by Eaves [4] and Saigal [16]. In particular, a concept of index can be defined, and this index can be shown to be invariant along the path.

In many applications including mathematical programming and economics, one needs to consider more general types of relations than functions. In particular, fixed point theorems dealing with point-to-set mappings, i.e., functions from R^n into subsets of R^n, are needed. The most celebrated of these follows.

Kakutani's Theorem [7]. Let S be an n-dimensional simplex, and let S* be the class of all nonempty convex subsets of S. Let f be a mapping from S into S*, and let it be upper semicontinuous. Then there is an x in S such that x is in f(x).

The algorithms developed in Refs. 5 and 12 are designed to compute fixed points of point-to-set mappings and are effective under conditions similar to those of the Leray-Schauder theorem.

Recent work of Saigal [15] has demonstrated that for a continuously differentiable function f whose derivative satisfies a Lipschitz condition, these algorithms can be made to converge quadratically to a fixed point, a desirable characteristic shared with the Newton's method [13, 7.1]. In the same work, smooth problems of up to 80 dimensions are solved.

The use of these algorithms can be made even when the conditions which guarantee success cannot be verified. In Ref. 12, many nonlinear programming problems are successfully solved, though the conditions can only be verified for a few. In case the algorithm fails on a problem, very little can be concluded. Ideally, a conclusion that there is no fixed point would be valuable, but no specific results that do this exist. To date, many fairly large problems of up to 30 dimensions for point-to-set mappings and 80 dimensions for smooth mappings have been successfully solved by these methods.

SOME STANDARD APPLICATIONS

Some applications of the fixed point computing methods include the economic equilibrium problem [18], nonlinear programming problem [12], nonlinear boundary value problem [1], solving systems of nonlinear equations [15], and finding roots of complex polynomials [9]. We will now discuss two of these applications.

The Nonlinear Programming Problem

The problem we consider is that of finding x to minimize

$$\theta(x)$$

subject to

$$g_i(x) \leq 0, \quad i = 1, \ldots, m$$

where θ and g_i are real valued functions of n variables $x_1, \ldots, x_n$. In addition, we shall assume that each of these functions is convex, and that in case the set $G = \{x: g_i(x) \leq 0 \ (i = 1, \ldots, m)\}$ is nonempty, there is an x such that $g_i(x) < 0$ for each $i = 1, \ldots, m$.

For a convex function θ, a point x^* in R^n is called a subgradient at x if and only if

$$\theta(y) \geq \theta(x) + \langle x^*, y - x\rangle, \quad \text{for all } y \text{ in } R^n$$

where

$$\langle x, y\rangle = x_1y_1 + \cdots + x_ny_n$$

The subdifferential of θ at x is the set $\partial\theta(x)$ of all subgradients of θ at x. A fact about convex functions is that $\partial\theta(x) \neq \phi$ for each x. Also $\partial\theta(x)$ is closed and <u>convex</u> for each x.

Now, define the mapping

$$s(x) = \max_{1 \leq i \leq m} g_i(x)$$

and note that s(x) is convex. Also, let

$$I(x) = \{i: s(x) = g_i(x)\} \neq \phi$$

for each x. We note that

$$\partial s(x) = \text{hull}\left\{\bigcup_{i \in I(x)} \partial g_i(x)\right\}$$

where hull (C) is the smallest convex set containing C.

Now, consider the following mapping from R^n into convex subsets of R^n:

$$f(x) = \begin{cases} x - \partial\theta(x), & s(x) < 0 \\ x - \text{hull}\ \{\partial\theta(x) \cup \partial s(x)\}, & s(x) = 0 \\ x - \partial s(x), & s(x) > 0 \end{cases}$$

It is readily confirmed that this mapping is upper semicontinuous. Now, let $\bar{x}$ be a fixed point of this mapping. We now show that $\bar{x}$ solves the nonlinear programming problem. Depending on the value of s(x), we have the following three cases.

Case 1: $s(\bar{x}) < 0$. Then, $0 \in \partial\theta(\bar{x})$ and thus $\bar{x}$ is a global minimizer of θ, and is also in G, and thus solves the problem.

Case 2: $s(\bar{x}) > 0$. Then $0 \in \partial s(\bar{x})$, and $\bar{x}$ is a global minimizer of s, and since $s(\bar{x}) > 0$, the set G is empty.

Case 3: $s(\bar{x}) = 0$. Then $0 \in \text{hull}\ \{\partial\theta(\bar{x}) \cup \partial s(\bar{x})\}$, and thus there exist nonnegative numbers ρ_0, ρ_i $[i \in I(\bar{x})]$, and vectors $y_0 \in \partial\theta(\bar{x})$ and $y_i \in \partial g_i(\bar{x})$, $i \in I(\bar{x})$ such that

$$\rho_0 y_0 + \sum_{i \in I(\bar{x})} \rho_i y_i = 0$$

Now, if $\rho_0 = 0$, then since $\bar{x}$ is a global minimizer of s, $g_i(x) < 0$ for each i is impossible. Since we have assumed the contrary, $\rho_0 \neq 0$, and we have $\bar{x}$ and ρ_i/ρ_0, $i \in I(x)$ satisfying the standard Karush-Kuhn-Tucker necessary and sufficient conditions for a solution to the problem.

Thus solving a nonlinear programming problem can be reduced to finding a fixed point of the mapping f. Extensions to the general (nonconvex) case appear in Ref. 12.

An an example, consider the following problem.

Minimize

$$x_1^2 + x_2^2 - 2x_1 - 3x_2$$

subject to

$$x_1 + x_2 \leq 1$$

Then the mapping

$$f(x_1, x_2) = \begin{cases} (-x_1 + 2,\ -x_2 + 3), & x_1 + x_2 < 1 \\ \text{hull}\ \{(-x_1 + 2,\ -x_2 + 3),\ (x_1 - 1,\ x_2 - 1)\}, & x_1 + x_2 = 1 \\ (x_1 - 1,\ x_2 - 1), & x_1 + x_2 > 1 \end{cases}$$

has $(\frac{1}{4}, \frac{3}{4})$ as a fixed point, and note that this also solves the nonlinear programming problem.

Computing Economic Equilibrium Prices

We consider an economy in which n commodities are exchanged, a typical vector of commodities being denoted by $x = (x_1, \ldots, x_n)$. We also assume that there are m traders participating in this exchange, each endowed with a utility function $U_i(x_1, \ldots, x_n)$ $(i = 1, \ldots, m)$ on the space of commodities which specifies the individual's preferences; that is, the individual chooses a vector of commodities from those available to him, which maximizes his utility function. In addition, each trader has an initial vector of assets

$$\omega^i = (\omega_1^i, \ldots, \omega_n^i)$$

reflecting his ownership of commodities prior to trading. The problem we then consider is to find a price vector $\pi = (\pi_1, \ldots, \pi_n)$, where π_i is the price of the commodity i, in the economy such that trading at these prices would result in no shortages of goods in the economy.

Now, given an arbitrary set of prices $\pi = (\pi_1, \ldots, \pi_n)$ of the commodities in the economy, the wealth W^i of the i-th trader is obtained by assuming that he disposes of his entire holdings at these prices, so

$$W^i = \sum_{j=1}^{n} \pi_j \omega_j$$

Also, given the wealth of each trader in the economy, his demand for the commodities is determined (as stated before) by maximizing his utility function subject to the constraint that his expenditure must not exceed his wealth; that is, by solving:

Maximize

$$U_i(x_1, \ldots, x_n)$$

subject to

$$\sum_{j=1}^{n} \pi_j x_j \le W^i$$

$$x_j \ge 0, \quad j = 1, \ldots, n$$

Thus a vector of demands $D^i(\pi) = (D_1^i(\pi), \ldots, D_n^i(\pi))$ which solves the above problem of the i-th trader would result.

As can be readily seen, these demands $D^i(\pi)$ as functions of π are homogeneous of degree 0. Thus we can normalize the price vector to satisfy

$$\sum_{j=1}^{n} \pi_j = 1$$

$$\pi_j \ge 0, \quad j = 1, \ldots, n$$

that is, lie in an (n - 1)-dimensional simplex, S.

As an example, if the utility function of a trader were

$$U(x_1, \ldots, x_n) = x_1^{a_1} x_2^{a_2} \ldots x_n^{a_n}$$

with $a_i \geq 0$ and $\sum_{i=1}^{n} a_i = 1$, then the demand vector would be

$$D(\pi) = \left(\frac{a_1 W}{\pi_1}, \frac{a_2 W}{\pi_2}, \ldots, \frac{a_n W}{\pi_n} \right)$$

where W is the wealth given in terms of the initial holding $(\omega_1, \ldots, \omega_n)$ by

$$W = \sum_{j=1}^{n} \pi_j \omega_j$$

Now, summing the individual demands $D_j^i(\pi)$, we obtain the market demands $D(\pi) = [D_1(\pi), \ldots, D_n(\pi)]$ where

$$D_j(\pi) = \sum_{i=1}^{m} D_j^i(\pi)$$

For each commodity, define the excess demand function

$$g_j(\pi) = D_j(\pi) - \omega_j$$

where $\omega_j = \omega_j^1 + \cdots + \omega_j^m$, the total commodity j in the economy shared among the m traders.

The demand functions, in addition, satisfy another structural condition known as the Walras law, which states that

$$\pi_1 g_1(\pi) + \cdots + \pi_n g_n(\pi) = 0$$

for all price vectors π, whether they are in equilibrium or not.

An equilibrium price vector $\bar{\pi}$ is one for which excess demands are less than or equal to zero, or

$$g_i(\bar{\pi}) \leq 0 \quad (i = 1, \ldots, n)$$

thus at these prices the total market demand can be satisfied by the initial stock of assets.

Now, let $k = (k_1, \ldots, k_n) > 0$ be fixed. Define

$$s_i(\pi) = \max\{-\pi_i, k_i g_i(\pi)\} \quad (i = 1, \ldots, n)$$

Clearly $s(\pi)$ is continuous when $g(\pi)$ is. Now define the mapping

$$f(\pi) = \frac{\pi + s(\pi)}{\langle \pi + s(\pi), e \rangle}$$

where e is the n-vector of all 1's.

It can be readily confirmed that $\pi + s(\pi) \geq 0$. It also follows that $\langle \pi + s(\pi), e \rangle > 0$ for all $\pi \in S$, which then implies that f is a continuous mapping from S into S, and by

Brouwer's theorem it has a fixed point $\bar{\pi}$. We now show that this fixed point is an equilibrium price vector. Since

$$\bar{\pi} = \frac{\bar{\pi} + s(\bar{\pi})}{\langle \bar{\pi} + s(\bar{\pi}),\ e\rangle}$$

$$s(\bar{\pi}) = \lambda\bar{\pi} \text{ (for some } \lambda)$$

or

$$\langle s(\bar{\pi}),\ g(\bar{\pi})\rangle = \lambda\langle \bar{\pi},\ g(\bar{\pi})\rangle = 0$$

Also $s_i(\bar{\pi}) \cdot g_i(\bar{\pi}) \geq 0$; and since $g_i(\bar{\pi}) > 0$ implies $s_i(\bar{\pi}) = k_i g_i(\bar{\pi}) > 0$, it must be that $g_i(\bar{\pi}) \leq 0$ for each $i = 1, \ldots, n$, and the result follows.

These equilibrium prices in such an economy are of interest since they reflect the relative scarcity of commodities in this economy.

THE BASICS OF THE ALGORITHM

We now illustrate the basics of the procedure by considering the problem of computing a fixed point of a continuous mapping f which maps the n-dimensional simplex

$$S = \left\{x:\ \sum_{i=1}^{n+1} x_i = 1,\ x_i \geq 0\right\} \subset R^{n+1}$$

into itself. The first step in the procedure is to subdivide the simplex S in such a way that each piece of the subdivision is an n-simplex, and such that any two n-simplexes that meet in this subdivision do so on a common face. A standard subdivision of S is obtained by choosing any integer $K \geq 1$, and considering all vectors in S of the form

$$\left(\frac{y_1}{K}, \ldots, \frac{y_{n+1}}{K}\right)$$

with $\sum_{i=1}^{n+1} y_i = K$ and y_i integer, as the vertices of the subdivision. An n-simplex of this subdivision is then uniquely obtained by a choice of a permutation π of $\{1, \ldots, n\}$ and a vertex v of this subdivision. The vertices $v^1, \ldots, v^{n+1}$ of the simplex (v, π) then are

$$v^1 = v$$

$$v^{i+1} = v^i + \frac{1}{K} Q_{\pi(i)} \qquad (i = 1, \ldots, n) \tag{2}$$

where $\pi(i)$ is the i-th component of the permutation π, and Q_i is the i-th column of the $(n + 1) \times n$ matrix.

$$\begin{bmatrix} -1 & & & & \\ 1 & -1 & & & \\ & 1 & \ddots & & \\ & & \ddots & \ddots & \\ & & & \ddots & -1 \\ & & & & 1 \end{bmatrix}$$

It can be verified that the size of any such simplex, called the <u>mesh</u> of the subdivision, is no greater than $\sqrt{n}/K$. A standard subdivision of a 2-simplex with K = 4 is shown in Fig. 2.

Given a subdivided simplex S and a continuous function f, one defines a labeling function ℓ which associates a number from $\{1, \ldots, n+1\}$ to each of the vertices of the subdivision. This labeling function is generally

$$\ell(x) = j, \quad \text{if } f_j(x) \le x_j \text{ and } x_j \ne 0 \tag{3}$$

and is well defined, since

$$\sum_{j=1}^{n+1} f_j(x) = \sum_{j=1}^{n+1} x_j = 1$$

implies that $f_j(x) > x_j$ for all j, with $x_j > 0$.

As an example to demonstrate the labeling, consider the mapping f: S → S where $S = \{x: x_1 + x_2 = 1,\ x_1 \ge 0,\ x_2 \ge 0\}$ and

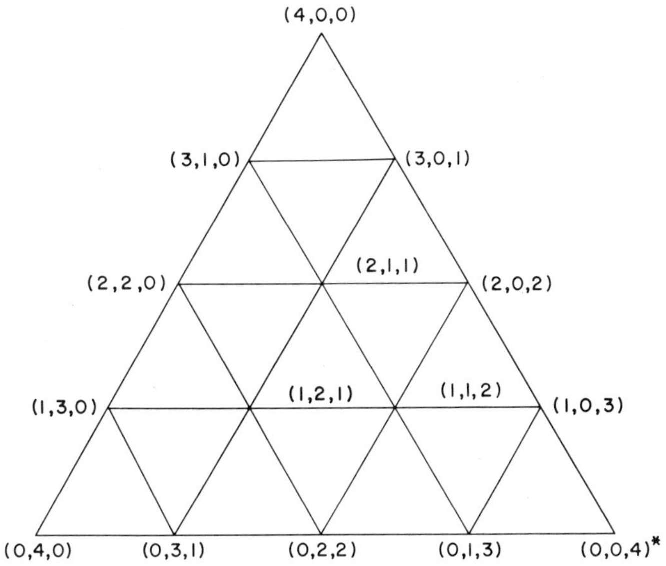

FIG. 2. A subdivided 2-simplex with K = 4. *The coordinates of the vertices should be divided by 4.

$$f(x_1, x_2) = \frac{1}{\theta(x_1)} \begin{bmatrix} \frac{1}{2} + \frac{x_1^2}{2} - \frac{x_2^2}{3} \\ \\ \frac{1}{2} + \frac{x_1^2}{3} - \frac{x_2^2}{2} \end{bmatrix}$$

where $\theta(x_1) = \frac{7}{6} - \frac{1}{3}x_1 + \frac{1}{3}x_1^2$. Also

$f(1, 0) = (\frac{6}{7}, \frac{1}{7})$, the label $\ell(1, 0) = 1$

$f(0, 1) = (\frac{1}{7}, \frac{6}{7})$, the label $\ell(0, 1) = 2$

On such an assignment, certain n-simplexes in the subdivision become completely labeled; that is, their vertices carry all the labels 1 through $n + 1$. Let σ be such a completely labeled simplex with vertices $v^1, \ldots, v^{n+1}$ with vertex v^i labeled i for each $i = 1, \ldots, n + 1$. Then it can be established that for any given $\epsilon > 0$, it the size of the simplex σ is $\delta > 0$ (determined by the uniform continuity of f), any point x in σ is such that $\|f(x) - x\| \leq n(\epsilon + \delta)$. Thus the task of computing approximate fixed points can be accomplished by searching, through the subdivision, for a completely labeled simplex. The algorithms are designed to do precisely this.

Figure 3 shows a subdivision of a 2-simplex with $K = 7$. It is in this setting that we shall describe the workings of the algorithm. Clearly, in view of (3), we need only specify the labeling of the vertices of the subdivision and not the function which led to such a labeling. Please note that the labeling rule implies a point x cannot be labeled i if $x_i = 0$. In the setting of the Sperner's lemma, such a labeling is called proper. The lemma then guarantees that there must be an odd number of completely labeled simplexes in the subdivision. As a by-product, we will also be proving this lemma.

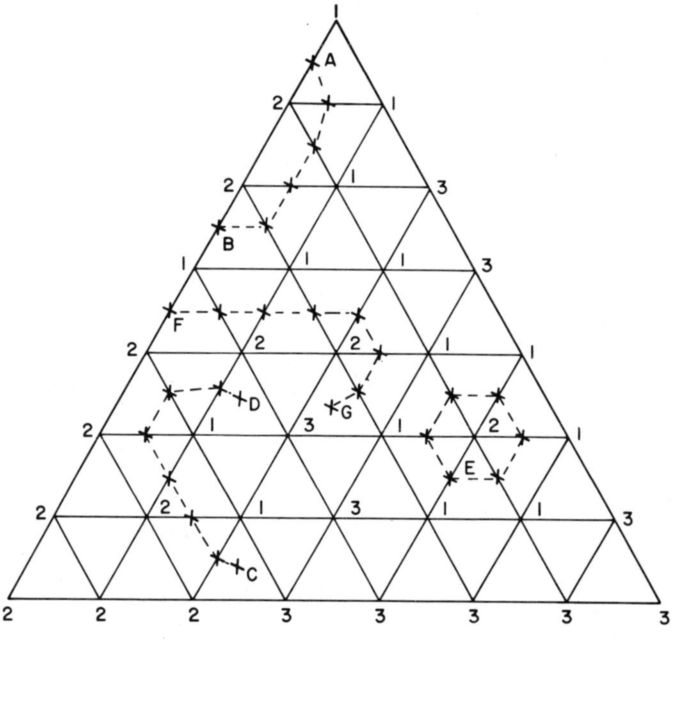

FIGURE 3.

Consider all line segments in the Fig. 3 carrying both the labels 1 and 2, and all the completely labeled triangles. Now, place a cross in the middle of each such line and in the center of each such triangle. By joining these crosses by dashed lines whenever they belong to the same triangle in the subdivision, we can create at most four types of paths as shown in the Fig. 3. Also, note that a path never returns to a triangle through which it has passed. These types can be enumerated as follows.

1. Paths starting at a line segment on the boundary, and ending at another line segment on the boundary (the path between A and B).
2. Paths starting inside a triangle of the subdivision and ending in a line segment on the boundary (the path between F and G).
3. Paths starting inside a triangle and ending inside another triangle (the path between C and D).
4. Paths ending as a loop (E). (The algorithm never reaches these paths.)

The paths of type 3 and 4 suggest an algorithm for finding a completely labeled triangle. The basic elements of such an algorithm are the following.

Step 0: Find a line segment L on the boundary that has both the labels 1 and 2. There is only one triangle T of the subdivision on this line. Let the vertex of the triangle not on this line segment be v.

Step 1: If the label on v is 3, stop. Otherwise go to Step 2.

Step 2: Since the label on v is either 1 or 2, one of the labels 1 or 2 is repeated in the triangle T. Hence, in T, there is exactly one more line segment which has both the labels 1 and 2. Let this line segment now be L.

Step 3: If L is on the boundary, stop. Otherwise, there is another triangle T on L (on the opposite side of the previous triangle). Make v the vertex on this triangle T not on L, and go to Step 1.

The sequence of steps described in the above algorithm are finite since no triangle is considered more than once, and there are only a finite number of triangles in the subdivision. This is so since for a triangle to appear more than once in the procedure, it must have more than two line segments labeled 1 and 2, which is not the case (as seen in Step 2). Thus, after a finite search, the algorithm will stop at Step 1 or Step 3. Stopping at Step 1 generates the path type 4, which then leads to a completely labeled triangle, while Step 3 leads to a type 3 path, which has failed to produce a completely labeled triangle. The termination in Step 3 is a real possibility, and in an implementable algorithm one guarantees that this will not happen by providing a unique starting line segment on the boundary.

The generalization of this algorithm to higher dimensions is now obvious. If we make the straightforward analogy between a triangle and an n-simplex, line segment and a (n - 1)-dimensional face of the n-simplex, it can be readily seen that all the elements of the algorithm are maintained. Thus, the algorithm would start with a (n - 1)-dimensional face in the boundary which has all the labels 1, 2 through n. If the label on the additional vertex on the unique n-simplex on this face is repeated, then there is exactly one more (n - 1)-face which carries all the labels

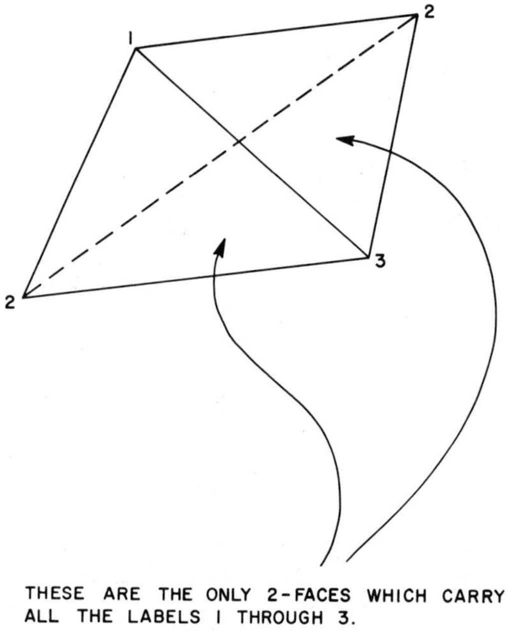
1
2
2
3
THESE ARE THE ONLY 2-FACES WHICH CARRY ALL THE LABELS 1 THROUGH 3.

FIGURE 4.

1, 2 through n. (As was the case in 2 dimensions. This is shown in Fig. 4 for 3 dimensions.) In addition, a subdivision in higher dimensions also shares the property that each (n - 1)-dimensional face on the boundary belongs to exactly one n-simplex, and otherwise to exactly two.

Thus, summarizing the basic steps in the development of an algorithm, one needs a procedure, called triangulation, for subdividing the space; a labeling method such that the distinguished simplexes called completely labeled can be related, in some manner, to fixed points of the mapping. Under the labeling method,

	$\bar{v}$	$\bar{\pi}$
$i = 1$	$v + \frac{1}{k} Q_{\pi_1}$	$(\pi_2, \ldots, \pi_n, \pi_1)$
$2 \le i \le n$	v	$(\pi_1, \ldots, \pi_i, \pi_{i-1}, \ldots, \pi_n)$
$i = n + 1$	$v - \frac{1}{k} Q_{\pi_{n+1}}$	$(\pi_n, \pi_1, \ldots, \pi_{n-1})$

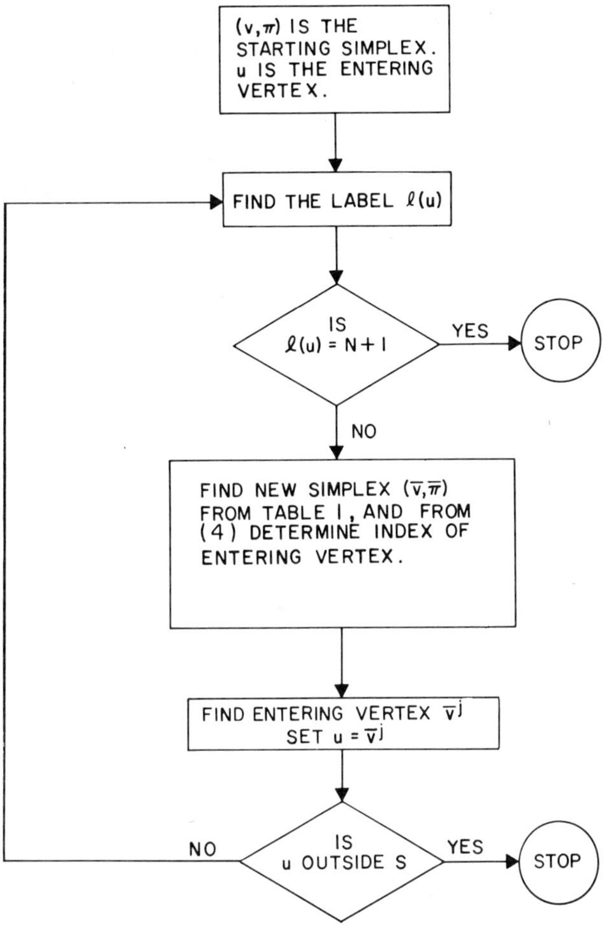

FIGURE 5.

certain facets (faces of dimension 1 less than the dimension of a simplex of the subdivision) became distinguished, and any simplex containing these distinguished facets is either completely labeled (thus contains exactly one such facet) or contains exactly two such facets. Finally, to avoid termination at Step 3, the boundary of the space under consideration must contain exactly one distinguished facet.

The computer implementation of such an algorithm involves the determination of the next simplex when a vertex with the repeated label has been dropped. For the subdivision procedure described in this section, this is readily obtained as follows.

Let the present simplex be (v, π), where $v = (v_1, \ldots, v_{n+1})$ and $\pi = (\pi_1, \ldots, \pi_n)$, and its vertices [generated recursively by (2)] be $v^1, \ldots, v^{n+1}$. Also, assume that the vertex whose label is repeated is v^i. Then the new simplex is $(\bar{v}, \bar{\pi})$, where $\bar{v}$ and $\bar{\pi}$ are obtained from Table 1.

Now, if the vertices of the new simplex $(\bar{v}, \bar{\pi})$ are $\bar{v}^1, \bar{v}^2, \ldots, \bar{v}^{n+1}$, the entering vertex is $\bar{v}^j$ where

$$j = \begin{cases} n+1 & i = 1 \\ i & 2 \leq i \leq n \\ 1 & i + n + 1 \end{cases} \tag{4}$$

The flow chart of the computer program implementing an algorithm on the subdivision of this section is given in Fig. 5.

AN ALGORITHM

One of the major disadvantages of the algorithm developed in the preceding section is that one is restricted to initiate the search at the boundary of the space. R^n has no such boundary, and thus the ability to initiate the algorithm at any point is required. This is achieved by considering the space $R^n \times (0, D]$ for the search where D is some positive real number. This space now has the boundary $R^n \times \{0\} \cup R^n \times \{D\}$. The algorithm will be developed on a subdivision of this space such that a unique distinguished facet is available in $R^n \times \{D\}$ to initiate the algorithm. We shall now describe the various details of the method. Note that we are considering computing a fixed point of the mapping f from R^n into R^n.

The Labeling

Let $r(x) = Ax - a$ be an affine mapping, with A an $n \times n$ nonsingular matrix. We will label a point (x, t) in $R^n \times (0, D]$ by the rule that

$$\ell(x, t) = \begin{cases} r(x) & \text{if } t = D \\ f(x) - x & \text{if } t < D \end{cases}$$

where $\ell(x, t)$ is the label on the point (x, t). As is evident, this labeling is quite different from the one used in the section entitled The Basics of the Algorithm.

Note that the simplexes in the subdivision of $R^n \times (0, D]$ are (n + 1)-dimensional, and their facets n-dimensional. We shall call a (n + 1)-simplex σ with vertices $(v^1, t_1), (v^2, t_2), \ldots, (v^{n+2}, t_{n+2})$ distinguished if the system of equations

$$\sum_{i=1}^{n+2} \lambda_i \ell(v^i, t_i) = 0 \tag{5}$$

$$\sum_{i=1}^{n+2} \lambda_i = 1 \tag{6}$$

$$\lambda_i \geq 0, \quad i = 1, \ldots, n + 2 \tag{7}$$

has a solution. This is a system with (n + 2) variables and (n + 1) equations, and, in view of (6), is bounded. In case all solutions are nondegenerate (in the usual linear programming sense, i.e., each solution has at least n + 1 positive variables), it can be shown that this system has two solutions with exactly n + 1 positive variables. These solutions correspond, in a unique way, with two facets of the simplex. We shall consider these two facets as the distinguished facets (as required for our labeling).

Let x^D be such that $Ax^D - a = 0$, and let (v^1, D), (v^2, D), ..., (v^{n+1}, D) be the vertices of a facet τ in $R^n \times \{D\}$. We note that this facet is distinguished if and only if (x^D, D) is in τ, and thus if we require (x^D, D) to lie in a unique facet of the subdivision, a unique starting facet on the boundary results.

We now confirm that distinguished facets under the labeling ℓ are related to fixed points of f. Let τ be a distinguished facet with vertices (v^1, t_1), ..., (v^{n+1}, t_{n+1}) with $t_i < D$, i = 1, ..., n + 1. Also, let $\epsilon > 0$ be arbitrary and given, and let $\delta > 0$ be determined by the uniform continuity of f, and let the size of τ be less than δ. Then there is a point $\bar{x}$ in τ such that $\|f(\bar{x}) - \bar{x}\| \leq \epsilon$. To see this, note that since $t_i < D$, (5)-(7) reduces to

$$\sum_{i=1}^{n+1} \lambda_i (f(v^i) - v^i) = 0$$

$$\sum_{i=1}^{n+1} \lambda_i \qquad\qquad = 1$$

$$\lambda_i \geq 0, \quad i = 1, \ldots, n+1$$

And since τ is distinguished, the above system has a solution $(\bar{\lambda}_i, \ldots, \bar{\lambda}_{n+1})$. Now, define

$$\bar{x} = \sum_{i=1}^{n+1} \bar{\lambda}_i v^i$$

and note that

$$\sum_{i=1}^{n+1} \bar{\lambda}_i f(v^i) = \bar{x}$$

Hence

$$f(\bar{x}) - \bar{x} = f(\bar{x}) - \sum_{i=0}^{n} \bar{\lambda}_i f(v^i)$$

and since $\|\bar{x} - v^i\| \leq \delta$, we have the result.

Another disadvantage of the basic algorithm is that it works with a fixed mesh. The algorithm we develop here will also have the property that the mesh of a facet τ in $R^n \times \left\{\frac{D}{2^k}\right\}$ goes to zero as k approaches ∞, and thus starting with a unique distinguished facet τ_0 in $R^n \times \{D\}$, one will find, progressively, distinguished facets τ_k in $R^n \times \left\{\frac{D}{2^k}\right\}$ for k approaching ∞. We now describe the triangulation procedure for obtaining the subdivision of $R^n \times (0, D]$ which, in addition, has the above property of the facets.

The Triangulation J3

The triangulation procedure, called J3 in the literature, for obtaining the subdivision of $R^n \times (0, D]$ was developed by Todd [19] and is related to the triangulation K3 of Ref. 5 on which the original algorithm was developed and described in 1971. We now describe this triangulation.

Given an arbitrary positive real number D, the vertices J_3^0 of this triangulation are all points v in R^{n+1} such that $v_{n+1} = D \cdot 2^{-k}$ for some integer $k = 0, 1, 2, \ldots$ and for each $i = 1, \ldots, n$, v_i/v_{n+1} are integers. In addition, a subset $\bar{J}_3^0$ of vertices in J_3^0 are called central vertices if for each $v \in \bar{J}_3^0$ and $i = 1, \ldots, n$, v_i/v_{n+1} is an odd integer.

A vertex v in J_3^0 is said to have depth k if $v_{n+1} = D \cdot 2^{-k}$. Note that to each central vertex v of depth $k \geq 1$, there is a unique nearest central vertex of depth $k - 1$ obtained as follows.

Define $\nu(v) = [\nu_1(v), \nu_2(v), \ldots, \nu_{n+1}(v)]$ where

$$\nu_i(v) = \begin{cases} -1 & \text{if } v_i/v_{n+1} \text{ is 1 mod 4} \\ +1 & \text{if } v_i/v_{n+1} \text{ is 3 mod 4} \end{cases}$$

Then the nearest central vertex to v is

$$y(v) = v - v_{n+1} \cdot \nu(v)$$

Now, each simplex σ in J3 has a unique representation by a triplet (v, π, s) where v is a central vertex, π a permutation of $\{1, \ldots, n + 1\}$, and $s = (s_1, \ldots, s_{n+1})$ with $s_i \in \{-1, +1\}$, $i = 1, \ldots, n + 1$. If $\nu = \nu(v)$, the vertices of this simplex are generated as follows [where $\pi)(j) = n + 1$]:

$$\begin{aligned} v^1 &= v \\ v^{i+1} &= v^i + v_{n+1} s_{\pi(i)} u_{\pi(i)} \qquad (i = 1, \ldots, j - 1) \\ v^{j+1} &= v^j - v_{n+1} \sum_{\ell=j+1}^{n+1} \nu_{\pi(\ell)} u_{\pi(\ell)} + v_{n+1} u_{n+1} \\ v^{k+1} &= v^k + 2 v_{n+1} \nu_{\pi(k)} u_{\pi(k)} \qquad (j + 1 \leq k \leq n + 1) \end{aligned} \tag{8}$$

The simplex σ is thus the convex hull of the vertices $v^1, \ldots, v^{n+2}$. That J3 is indeed a triangulation is proved in Ref. 19. The triangulation J3 of $R \times (0, 2]$ is shown in Fig. 6.

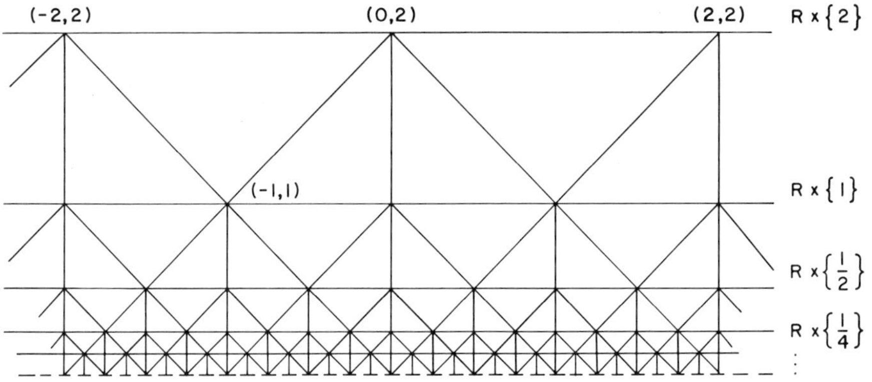

FIG. 6. Triangulation J3 of R × (0, 2].

Now, let (v, π, s) be a simplex in J3, and let $\pi(j) = n + 1$. Also, let $v^1, \ldots, v^{n+2}$ be the vertices of this simplex generated by the recursive relations (8), and assume that the vertex which has a repeated label (and is hence dropped) is v^i. The new simplex ($\bar{v}$, $\bar{\pi}$, $\bar{s}$) is then obtained by the rules given in Table 2. (The recursive relations can be simplified by defining a n-dimension vector b such that $b_i \in \{-1, +1\}$ and substituting $b_{\pi(i)}$ for $s_{\pi(i)}$ whenever $1 \leq i \leq j - 1$, and for $\nu_{\pi(i)}$ for $j + 1 \leq i \leq n + 1$. This, in effect, combines ν and s into one vector. Table 2 updates b rather than ν and s separately.)

Pivoting

Implementation of the method of this section requires the determination of the vertex that has a "repeated label." This determination was simple in the basic algorithm, since it was easily recognizable. In the method under consideration, this determination requires a pivot operation, identical to one in the simplex method of the linear programming theory. This is now discussed in some detail.

As we have seen, the method is initiated with a distinguished facet τ on the boundary $R^n \times \{D\}$ of the space. Let the vertices of this facet be (v^1, D), (v^2, D), $\ldots$, (v^{n+1}, D). Define the matrix

$$B = \begin{bmatrix} r(v^1) & r(v^2) & \cdots & r(v^{n+1}) \\ 1 & 1 & & 1 \end{bmatrix}$$

and note that this is a $(n + 1) \times (n + 1)$ nonsingular matrix. Also, since the facet τ is distinguished, there is a solution $\bar{\lambda}$ to the system

$$B\lambda = b$$

$$\lambda \geq 0$$

where $b = \begin{bmatrix} \vec{0} \\ 1 \end{bmatrix}$ and $\vec{0}$ is the n-dimensional null vector.

TABLE 2

		$\bar{v}$	$\bar{\pi}$	$\bar{b}$
$i = 1$	$j = 1$	$v - v_{n+1}(-1, b)$	$[\pi(2), \ldots, \pi(n+1), \pi(1)]$	b
	$j > 1$	$v + 2v_{n+1}b_{\pi(1)}u^{\pi(1)*}$	π	$b - 2b_{\pi(1)}u^{\pi(1)*}$
$2 \leq i \leq j - 1$		v	$[\pi(1), \ldots, \pi(i+1), \pi(i), \ldots, \pi(n+1)]$	b
$i = j$	$b_{\pi(j-1)} = \nu_{\pi(j-1)}$	v	$[\pi(1), \ldots, \pi(j), \pi(j-1), \ldots, \pi(n+1)]$	b
	$b_{\pi(j-1)} = -\nu_{\pi(j-1)}$	v	$[\pi(1), \ldots, \pi(j-2), \pi(j), \ldots, \pi(n+1), \pi(j-1)]$	$b - 2b_{\pi(j-1)}u^{\pi(j-1)}$
$j + 1 \leq i \leq n + 1$		v	$[\pi(1), \ldots, \pi(i+1), \pi(i), \ldots, \pi(n+1)]$	b
$i = n + 2$	$j = n + 1$	$v + \frac{1}{2}v_{n+1}(-1, b)$	$[\pi(n+1), \pi(1), \ldots, \pi(n)]$	b
	$j < n + 1$	v	$[\pi(1), \ldots, \pi(j-1), \pi(n+1), \pi(j), \ldots, \pi(n)]$	$b - 2b_{\pi(n+1)}u^{\pi(n+1)}$

* u^i has been generically used for the i-th unit vector of the correct dimension.

Now, let σ be the unique simplex on this facet, let (v, t) be the vertex of this simplex not in τ, and let $\ell(v, t) \equiv c$ be the label on this vertex. The "repeated label" is now determined by that column of B which is replaced by $\binom{c}{1}$ such that for the resulting matrix $\bar{B}$ the system

$$\bar{B}\lambda = b$$

$$\lambda \geq 0$$

has a solution. This determination is carried out in precisely the same manner as in the simplex method with the lexicographic pivot rule to guarantee that a distinguished simplex will have exactly two distinguished facets. The lexicographic pivot also eliminated the need for such a nondegeneracy assumption.

The Geometry of thc Algorithm

The algorithm just described has a very appealing geometric interpretation. We shall now describe this geometry.

Define a piecewise linear mapping H from $R^n \times (0, D]$ to R^n such that on any vertex (x, t) of the triangulation J3

$$H(x, t) = \ell(x, t)$$

and is linear on each simplex σ of the triangulation; that is, if the vertices of σ are $(v^1, t_1), \ldots, (v^{n+2}, t_{n+2})$ and (x, t) in σ is such that

$$(x, t) = \sum_{i=1}^{n+2} \lambda_i(v^i, t_i) \tag{9}$$

then

$$H(x, t) = \sum_{i=1}^{n+2} \lambda_i H(v^i, t_i) \tag{10}$$

As can be readily confirmed, $H(x, D) = r(x)$ and $H\left(\cdot, \frac{D}{2^k}\right)$ is a piecewise linear approximation to $f(x) - x$ on a triangulation $R^n \times \left\{\frac{D}{2^k}\right\}$. Also, as the size of the facets in this space approaches 0 as k approaches ∞, this approximation approaches $f(x) - x$ uniformly.

Now, let τ be a distinguished facet with vertices $(v^1, t_1), \ldots, (v^{n+1}, t_{n+1})$. Then there is a point (x, t) in τ such that $H(x, t) = 0$. This follows since (5)-(7) holds for distinguished facets, and (9)-(10) then gives the result. Also, since H is linear on any simplex σ, if σ is distinguished, there is a line segment $[x_h(\sigma), t_h(\sigma)]h \in [0, 1]$, in σ, such that $[x_0(\sigma), t_0(\sigma)]$ and $[x_1(\sigma), t_1(\sigma)]$ lie in the boundary of σ and $H[x_h(\sigma), t_h(\sigma)] = 0$ for each $h \in [0, 1]$.

Since r(x) is one-to-one and linear, $H(x, D) = 0$ has a unique solution (x^D, D). Thus, starting with the point (x^D, D) in $R^n \times \{D\}$, the algorithm generates a piecewise linear path (x_h, t_h), h going from 0 to infinity such that

$$(x_0, t_0) = (x^D, D)$$

and

$$H(x_h, t_h) = 0$$

for each h. (Such a path is obtained by "joining" the linear pieces in each distinguished simplex generated by the algorithm.) In addition, if $t_h \to 0$, since $H(\cdot, t_h)$ approaches $f(x) - x$, all cluster points of x_h are fixed points of f. (In Ref. 5 a modification of the labeling is given which insures that, in every case, t_h will approach 0.)

Conditions for Success

We now state some conditions on the mapping f under which the piecewise linear path generated by the algorithm will converge to a fixed point of f. We will assume that f is uniformly continuous and r(x) is one-to-one affine mapping with $r(x_0) = 0$. Our first result follows.

Theorem 1. Let there exist an open, bounded set C containing x_0, and a $\delta > 0$ such that any simplex of size less than δ which intersects the boundary of C is not distinguished. Then there is a $D > 0$ such that the path starting at x_0 never leaves C, and thus converges to a fixed point of f.

Proof: Let $D = \delta/\sqrt{n+1}$. The algorithm generates simplexes with diameter less than δ and starts inside C. Since none of these simplexes intersect the boundary of C, they all lie inside C. Also, as for each $t > 0$, there are a finite number of simplexes of J3 inside $C \times [t, D]$; in the path (x_h, t_h) thus generated, t_h must approach 0 as h approaches infinity. Hence x_h has a cluster point, and we have the result.

The next result can be considered as a generalization of the Leray-Schauder theorem. Here $h(x) = f(x) - x$.

Theorem 2. Let C be an open bounded set containing x_0 such that $h(x) + \rho r(x) \neq 0$ for all $\rho \geq 0$ and x in the boundary of C. Then there is a $D > 0$ such that the algorithm will compute a fixed point of f.

Proof: Assume the contrary. Then, for each D_k, $k = 1, 2, \ldots$, such that $D_k \to 0$ as $k \to \infty$, there is a distinguished simplex σ^k which intersects the boundary of C. Let x^k be a point in this intersection. Since x^k lies in a compact region, there is a subsequence $x^{k'}$ which converges to a point $\bar{x}$ on the boundary of C. In addition, $\sigma^{k'}$ also converge to $\bar{x}$ (since $D_{k'} \to 0$). Since σ^k is distinguished, there is an m_k such that if the vertices of σ^k are $v_1^k, \ldots, v_{n+2}^k$

$$\sum_{i=1}^{m_k} \lambda_i^k h\left(v_i^k\right) + \sum_{i=m_k+1}^{n+2} \lambda_i^k r\left(v_i^k\right) = 0$$

where λ_i^k are nonnegative and sum to 1. Since $m_{k'}$ lie in the set $\{1, \ldots, n+2\}$ and $\lambda_i^{k'}$, $i = 1, \ldots, n+2$ lie in the compact set $[0, 1]$, there is a subsequence k'' of k' such that $m_{k''} = m$ and $\lambda_i^{k''} \to \lambda_i$ for each $i = 1, \ldots, n+2$. Also, on this subsequence, $v_i^{k''} \to \bar{x}$. Thus, using the continuity of h and r, we obtain

$$h(\bar{x}) \cdot \sum_{i=1}^{m} \lambda_i + r(\bar{x}) \cdot \sum_{i=m+1}^{n+2} \lambda_i = 0$$

Since $r(x) \neq 0$ for all x in the boundary of C, and $\bar{x}$ is such a point, we conclude that $h(\bar{x}) + \rho r(x) = 0$ with

$$\rho = \frac{\sum_{i=m+1}^{n+2} \lambda_i}{\sum_{i=1}^{m} \lambda_i} > 0$$

and we have a contradiction.

To see how the Leray-Schauder theorem follows from Theorem 2, define $r(x) = -x$ and $h(x) = f(x) - x$. Then $f(x) \neq \lambda x$ for all $\lambda \geq 1$ if and only if $h(x) + \rho r(x) \neq 0$ for all $\rho \geq 0$. Thus, starting the algorithm at $x_0 = 0$, we have the proof.

COMPUTATIONAL CONSIDERATIONS

In some of these algorithms, especially in economics, fixed points are known to exist, and a reasonable choice of the initial affine mapping r can be made so that the algorithms will compute a fixed point. Saigal [16] has demonstrated that even when the mapping is affine [i.e., $f(x) = Ax - a$] and there is a unique fixed point, a proper choice of r has to be made, otherwise the algorithm fails to compute it. Thus, since in a practical problem choosing r and D to satisfy the convergence conditions of the previous section may be very time consuming, a reasonable approach would be to make an arbitrary choice and apply the algorithm. If failure occurs, hopefully enough will be learnt about the mapping to modify the choice of r. At present there is no theory which lays the guidelines for such a modification, but the results of Ref. 16 are potentially useful. Unfortunately, if the failure is due to the nonexistence of a fixed point, such an approach could be very frustrating.

A reasonable approach for detecting failure may be to keep track of the entering vertex v, and terminate when $\sum_{i=1}^{n} |v_i| > B$ for some predefined positive number B. In practical problems, B can be generally defined.

Some attention has been paid to the study of growth of computational effort as a function of the dimension of the problem. By making reasonable hypotheses on the behavior of the algorithm, Saigal [14] and Todd [20] have associated measures on triangulations which would give an indication of this growth. Theoretically, one shows that these measures, which count the number of simplexes the algorithm would generate to pass through a unit region of the space, grow approximately as the square of the dimension n. Since it takes $O(n^2)$ multiplications to pass through each simplex, it is expected that computational work would grow as $O(n^4)$. This seems to be supported by a growing body of computational experience.

REFERENCES

1. Allgower, E. L., and M. M. Jeppson, The approximation of solutions of nonlinear elliptic boundary value problems with several solutions, Springer Lect. Notes 333, 1-20 (1973).
2. Brouwer, L. E. J., Über eineindentige, stetige Transformationen von Flächen in Sich, Math. Ann. 67, 176-180 (1910).
3. Eaves, B. C., Homotopies for computation of fixed points, Math. Programming 3, 1-22 (1972).
4. Eaves, B. C., A short course in solving equations with PL homotopies, in Proceedings of the Ninth SIAM-AMS Symposium in Applied Mathematics, Vol. 9 (R. W. Cottle and C. E. Lemke, eds.), 1976.
5. Eaves, B. C., and R. Saigal, Homotopies for computation of fixed points on unbounded regions, Math. Programming 3, 225-237 (1972).
6. Hansen, T., On the Approximation of a Competitive Equilibrium, Ph.D. Thesis, Yale University, 1968.
7. Kakutani, S., A generalization of the Brouwer's fixed point theorem, Duke Math. J. 8, 457-459 (1941).
8. Kuhn, H. W., Simplicial approximation of fixed points, Proc. Natl. Acad. Sci. U.S.A. 61, 1238-1242 (1968).
9. Kuhn, H. W., A new proof of the fundamental theorem of algebra, Math. Programming Stud. 1, 148-158 (1974).
10. Lemke, C. E., Bimatrix equilibrium points and mathematical programming, Manage. Sci. 11, 681-689 (1965).
11. Lemke, C. E., and J. T. Howson, Jr., Equilibrium points of bimatrix games, SIAM J. Appl. Math. 12, 413-423 (1964).
12. Merrill, O. H., Applications and Extensions of an Algorithm That Computes Fixed Points of Certain Upper Semi-Continuous Point to Set Mappings, Ph.D. Dissertation, University of Michigan, 1972, 228 pp.
13. Ortega, J. M., and W. C. Rheinbolt, Iterative Solutions of Nonlinear Equations in Several Variables, Academic, New York, 1970.
14. Saigal, R., Investigations into the efficiency of fixed point algorithms, in Fixed Points—Algorithms and Applications (S. Karamardian, ed.), Academic, New York, 1977.
15. Saigal, R., On the convergence rate of algorithms for solving equations that are based on methods of complementary pivoting, Math. Oper. Res., to appear.
16. Saigal, R., On paths generated by fixed point algorithms, Math. Oper. Res., to appear.

17. Scarf, H., The approximation of fixed points of a continuous mapping, SIAM J. Appl. Math. 15, 1328-1343 (1967).
18. Scarf, H., The Computation of Economic Equilibrium (in collaboration with T. Hansen), Yale University Press, New Haven, Connecticut, 1973, 249 pp.
19. Todd, M. J., Union Jack triangulations, in Fixed Points—Algorithms and Applications (S. Karamardian, ed.), Academic, New York, 1977.
20. Todd, M. J., On triangulations for computing fixed points, Math. Programming 10, 322-346 (1976).

10
CLASSICAL OPTIMIZATION

LEON COOPER

Department of Industrial Engineering
and Engineering Management
Southern Methodist University
Dallas, Texas

ORIGINS OF OPTIMIZATION

The concept of the optimum (greatest or least) value of a mathematical function is one that was formulated in a precise way rather late in the history of mathematics. However, its origins are steeped in antiquity.

Virgil tells us of the founding of the city of Carthage by Queen Dido, who was allowed to have the largest area of land that could be surrounded by the hide of a bull. Queen Dido prepared a rope of unspecified thickness (a nontrivial detail omitted in Virgil) and then, perhaps guided by the gods, hit upon the optimal solution. With the sea as a diameter, she arranged the rope of finite length in the form of a semicircle. Indeed, this half-circle has the largest possible area for a fixed perimeter. Archimedes (287–212 B.C.) conjectured, but did not prove, that this was the correct solution. This was not proven until the development of the calculus of variations in the nineteenth century.

A great deal of thought about maximization and minimization is found in some of the work of the ancient Greek geometers. For example, in Book V of his treatise on *Conic Sections*, Appolonius (ca. 262–190 B.C.) deals with the problem on the maximum and minimum lengths that can be drawn from various points to a conic section. In work of great originality, he investigated maximum and minimum length distances for certain points on the major axis of a central conic or on the axis of a parabola. He does the same for points on the minor axis of an ellipse. He also proved that a line from a point within a conic to a point on

Now deceased

the conic which is a maximum or minimum is perpendicular to a tangent line through the point, a result of great significance. In short, he proved that the maximum and minimum lines to the point are normal to a tangent at the point of tangency.

Intuitively, the solution to another geometric optimization problem has been known for a long time, viz., that the shortest distance between two points on a plane is a straight line. The early Greeks used this principle in thinking about the behavior of light. Heron of Alexandria thought of light traveling between two points by the shortest path. Later Fermat, in the seventeenth century, formulated a principle of least time which generalized the earlier principle. Virtually all the principles of classical mechanics and optics down to the present day in wave mechanics have been or can be formulated in terms of various minimum principles.

The creation of the calculus, which led to the existence of the subject of this article, was, at least in part, motivated by the problem of finding the maximum or minimum value of a function. If a cannonball is shot from a cannon, the distance it will travel horizontally (the range) depends on the angle of inclination of the cannon to the ground. An early problem was to find the angle that would maximize the range. In the seventeenth century, Galileo determined that the maximum range (in a vacuum) is obtained for an angle of inclination of 45°. He was also able to determine the maximum heights reached by projectiles for various angles. Another early scientific influence on the development of the calculus was the study of the motion of the planets. Problems of optimization were involved in such determinations as the greatest and least distances of a planet from the sun.

Again the pervasive influence of geometry in leading to the development of calculus, one of the two greatest mathematical developments of all time, and hence to classical analysis, should be noted. Kepler made an early and crucial observation in his *Stereometria Doliorum* in 1615. He was interested in the optimal shape of casks for wine. He showed that, of all right parallelepipeds inscribed in a sphere and having square bases, the cube is the largest. He proceeded by calculating the volume for various choices of the dimensions. He then made an exceedingly important observation, viz., that as the maximum volume was approached, the *change* in volume for a *fixed* change in dimensions grew smaller and smaller. In the language of the calculus, the first derivative approached zero.

In a similar vein, Fermat, in his *Methodus ad Disquirendam Maximam and Minimam* in 1637, gave his method for finding maxima and minima by using the following as an example. Given a straight line segment, it is required to find a point on the line segment such that the rectangle contained by the two segments is a maximum. If the length of the whole line segment is L and the point marks off a part of length P, then the rectangle has area $P(L - P) = PL - P^2$. Fermat then replaces P by $P + E$. The remaining part is $L - (P + E)$ and the rectangular area is now $(P + E)(L - P - E)$. He then maintains, indicating complete insight into the principles we now understand, that the two areas should be equated, resulting in

$$PL + EL - P^2 - 2EP - E^2 = PL - P^2 \tag{1}$$

He then subtracts $PL - P^2$ from both sides and divides by E to obtain

$$L = 2P + E \tag{2}$$

He then argues that $E = 0$ at a maximum and so obtains $L = 2P$. Therefore, the rectangle is a square. He also generalizes the argument for any function, but of course does not justify dividing by E and setting E to zero.

It is interesting to note that at least one of the influences that led to the development of the calculus was the problem of finding the maxima and minima of a function. In turn, the calculus became a splendid instrument for examining such problems in a general setting. This is the subject matter of this article.

MATHEMATICAL BACKGROUND

We present here some of the mathematical concepts, either as definitions or theorems, that are required for the subsequent discussion. For proofs of the theorems that are omitted, the reader may consult Apostol's *Mathematical Analysis* [1].

Classical optimization theory is restricted to a consideration of finding maxima and minima of *continuous* and *differentiable* functions. Hence we need to define these terms.

Continuity. A function of n variables $x_1, x_2, \ldots, x_n$ is continuous at the point[1] $\mathbf{x}^0 = (x_1{}^0, x_2{}^0, \ldots, x_n{}^0)$ if for every $\epsilon > 0$ there exists a set of corresponding δ_j, $j = 1, 2, \ldots, n$ such that for $|h_j| < \delta_j$, $j = 1, 2, \ldots, n$ and $\delta_j > 0$, $j = 1, 2, \ldots, n$

$$|f(x_1{}^0 + h_1, x_2{}^0 + h_2, \ldots, x_n{}^0 + h_n) - f(x_1{}^0, x_2{}^0, \ldots, x_n{}^0)| \leq \epsilon \tag{3}$$

In vector notation we would write (3) as:

$$|f(\mathbf{x}^0 + \mathbf{h}) - f(\mathbf{x}^0)| \leq \epsilon \tag{4}$$

That not all functions are continuous should be obvious to the reader. For example, the function

$$f(x) = \begin{cases} 0, & -\infty \leq x \leq 0 \\ x, & 0 \leq x \leq 4 \\ \dfrac{x}{2}, & 4 \leq x \leq \infty \end{cases} \tag{5}$$

is an example of a discontinuous function of one variable.

Differentiability. A function $f(\cdot)$ of n variables is differentiable at a point $\mathbf{x}^0$ if the derivative of $f(\cdot)$ with respect to each of the independent variables exists. The derivative with respect to each variable is defined as:

$$\frac{\partial f(\mathbf{x}^0)}{\partial x_j} = \lim_{h_j \to 0} \frac{f(x_1{}^0, x_2{}^0, \ldots, x_j{}^0 + h_j, \ldots, x_n{}^0) - f(\mathbf{x}^0)}{h_j} \quad (j = 1, 2, \ldots, n) \tag{6}$$

[1] The boldface quantities such as $\mathbf{x}$ are to be regarded as points in an n-dimensional Euclidean space or simply as n-component vectors.

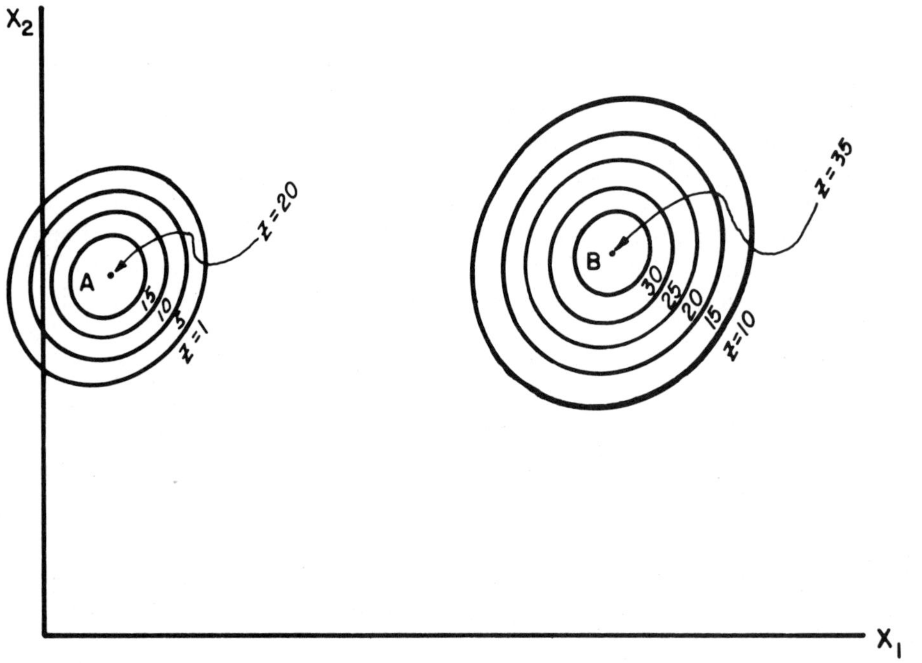

Fig. 1. Local and global maxima.

(The notation $\partial f(\mathbf{x}^0)/\partial x_j$ indicates the derivative evaluated at the point $\mathbf{x}^0$. This is sometimes noted as $\partial f(\mathbf{x})/\partial x_j|_{\mathbf{x}=\mathbf{x}^0}$.)

A function may be continuous without being differentiable. For example, $f(x) = x^{2/3}$ has no derivative at the point $x = 0$.

Absolute (Global) Maximum. A function $f(\cdot)$ takes on its absolute (or global) maximum at a point $\mathbf{x}^*$ if $f(\mathbf{x}) \leq f(\mathbf{x}^*)$ for all values of $\mathbf{x}$ over which the function f is defined. We will assume, in order to rule out certain anomalous cases, that the values of $\mathbf{x}$ at which $f(\cdot)$ attains its maximum are actually in the set of values over which $\mathbf{x}$ is defined.

The definition of absolute (or global) minimum can be obtained from the preceding definition by reversing the sense of the inequality between $f(\mathbf{x})$ and $f(\mathbf{x}^*)$. For example, the function $f(x_1, x_2) = 2x_1^2 + 3x_2^2 - 8x_1 - 12x_2 + 40$ has a global minimum at $x^* = (2, 2)$. We will see, subsequently, how this can be established.

Strong Relative (or Local) Maximum. A function $f(\cdot)$ takes on a strong relative (or local) maximum at a point $\mathbf{x}^0$ if there exists an ϵ, $0 < \epsilon < \delta$, such that for all $\mathbf{x}$ satisfying $0 < \|\mathbf{x} - \mathbf{x}^0\| < \epsilon$, it is the case that $f(\mathbf{x}) < f(\mathbf{x}^0)$.

In geometric language, the above definition states that if a function $f(\cdot)$ has a strong local maximum at some point $\mathbf{x}^0$ in an n-dimensional Euclidean space E^n, then there is a hypersphere (neighborhood) about $\mathbf{x}^0$ of radius ϵ, such that for every point $\mathbf{x}$ in the interior of the hypersphere, $f(\mathbf{x})$ is strictly less than $f(\mathbf{x}^0)$. A strong relative minimum is defined by reversing the inequality between $f(\mathbf{x})$ and $f(\mathbf{x}^0)$ in the preceding definition.

Weak Relative (or Local) Maximum. A function $f(\cdot)$ takes on a weak relative (or local) maximum at a point $\mathbf{x}^0$ if there exists an ϵ, $0 < \epsilon < \delta$, such that for all $\mathbf{x}$ satisfying $0 < \|\mathbf{x} - \mathbf{x}^0\| < \epsilon$, it is the case that $f(\mathbf{x}) \leq f(\mathbf{x}^0)$ and there is at least one point $\mathbf{x}$ in the interior of the hypersphere $\|\mathbf{x} - \mathbf{x}^0\| < \epsilon$ such that $f(\mathbf{x}) = f(\mathbf{x}^0)$.

In general, we will not distinguish between strong and weak local maxima. They will simply be called local maxima. It should be clear that a weak relative (or local) minimum is defined by reversing the inequality between $f(\mathbf{x})$ and $f(\mathbf{x}^0)$ in the preceding definition.

In connection with the definition of maxima and minima, it should be noted that if $f(\cdot)$ has an absolute maximum at a point $\mathbf{x}^*$, then $-f(\cdot)$ has an absolute minimum at $\mathbf{x}^*$, and vice versa. Similarly, if $f(\cdot)$ has an absolute maximum at a point $\mathbf{x}^0$, then $-f(\cdot)$ has an absolute minimum at $\mathbf{x}^0$, and vice versa.

In Fig. 1 we have shown graphically a function $z = f(\mathbf{x}) = f(x_1, x_2)$. What is plotted are the contours of z in the two-dimensional plane, E^2. At the points marked A and B, two relative maxima are shown. Assuming that $f(x_1, x_2)$ goes to $-\infty$ as $x_1, x_2 \to \infty$, we see that the point B is also the global maximum.

Convex and Concave Functions. A function $f(\cdot)$ is convex over some convex set X in E^n if for any two points $\mathbf{x}_1$ and $\mathbf{x}_2$ in X and for all λ, $0 \leq \lambda \leq 1$, $f[\lambda\mathbf{x}_1 + (1 - \lambda)\mathbf{x}_2] \leq \lambda f(\mathbf{x}_1) + (1 - \lambda)f(\mathbf{x}_2)$. If $f[\lambda\mathbf{x}_1 + (1 - \lambda)\mathbf{x}_2] \geq \lambda f(\mathbf{x}_1) + (1 - \lambda)f(\mathbf{x}_2)$, then the function is concave. If in the aforementioned expressions the inequalities are strict, then the function $f(\cdot)$ is said to be strictly convex in the first case and strictly concave in the second.

Consider the example of Fig. 2. The function $f(\cdot)$ is defined over the convex set X equal to the real line. Any point on the curve of $f(\cdot)$ between x_1 and x_2 is equal to $f[\lambda x_1 + (1 - \lambda)x_2]$ for $0 \leq \lambda \leq 1$, $\lambda f(x_1 + (1 - \lambda)f(x_2)$ will be a point on the straight line segment shown in Fig. 2. Hence a convex function is one that lies on or below a line segment drawn between two points on its curve. In general, a function $z = f(\mathbf{x})$ is a hypersurface in n-dimensional space. It is convex if the line segment which connects any two points $[\mathbf{x}_1, z_1]$ and $[\mathbf{x}_2, z_2]$ on the surface of $f(\mathbf{x})$ lies entirely on or above the hypersurface. The reverse holds true for concave functions.

We will now state without proof some important results which relate to the use of convex and concave functions. Proofs of these theorems can be found in Ref. 2.

Theorem 1. Let the functions $f_k(\cdot)$, $k = 1, 2, \ldots, p$ be convex (concave) functions over some convex set[2] X in E^n. Then the function $f(\mathbf{x}) = \sum_{k=1}^{p} f_k(\mathbf{x})$ is also a convex (concave) function over X.

Theorem 1 says that the sum of convex functions is a convex function and the sum of concave functions is a concave function.

[2] A convex set is one such that a straight line between any two points in the set is also in the set. More precisely, for any two points $\mathbf{x}_1$ and $\mathbf{x}_2$ in the set, the convex combination, $\lambda\mathbf{x}_1 + (1 - \lambda)\mathbf{x}_2$, $0 \leq \lambda \leq 1$, is also in the set.

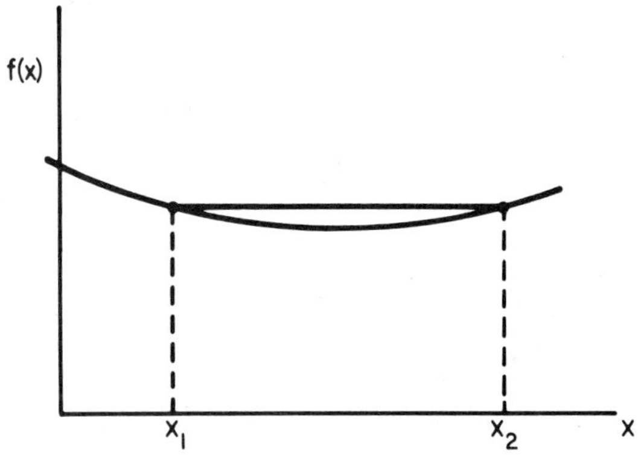

Fig. 2. A convex function.

Theorem 2. If $f(\cdot)$ is a convex function over the nonnegative orthant of E^n, then if $W = \{\mathbf{x} \mid f(\mathbf{x}) \le b, \mathbf{x} \ge \mathbf{0}\}$ is not empty, W is a convex set.

We will now state some general mathematical results from the calculus for purposes of reference subsequently. Proofs of these results can be found in Ref. 1.

Theorem 3 **(Mean Value Theorem).** If $f(\cdot)$ is continuous in the closed interval $x_0 \le x \le x_0 + h$ and differentiable at every point in the open interval $x_0 < x < x_0 + h$, then there exists at least one point x_m in the open interval specified where

$$\frac{f(x_0 + h) - f(x_0)}{h} = f'(x_m) \tag{7}$$

By $f'(x_m)$ we mean $df(x)/dx \,|_{x=x_m}$

The point x_m is not specified by this theorem. Its existence is merely guaranteed. An alternative way of expressing this theorem is to write:

$$\frac{f(x_0 + h) - f(x_0)}{h} = f'(x_0 + \theta h), \qquad 0 < \theta < 1 \tag{8}$$

This follows from the fact that x_m lies in the open interval $(x_0, x_0 + h)$ and any x in this interval can be expressed as $x = x_0 + \theta h$ where $0 < \theta < 1$. Therefore $x_m = x_0 + \theta h$ from which (8) follows.

A result which is often of interest and use for functions of a single variable but which is not easily generalized for functions of more than one variable is the following. It gives a *necessary and sufficient* condition for a function to have a maximum or minimum at a point and includes the case where the derivative may not exist at the maximal or minimal point—hence its value.

Theorem 4. Given a function $f(\cdot)$ defined on an interval, containing a point x_0 and that $f(\cdot)$ is continuously differentiable everywhere in the interval (with the possible exception of x_0) and further that $f'(\cdot)$

vanishes at a finite number of points. Then $f(\cdot)$ has a maximum or minimum at x_0 *if and only if* the point x_0 divides the interval over which $f(\cdot)$ is defined into two subintervals in which $f'(\cdot)$ has different signs. More precisely, the function has a maximum if the derivative is positive to the left of x_0 and negative to the right. It has a minimum if the reverse holds.

In Fig. 3 we see a function which has a minimum at a point x_0, for which the derivative is not defined at x_0. The theorem, of course, is still true if the derivative is defined everywhere.

One of the most important results in analysis, and of major significance in optimization, is what is known as Taylor's theorem. We shall give the theorem in its multidimensional form. In order to do so we shall require some special notation. Let

$$D^j f(\mathbf{x}) = \sum_{i_1=1}^{n} \sum_{i_2=1}^{n} \cdots \sum_{i_j=1}^{n} h_{i_1} h_{i_2} \cdots h_{i_j} \frac{\partial^j f(\mathbf{x})}{\partial x_{i_1} \partial x_{i_2} \cdots \partial x_{i_j}} \tag{9}$$

$$S_N(\mathbf{x}) = \sum_{j=1}^{n} \frac{1}{j!} D^j f(\mathbf{x}) \tag{10}$$

Further, let $\mathbf{x}_2 = \mathbf{x}_1 + \mathbf{h}$, where $\mathbf{h} = (h_1, h_2, \ldots, h_n)$. Then we define a function, usually called the *remainder term*, as

$$R_N(\mathbf{x} + \theta\mathbf{h}) = \frac{1}{(N+1)!} D^{N+1} f(\mathbf{x} + \theta\mathbf{h}), \qquad 0 < \theta < 1 \tag{11}$$

We now state Taylor's theorem.

Theorem 5 **(Taylor's Theorem).** A function $(\cdot)$ which is continuous and which has continuous partial derivatives of required order may be represented at point $\mathbf{x}_2 = \mathbf{x}_1 + \mathbf{h}$ in terms of its value at a point $\mathbf{x}_1$ by

$$f(\mathbf{x}_2) = f(\mathbf{x}_1) + S_N(\mathbf{x}) + R_N(\mathbf{x}_1 + \theta\mathbf{h}) \tag{12}$$

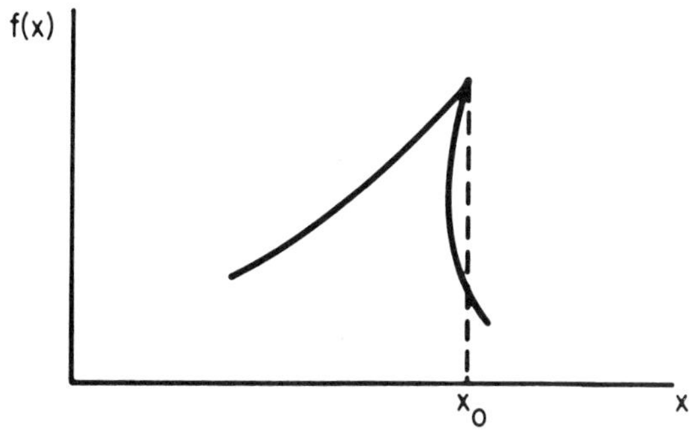

Fig. 3. Nondifferentiable maximum.

In particular, a value of $0 < \theta < 1$ exists to make (12) hold true. In simple language what Taylor's theorem states is that we may approximate any given function $f(\cdot)$ by polynomial of order N if the first $N + 1$ partial derivatives of the function are continuous. The two most commonly used forms of Taylor's theorem, which we shall use subsequently, are as follows.

$$f(\mathbf{x}_2) = f(\mathbf{x}_1) + \nabla f[\theta\mathbf{x}_1 + (1 - \theta)\mathbf{x}_2]\mathbf{h} \tag{13}$$

$$f(\mathbf{x}_2) = f(\mathbf{x}_1) + \nabla f(\mathbf{x}_1)\mathbf{h} + \tfrac{1}{2}\mathbf{h}' H[\theta\mathbf{x}_1 + (1 - \theta)\mathbf{x}_2]\mathbf{h} \tag{14}$$

Equation (13) is a first-order approximation and ∇f is the gradient vector and is defined by

$$\nabla f = \left(\frac{\partial f}{\partial x_1}, \frac{\partial f}{\partial x_2}, \ldots, \frac{\partial f}{\partial x_n}\right) \tag{15}$$

The notation $\nabla f[\theta\mathbf{x}_1 + (1 - \theta)\mathbf{x}_2]$ indicates that the gradient vector is to be evaluated at the point $\mathbf{x} = \theta\mathbf{x}_1 + (1 - \theta)\mathbf{x}_2$ in E^n. Taylor's theorem assures us that there exists a θ such that (13) holds.

In Eq. (14) we have written the theorem in terms of second partial derivatives as well as first partial derivatives. H in Eq. (14) is the Hessian matrix of $f(\cdot)$ and is defined as a matrix of the n^2 second partial derivatives of $f(\cdot)$. In other words, it is the $n \times n$ matrix $H = \|\partial^2 f/\partial x_u \partial x_v\|$ where $u, v = 1, 2, \ldots, n$. The notation $H[\theta\mathbf{x}_1 + (1 - \theta)\mathbf{x}_2]$ indicates that the Hessian matrix is evaluated at the point $\mathbf{x} = \theta\mathbf{x}_1 + (1 - \theta)\mathbf{x}_2$.

We require one last result from analysis. It is known as the *implicit function theorem*. Consider a set of m equations in n variables where $m < n$:

$$g_i(\mathbf{x}) = 0, \qquad i = 1, 2, \ldots, m$$

and

$$\mathbf{x} = (x_1, x_2, \ldots, x_n) \tag{16}$$

There may be situations in which we may wish to use these m equations to eliminate some subset m of the n variables in some expression. Suppose we wish to do this at some specific point $\mathbf{x}^0$. Without loss of generality, suppose we wish to eliminate the first m variables. Therefore, what we would like to know is under what circumstances there exist a set of m functions ϕ_i, $i = 1, 2, \ldots, m$, such that

$$x_i = \phi_i(x_{m+1}, x_{m+2}, \ldots, x_n), \qquad i = 1, 2, \ldots, m \tag{17}$$

at $\mathbf{x}^0$ or in a neighborhood of $\mathbf{x}^0$. The following theorem supplies this information.

Theorem 6 **(Implicit Function Theorem).** If the rank of the Jacobian matrix J_m evaluated at the point $\mathbf{x}^0$ is equal to m, then this is a necessary and sufficient condition for the existence of a set of m functions ϕ_i, $i = 1, 2, \ldots, m$, which are unique, continuous, and differentiable in some neighborhood of $\mathbf{x}^0$.

The Jacobian matrix J_m is defined as

$$J_m = \begin{bmatrix} \dfrac{\partial g_1}{\partial x_1} & \dfrac{\partial g_1}{\partial x_2} & \cdots & \dfrac{\partial g_1}{\partial x_m} \\ \dfrac{\partial g_2}{\partial x_1} & \dfrac{\partial g_2}{\partial x_2} & \cdots & \dfrac{\partial g_2}{\partial x_m} \\ \vdots & \vdots & & \vdots \\ \dfrac{\partial g_m}{\partial x_1} & \dfrac{\partial g_m}{\partial x_2} & \cdots & \dfrac{\partial g_m}{\partial x_m} \end{bmatrix} \tag{18}$$

The statement that the Jacobian matrix is evaluated at $\mathbf{x}_0$ means that each of the elements of J_m is evaluated at the point $\mathbf{x}_0$.

For simplicity in the statement of Theorem 6 we assumed that the first m variables were to be expressed in terms of the remaining $n - m$ variables. A more exact statement is as follows. If one selects any m column vectors of the form

$$\begin{bmatrix} \dfrac{\partial g_1}{\partial x_j} \\ \dfrac{\partial g_2}{\partial x_j} \\ \vdots \\ \dfrac{\partial g_m}{\partial x_j} \end{bmatrix}$$

Then, if the resulting Jacobian matrix has rank m, these particular variables may be eliminated, i.e., the required set of m functions ϕ_i exist. There are $\binom{m}{n} = n!/m!(n - m)!$ possible combinations of m columns out of a total set of n and hence that number of Jacobians. Every Jacobian that has rank m represents a set of variables that can be eliminated. An equivalent statement to having rank m is that the Jacobian matrix be nonsingular, i.e., that its determinant is not zero.

UNCONSTRAINED OPTIMIZATION

We shall divide the subject of unconstrained optimization, i.e., maximizing or minimizing a function under conditions where the variables are not constrained in any way, into two major parts. The first is the study of functions of a single variable and the second treats functions of several variables.

There are two basic approaches that one can take to find the optimum of a function. The first approach, which is often called a *direct method*, consists of evaluating the given function $f(\cdot)$ at some point $\mathbf{x}^1$ and then seeking, by some method, to find another point $\mathbf{x}^2$, where $\mathbf{x}^2 = g(\mathbf{x}^1)$, some well-defined function of $\mathbf{x}^1$, such that $f(\mathbf{x}^2) > f(\mathbf{x}^1)$ if we seek a maximum or $f(\mathbf{x}^2) < f(\mathbf{x}^1)$ if we seek a minimum. We repeat this process until no further change is possible. We shall *not* be concerned with this general approach in this article.

The second approach is what is referred to as an *indirect method* and is the approach of calculus and classical optimization. What is involved here is the development and use of necessary and sufficient conditions that an optimal point, a local maximum or minimum, must satisfy. In certain cases it is possible to determine whether or not a global optimum has been found.

We first consider the derivation and proof of a *necessary* condition for the existence of a local or relative optimum of a function. This is provided in the following theorem.

Theorem 7. A necessary condition that a continuous function $f(\cdot)$, whose first derivative is continuous over E^1, has a local minimum or maximum at a point x_0 is that $df(x_0)/dx = 0$.

Proof: We recall from the definition that if $f(\cdot)$ has a relative minimum at a point x_0, then there exists an $\epsilon > 0$ such that for some interval about x_0, $f(x) \geq f(x_0)$ for $|x - x_0| < \epsilon$. Therefore, let us consider points about x_0 of the form

$$x = x_0 + h, \qquad 0 < |h| < \epsilon$$

Then we may write

$$f(x_0 + h) - f(x) \geq 0, \qquad 0 < |h| < \epsilon \tag{19}$$

We may then divide Eq. (19) by h to obtain

$$\frac{f(x_0 + h) - f(x)}{h} \geq 0, \qquad h > 0 \tag{20}$$

$$\frac{f(x_0 + h) - f(x)}{h} \leq 0, \qquad h < 0 \tag{21}$$

If we take the limit of the expressions in (20) and (21), we obtain

$$\lim_{h \to 0} \frac{f(x_0 + h) - f(x)}{h} = \frac{df(x_0)}{dx} \geq 0 \tag{22}$$

$$\lim_{h \to 0} \frac{f(x_0 + h) - f(x)}{h} = \frac{df(x_0)}{dx} \leq 0 \tag{23}$$

Together, Eqs. (22) and (23) imply that

$$\frac{df(x_0)}{dx} = 0 \tag{24}$$

A corresponding argument with the inequalities reversed can be made for a relative maximum. Hence the theorem is proved.

A point at which the first derivative is zero is called a stationary point. Therefore, we have proven that a *necessary* condition for x_0 to be a local maximum or minimum is that x_0 be a stationary point. Since Theorem 7 provides only a necessary condition, a stationary point may be either a maximum, a minimum, or neither. The following obvious examples illustrate this:

$f_1(x) = 4x^2$	has a minimum
$f_2(x) = -3x^2$	has a maximum
$f_3(x) = 7x^3$	has neither a maximum nor a minimum

In order to discriminate between maxima and minima and other stationary points we require a sufficient condition. This is provided by the following theorem.

Theorem 8. Given a continuous function $f(\cdot)$ whose first two derivatives are continuous at x_0. Then if $f'(x_0) = 0$, a sufficient condition for $f(\cdot)$ to have a minimum at x_0 is that $f''(x_0) > 0$ and a sufficient condition for $f(\cdot)$ to have a maximum at x_0 is that $f''(x_0) < 0$.

Proof: Using Taylor's theorem we may write

$$f(x_0 + h) = f(x_0) + hf'(x_0) + \frac{h^2}{2} f''[\theta x_0 + (1 - \theta)(x_0 + h)], \qquad 0 \leq \theta \leq 1 \quad (25)$$

If $f(\cdot)$ has a relative minimum at x_0, then we know from Theorem 7 that $f'(x_0) = 0$. Therefore we may rewrite Eq. (25) as

$$f(x_0 + h) - f(x_0) = \frac{h^2}{2} f''[\theta x_0 + (1 - \theta)(x_0 + h)], \qquad 0 \leq \theta \leq 1 \quad (26)$$

If $f(\cdot)$ is to have a minimum at x_0, then it follows from Eq. (26) that

$$f(x_0 + h) - f(x_0) = \frac{h^2}{2} f''[\theta x_0 + (1 - \theta)(x_0 + h)] > 0, \qquad 0 \leq \theta \leq 1 \quad (27)$$

Suppose that $f''(x_0) < 0$. Then it follows from the continuity of the second derivative that $f''[\theta x_0 + (1 - \theta)(x_0 + h)] < 0$ and therefore that

$$f(x_0 + h) - f(x_0) = \frac{h^2}{2} f''(\theta x_0 + (1 - \theta)(x_0 + h)] < 0 \quad (28)$$

and therefore x_0 cannot be a minimum point. Conversely, if $f''(x_0) > 0$ at $f'(x_0) = 0$, $f(x_0)$ is clearly a minimum. We can repeat a corresponding argument for the case of a maximum and arrive at the conclusion that if $f''(x_0) < 0$ at $f'(x_0) = 0$, $f(x_0)$ is a maximum.

While Theorem 8 gives sufficient conditions for functions for which the second derivative does not vanish, it is certainly possible that at some point x_0, both first and second derivatives will vanish. The following theorem is more general and gives sufficient conditions for any case.

Theorem 9. Assume that $f(\cdot)$ and its first n derivatives are continuous. Then

> $f(\cdot)$ has a relative maximum or minimum at x_0 if and only if n is even, where n is the order of the first nonvanishing derivative at x_0. The function $f(\cdot)$ has a maximum at x_0 if $f^{(n)}(x_0) < 0$ and a minimum if $f^{(n)}(x_0) > 0$.

Proof: By hypothesis the first $n - 1$ derivatives of $f(\cdot)$ vanish at x_0, i.e.,

$$f'(x_0) = f''(x_0) = \cdots = f^{(n-1)}(x_0) = 0 \tag{29}$$

We have also assumed that $f^{(n)}(x_0) \neq 0$ and that n is an even number. From Taylor's theorem and Eq. (29) we then have that

$$f(x_0 + h) - f(x_0) = \frac{h^n}{n!} f^{(n)}[\theta x_0 + (1 - \theta)(x_0 + h)], \qquad 0 \leq \theta \leq 1 \tag{30}$$

Since $f^{(n)}$ is continuous, we are assured that $f^{(n)}[\theta(x_0) + (1 - \theta)(x_0 + h)]$ will have the same sign as $f^{(n)}(x_0)$. Hence from Eq. (30) this implies that $f(x_0 + h) - f(x_0)$ will have the same sign as $f^{(n)}(x_0)$ since n is even. This means that $f(x_0 + h) - f(x_0)$ will be positive whenever $f^{(n)}(x_0)$ is positive and negative whenever $f^{(n)}(x_0)$ is negative. Therefore x_0 will be a relative maximum or minimum depending on the sign of $f^{(n)}(x_0)$.

The foregoing argument disposes of the "if" part of the theorem. We shall prove the "only if" part by contradiction. Let us assume, contrary to our hypothesis, that the order of the first nonvanishing derivative is odd. Using Taylor's theorem as before, we again write

$$f(x_0 + h) - f(x_0) = \frac{h^n}{n!} f^{(n)}[\theta x_0 + (1 - \theta)(x_0 + h)], \qquad 0 \leq \theta \leq 1 \tag{31}$$

If x_0 is to be a maximum or minimum, then the right-hand side of Eq. (31) has to be nonpositive or nonnegative. However, when n is odd, h^n is positive or negative as h is positive or negative. Since $f(\cdot)$ is continuous, the sign of $f^{(n)}[\theta x_0 + (1 - \theta)(x_0 + h)]$ does not change sign as h goes from positive to negative. Therefore the term on the right-hand side of Eq. (31) changes its sign as $h^{(n)}$ changes its sign. Hence, from Eq. (31) we see that $f(x_0 + h) - f(x_0)$ will have different signs depending upon whether h is positive or negative. This leads to a contradiction, since under these conditions x_0 could not be a point at which $f(\cdot)$ has a maximum or minimum. Such a condition is a direct contradiction of the definition of a relative maximum or minimum. This concludes the proof.

Example. Consider the following function and let us examine its derivatives:

$$f(x) = (2x - 4)^4$$

$$\frac{df}{dx} = 8(2x - 4)^3$$

$$\frac{d^2f}{dx^2} = 48(2x - 4)^2$$

$$\frac{d^3f}{dx^3} = 192(2x - 4)$$

$$\frac{d^4f}{dx^4} = 384$$

It is easily seen that the first three derivatives vanish when $x = 2$. The first nonvanishing derivative is d^4f/dx^4. Therefore f has a minimum at $x = 2$ since the first nonvanishing derivative is even and positive.

As another example, consider the following function and its derivatives:

$$f(x) = (2x - 6)^3$$
$$\frac{df}{dx} = 6(2x - 6)^2$$
$$\frac{d^2f}{dx^2} = 24(2x - 6)$$
$$\frac{d^3f}{dx^3} = 48$$

The first two derivatives vanish at $x = 3$. The first nonvanishing derivative is d^3f/dx^3. Even though $df/dx = 0$ at $x = 3$, f has neither a maximum nor a minimum at $x = 3$ since the order of the first nonvanishing derivative is odd rather than even. The point $x = 3$ is a stationary point, i.e., $df/dx = 0$ but it is a point of inflection rather than a local maximum or minimum.

The foregoing results tell us a good deal about the conditions that must hold for a function f to have a local maximum or minimum. However, being existence theorems, they tell us little about how to find such local maxima or minima. Let us consider this. In practice, for a given function $f(x)$, we invoke the necessary condition (Theorem 7) and we have that the point x_0 we seek is a solution to

$$\frac{df(x_0)}{dx} = 0 \tag{32}$$

Whether or not it is a simple matter to solve Eq. (32) for the relative optimum (or optima) x_0 depends entirely on the nature of the function $f(\cdot)$. In general this may involve finding the roots of nonlinear or transcendental equations for which no computational procedure in closed form may exist. For example, suppose one was asked to find all maxima of $f(x) = x \sin x - x^2$. The necessary condition for a maximum is

$$f'(x) = x \cos x + \sin x - 2x = 0 \tag{33}$$

One cannot solve Eq. (33) explicitly for x. Hence it is necessary to resort to numerical methods. There are many methods which have been proposed for finding roots of general or special nonlinear equations. They are outside the scope of this article. They are properly the subject of numerical analysis and are described in such texts as Refs. 3 and 4. However, one such method deserves to be described. It is called Regula Falsi or the method of false position.

We assume that the function $f'(x)$ is both continuous and computable. The derivation of the numerical method can be made clear by referring to Fig. 4.

In Figure 4 we use the notation $F(x) \equiv f'(x)$ and hence we seek roots of $F(x) = 0$. We assume that at any iteration we have available two points x_{i-1}, x_i and their corresponding function values $F(x_{i-1})$, $F(x_i)$. If a secant if passed through the points $(x_{i-1}, F(x_{i-1}))$ and $(x_i, F(x_i))$. The intersection of the secant with the x-axis yields the new approximation, x_{i+1}, to the root we seek. It can be seen from Fig. 4

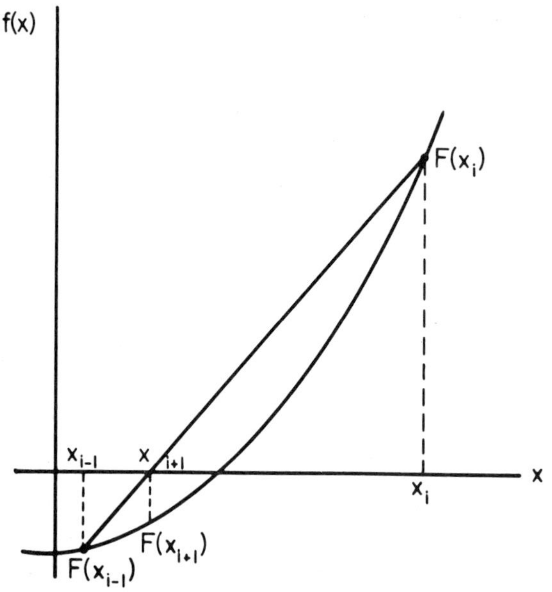

Fig. 4. Method of false position.

that the slope $\Delta F/\Delta x$ of the secant is given by

$$\frac{\Delta F}{\Delta x} = \frac{F(x_i) - F(x_{i-1})}{x_i - x_{i-1}} = \frac{F(x_i) - F(x_{i+1})}{x_i - x_{i+1}} \approx \frac{F(x_i)}{x_i - x_{i+1}} \tag{34}$$

Solving Eq. (34) for x_{i+1} yields the approximate iteration formula of the method of false position:

$$x_{i+1} = x_i - F(x_i)\frac{x_i - x_{i-1}}{F(x_i) - F(x_{i+1})} \tag{35}$$

Once the third point, x_{i+1}, has been found, either x_{i-1} or x_i is dropped. The rule that is used is to drop whichever point will allow us to bracket the root we seek. This is assured by keeping the points which have values of F of opposite sign. If we do this at each stage of the calculation, including the choice of the two initial points, then the convergence of this numerical scheme can be assured. An example of this method follows.

Example. Find a root of $F(x) = x^3 - 6x - 8 = 0$. We note that $F(2) = -12$, $F(3) = 1$. Hence there is a root between $x = 2$ and $x = 3$. If $x_1 = 2$ and $x_2 = 3$, then

$$x_3 = 3 - 1\left(\frac{3-2}{1+12}\right) = 3 - 1\left(\frac{1}{13}\right) = 2.9231$$

We then have $F(2.9231) = -0.56213$. We retain $F(3) = 1$ and continue:

$$x_4 = 2.9231 - (-0.56213)\frac{2.9231 - 3}{-0.56213 - 1} = 2.8954$$

We then have $F(2.9508) = -0.01153$. It can be seen $F(x) \to 0$. This process can be continued to any desired degree of accuracy.

We previously defined convex and concave functions because of their special significance in optimization. The convex function shown in Fig. 2 has a single minimum over the interval shown. We shall now prove that this is always true.

Theorem 10. Let $f(\cdot)$ be a convex function over a closed interval $a \le x \le b$. Then any local minimum of $f(\cdot)$ in this interval is also the global minimum over the interval.

Proof: This proof is by contradiction. Assume that $f(\cdot)$ has a local minimum at x_0 in the interval $[a, b]$. Assume further that the global minimum is taken on at a different point, $\hat{x}$, such that $f(x_0) > f(\hat{x})$. We defined a convex function as one for which $f[\lambda x_1 + (1 - \lambda)x_2] \le \lambda f(x_1) + (1 - \lambda)f(x_2)$ for any x_1, x_2 in $[a, b]$ and all λ, $0 \le \lambda \le 1$. Therefore x_0 and $\hat{x}$ must satisfy this convexity property. Hence it must be true that

$$f[\lambda\hat{x} + (1 - \lambda)x_0] \le \lambda f(\hat{x}) + (1 - \lambda)f(x_0) \tag{36}$$

However, by hypothesis, $f(\hat{x}) < f(x_0)$. Therefore, if we substitute $f(x_0)$ for $f(\hat{x})$ on the right-hand side of Eq. (36), we strengthen the inequality and so we may write

$$f[\lambda\hat{x} + (1 - \lambda)x_0] < \lambda f(x_0) + (1 - \lambda)f(x_0) = f(x_0) \tag{37}$$

Now consider any neighborhood ϵ of x_0 such that $|\hat{x} - x_0| < \epsilon$. If λ is chosen so that $0 < \lambda < \epsilon/|\hat{x} - x_0|$, then we see that $x = \lambda\hat{x} + (1 - \lambda)x_0$ is clearly in this neighborhood ϵ of x_0. Therefore $f(x) = f[\lambda\hat{x} + (1 - \lambda)x_0] < f(x_0)$. This contradicts the fact that x_0 is a local minimum of $f(\cdot)$. Hence our original assumption that $f(\hat{x}) < f(x_0)$ is false, and we have shown that any local minimum is also the global minimum.

In a very similar proof, by reversing inequalities one can show that any local maximum of a concave function is the global maximum of the function over a given interval.

We now consider the reverse situation, viz., maximizing a convex function. An examination of Fig. 2 strongly suggests that if the function is not constrained to a finite interval, it may have no finite maximum. If a finite interval on x is considered, it appears as though the maximum would be at one or both of the end points of the interval. The following proof establishes this.

Theorem 11. The global maximum of a convex function $f(\cdot)$ over a closed interval $a \le x \le b$ will be taken on at either $x = a$ or $x = b$ or both.

Proof: Any point in the interval $[a, b]$ is given by $x = \lambda a + (1 - \lambda)b$, $0 \leq \lambda \leq 1$. Therefore from the definition of a convex function:

$$f[\lambda a + (1 - \lambda)b] \leq \lambda f(a) + (1 - \lambda)f(b) \tag{38}$$

It is clear from Eq. (38) that $f(\cdot)$ either has a minimum somewhere in the interval or it is a linear function over the interval. If the latter is true, then clearly the theorem is true. Suppose now it has a minimum in the interval. Either the minimum is at one end point or in the interior of the interval. Suppose the minimum is at $x = a$. Then by Eq. (38) we have that $f(a) \leq f(a)$ or $f(a) = f(a)$. As we increase x we cannot pass through a maximum at any point $x < b$, for by convexity and Eq. (38) $f(x) \leq \lambda f(a) + (1 - \lambda)f(b)$. Hence the only possibility would be for the maximum to occur at $x = b$. The same argument would hold for an interior minimum, except that we can either decrease or increase x to reach one or both end joints. Hence all maxima will be at end points and, therefore, one or both of the end points will be global maxima.

We turn now to consider the optimization of functions of several variables. It will be seen that the results of the foregoing sections for the function of a single variable can be generalized but with some attendant increase in complexity. The first result, Theorem 12, provides a necessary condition for a relative (local) maximum or minimum.

Theorem 12. If $\mathbf{x}^0$ is a point in E^n where $f(\cdot)$ takes on a local maximum or minimum, where $f(\cdot)$ is a continuous function with continuous first partial derivatives, then the first partial derivatives of $f(\cdot)$ with respect to each of the n variables must vanish at the point x^0.

Proof: If $f(\cdot)$ has a relative minimum at a point $\mathbf{x}^0$, then we know from the definition of relative minimum that there exists an $\epsilon > 0$ such that for all points $\mathbf{x}$ in an ϵ-neighborhood of $\mathbf{x}^0$, $f(\mathbf{x}) \geq f(\mathbf{x}^0)$. In particular, consider points in this neighborhood of the form

$$\mathbf{x} = \mathbf{x}^0 + h\mathbf{e}_j, \qquad 0 < |h| < \epsilon \tag{39}$$

In Eq. (39) $\mathbf{e}_j$ is a unit vector with n components and a "1" for the jth component. We may therefore write

$$f(\mathbf{x}^0 + h\mathbf{e}_j) - f(\mathbf{x}^0) \geq 0 \tag{40}$$

for all h, $0 < |h| < \epsilon$. Equation (40) may be divided by h to obtain

$$\frac{f(\mathbf{x}^0 + h\mathbf{e}_j) - f(\mathbf{x}^0)}{h} \geq 0, \qquad h > 0 \tag{41}$$

$$\frac{f(\mathbf{x}^0 + h\mathbf{e}_j) - f(\mathbf{x}^0)}{h} \leq 0, \qquad h < 0 \tag{42}$$

If we take the limit in Eqs. (41) and (42) as $h \to 0$, we obtain that

$$\lim_{h\to 0} \frac{f(\mathbf{x}^0 + h\mathbf{e}_j) - f(\mathbf{x}^0)}{h} = \frac{\partial f(\mathbf{x}^0)}{\partial x_j} \geq 0 \tag{43}$$

$$\lim_{h\to 0} \frac{f(\mathbf{x}^0 + h\mathbf{e}_j) - f(\mathbf{x}^0)}{h} = \frac{\partial f(\mathbf{x}^0)}{\partial x_j} \leq 0 \tag{44}$$

Equations (43) and (44) together imply that

$$\frac{\partial f(\mathbf{x}^0)}{\partial x_j} = 0, \qquad j = 1, 2, \ldots, n \tag{45}$$

A corresponding argument, with the inequalities reversed, can be made for a local maximum.

Just as in the case of a function of a single variable, what we have shown is that a necessary condition for $\mathbf{x}^0$ to be a local maximum or minimum is that $\mathbf{x}^0$ be a stationary point, i.e., that Eq. (45) hold at the point $\mathbf{x}^0$.

Let us now consider the derivation of sufficient conditions for optima of functions of several variables. We shall see that this case is considerably more complicated than was the case of a function of a single variable.

Theorem 13. If $f(\cdot)$ and its first two partial derivatives are continuous, then a sufficient condition for $f(\mathbf{x})$ to have a relative minimum (or maximum) at $\mathbf{x}^0$, when $\partial f(\mathbf{x}^0)/\partial x_j = 0, j = 1, 2, \ldots, n$, is that the Hessian matrix be positive definite[3] (negative definite).

Proof: From Taylor's theorem we know that

$$f(\mathbf{x}^0 + \mathbf{h}) = f(\mathbf{x}^0) + \nabla f(\mathbf{x}^0)\mathbf{h} + \tfrac{1}{2}\mathbf{h}'H[\theta\mathbf{x}^0 + (1 - \theta)(\mathbf{x}^0 + \mathbf{h})]\mathbf{h} \tag{46}$$

By hypothesis, the necessary conditions are satisfied, i.e., $\partial f(\mathbf{x}^0)/\partial x_j = 0, j = 1, 2, \ldots, n$, and therefore $\nabla f(\mathbf{x}^0) = \mathbf{0}$. Hence we can rewrite Eq. (46) as

$$f(\mathbf{x}^0 + \mathbf{h}) - f(\mathbf{x}^0) = \tfrac{1}{2}\mathbf{h}H[\theta\mathbf{x}_0 + (1 - \theta)(\mathbf{x}^0 + \mathbf{h})]\mathbf{h} \tag{47}$$

From Eq. (47) it is clear that $f(\mathbf{x}^0 + \mathbf{h}) - f(\mathbf{x}^0)$ will have the same sign as $\mathbf{h}'H[\theta\mathbf{x}^0 + (1 - \theta)(\mathbf{x}^0 + \mathbf{h})\mathbf{h}$. Since we have assumed the existence and continuity of the second partial derivative of $f(\cdot)$, it is clear that $\partial^2 f(\mathbf{x}^0)/\partial x_i \partial x_j$ will have the same sign as

$$\frac{\partial^2 f}{\partial x_i \partial x_j}[\theta\mathbf{x}^0 + (1 - \theta)(\mathbf{x}^0 + \mathbf{h})]$$

which are the components of the Hessian matrix, providing we are in an ϵ-

[3] A matrix P is said to be *positive definite* if the associated quadratic form $\mathbf{y}'P\mathbf{y}$ is always positive for any $\mathbf{y}$ except $\mathbf{y} = 0$. A matrix is *negative definite* if the quadratic form is always negative except for $\mathbf{y} = \mathbf{0}$. It is said to be *indefinite* if it is positive for some $\mathbf{y}$ and negative for others.

neighborhood of $\mathbf{x}^0$, $0 < \| \mathbf{h} \| < \epsilon$. This being the case, if $\mathbf{h}'H[\mathbf{x}^0]\mathbf{h}$ is positive, then $f(\mathbf{x}^0 + \mathbf{h}) - f(\mathbf{x}^0)$ will also be positive for $0 < \| \mathbf{h} \| < \epsilon$, and therefore $f(\mathbf{x})$ will be a minimum at $\mathbf{x}^0$. If H is positive definite, then this will be the case. A similar argument can be made for the negative definite case and a maximum.

Example. Find the minimum of

$$f(\mathbf{x}) = 3x_1^2 + x_2^2 + 2x_3^2 - 12x_1 - 6x_2 - 8x_3 + 140$$

$$\frac{\partial f}{\partial x_1} = 6x_1 - 12 = 0, \qquad x_1 = 2$$

$$\frac{\partial f}{\partial x_2} = 2x_2 - 6 = 0, \qquad x_2 = 3$$

$$\frac{\partial f}{\partial x_3} = 4x_3 - 8 = 0, \qquad x_3 = 2$$

There is one solution and it is $\mathbf{x}^0 = (x_1{}^0, x_2{}^0, x_3{}^0) = (2, 3, 2)$. Is this point a maximum or minimum or neither? We evaluate the second derivatives at $\mathbf{x}^0$:

$$\frac{\partial^2 f}{\partial x_1{}^2} = 6, \qquad \frac{\partial^2 f}{\partial x_1 \partial x_2} = 0, \qquad \frac{\partial^2 f}{\partial x_1 \partial x_3} = 0$$

$$\frac{\partial^2 f}{\partial x_2{}^2} = 2, \qquad \frac{\partial^2 f}{\partial x_2 \partial x_3} = 0, \qquad \frac{\partial^2 f}{\partial x_3{}^2} = 4$$

Therefore the Hessian matrix at $\mathbf{x}^0 = (2, 3, 2)$ is

$$H[\mathbf{x}^0] = \begin{bmatrix} 6 & 0 & 0 \\ 0 & 2 & 0 \\ 0 & 0 & 4 \end{bmatrix}$$

which is clearly a positive definite matrix. Therefore (2, 3, 2) is a minimum point.

Unfortunately, matters are a good deal more complicated than the example and Theorem 13 indicate. The case we have not yet considered is when the Hessian is semidefinite. By a positive semidefinite matrix P we mean that the associated quadratic form $\mathbf{y}'P\mathbf{y} \geq 0$ and at least one $\mathbf{y} \neq \mathbf{0}$ exists such that $\mathbf{y}'P\mathbf{y} = 0$. The inequality is reversed for a negative semidefinite matrix. There is a long involved history of attempts to resolve this problem involving such mathematicians as Lagrange, Peano, and Serret. The problem was finally resolved by Sheffer. A complete account of this theory can be found in Ref. 5. It will not be discussed here because it is of no computational importance. Indeed, even the sufficient conditions for the case where the Hessian is positive or negative definite are seldom employed in practice because of the complexity of the calculations. In applications, what one does is calculate all solutions to the necessary conditions and then evaluate the function for each solution. The least is the unconstrained minimum and the greatest is the unconstrained maximum.

It is not a simple matter in general to solve sets of simultaneous nonlinear equations, which is what is required to find a point that satisfies the necessary conditions

$$\frac{\partial f(\mathbf{x}^0)}{\partial x_j} = 0, \qquad j = 1, 2, \ldots, n \tag{48}$$

Methods for solving simultaneous nonlinear equations are described in standard texts on numerical analysis such as Refs. 3 and 4.

We turn now to considering optima of convex and concave functions of several variables. The following theorems are of use in aiding us in searching for global optima.

Theorem 14. Let $f(\cdot)$ be a convex function over a closed convex set X in E^n. Then if $x^0 \in X$ is a local minimum of $f(\cdot)$, it is also the global minimum of $f(\cdot)$.

Proof: This proof is by contradiction. Let us assume that $f(\cdot)$ has a local minimum at $\mathbf{x}^0 \in X$. We assume further that the global minimum over X occurs at a different point, $\mathbf{x}^* \neq \mathbf{x}^0$, and that $f(\mathbf{x}^0) > f(\mathbf{x}^*)$. Since $f(\cdot)$ is a convex function over X, and $\mathbf{x}^0$ and $\mathbf{x}^*$ are in X, then it must be true that

$$f[\lambda \mathbf{x}^* + (1 - \lambda)\mathbf{x}^0] \leq \lambda f(\mathbf{x}^*) + (1 - \lambda)f(\mathbf{x}^0), \qquad 0 \leq \lambda \leq 1 \tag{49}$$

By hypothesis, $f(\mathbf{x}^0) > f(\mathbf{x}^*)$. Therefore, if we substitute $f(\mathbf{x}^0)$ for $f(\mathbf{x}^*)$ on the right-hand side of Eq. (49), we strengthen the inequality and we have

$$f[\lambda \mathbf{x}^* + (1 - \lambda)\mathbf{x}^0] < \lambda f(\mathbf{x}^0) + (1 - \lambda)f(\mathbf{x}^0) = f(\mathbf{x}^0) \tag{50}$$

Now consider any ϵ-neighborhood of $\mathbf{x}^0$ such that $\| \mathbf{x}^* - \mathbf{x}^0 \| < \epsilon$. If λ is chosen so that $0 < \lambda < \epsilon / \| \mathbf{x}_0^* - \mathbf{x}^0 \|$, then we see that $\mathbf{x} = \lambda \mathbf{x}^* + (1 - \lambda)\mathbf{x}^0$ is in this ϵ-neighborhood of $\mathbf{x}^0$ and from Eq. (50):

$$f(\mathbf{x}) = f[\lambda \mathbf{x}^* + (1 - \lambda)\mathbf{x}^0] < f(\mathbf{x}^0) \tag{51}$$

Equation (50) contradicts the hypothesis that $\mathbf{x}^0$ is a local minimum. Hence $f(\mathbf{x}^*) < f(\mathbf{x}^0)$ is false and we have shown that any local minimum is a global minimum.

In a completely analogous fashion to Theorem 14 we can show that any local maximum of a concave function $f(\cdot)$ over a closed convex set X in E^n is also the global maximum over X.

We turn now to a consideration of the reverse problem, viz., finding the global maximum of a convex function over a convex set.

Theorem 15. Let X be a closed convex set bounded from below and let $f(\cdot)$ be a convex function over X in E^n. If $f(\cdot)$ has global maxima at points $\mathbf{x}^0$ with $\| \mathbf{x}^0 \|$ finite, then one or more of the $\mathbf{x}^0$ are extreme points[4] of X.

[4] An extreme point $\mathbf{x}^E$ of a convex set is a point that *cannot* be represented as a convex combination of two other points in the set, i.e., $\mathbf{x}^E \neq \lambda \mathbf{x}^1 + (1 - \lambda)\mathbf{x}^2$, $0 < \lambda < 1$. The vertices of a triangle in E^2 are examples of extreme points of the set consisting of the triangle and its interior.

Proof: Let $\mathbf{x}^0$ be a point in X at which $f(\mathbf{x})$ takes on its global maximum. We shall consider two cases.

Case 1: Suppose $\mathbf{x}^0$ is in the interior of X. For this case we claim that $f(\mathbf{x})$ is constant over all of X. We prove this by contradiction. Suppose it is not true. Then let $\hat{\mathbf{x}} \in X$ by any point such that $f(\hat{\mathbf{x}}) \neq f(\mathbf{x}^0)$. Since $\mathbf{x}^0$ is the global maximum, then $f(\hat{\mathbf{x}}) < f(\mathbf{x}^0)$. Since X is a convex set and $\mathbf{x}^0$ is in the interior of X, there must exist some point, say $\mathbf{x}^1 \in X$ and $0 < \lambda < 1$ such that

$$\mathbf{x}^0 = \lambda \mathbf{x}^1 + (1 - \lambda)\hat{\mathbf{x}} \tag{52}$$

However, since $f(\cdot)$ is a convex function, we have

$$f(\mathbf{x}^0) = f[\lambda \mathbf{x}^1 + (1 - \lambda)\hat{\mathbf{x}}] \leq \lambda f(\mathbf{x}^1) + (1 - \lambda) f(\hat{\mathbf{x}}) \tag{53}$$

Since $f(\mathbf{x}^0) > f(\hat{\mathbf{x}})$, if we substitute $f(\mathbf{x}^0)$ for $f(\hat{\mathbf{x}})$ in Eq. (53), we strengthen the inequality, and rearranging we have

$$\lambda f(\mathbf{x}^1) + (1 - \lambda) f(\mathbf{x}^0) > f(\mathbf{x}^0) \tag{54}$$

Simplifying (54) and noting that $\lambda > 0$, we have

$$f(\mathbf{x}^1) > f(\mathbf{x}^0) \tag{55}$$

which is a contradiction since $\mathbf{x}^0$ is the global maximum. Therefore if $\mathbf{x}^0$ is in the interior of X, $f(\mathbf{x})$ is constant over X and the theorem is trivially true for this case.

Case 2: Suppose $\mathbf{x}^0$ is on the boundary of X. We must show that $\mathbf{x}^0$ is an extreme point. If $\mathbf{x}^0$ is on the boundary of X, which is a convex set, then there is a well-known result of convex set theory which states that if $\mathbf{x}^0$ is not an extreme point, then there exists a supporting hyperplane to X at $\mathbf{x}^0$. (See p. 80 of Ref. 2.) Let S_1 be that supporting hyperplane and consider the intersection of S and S_1, i.e., $T_1 = S_1 \cap X$. T_1 is a closed convex set since the intersection of closed convex sets is a closed convex set. Furthermore, T_1 is not empty and contains at least one extreme point of X since X is bounded from below. Now consider T_1. T_1 lies in a space of dimension one lower than that of the space in which S_1 lies. Then T_1 is a set in a space of dimension $n - 1$. If $\mathbf{x}^0$ is in the interior of T_1, then we apply Case 1 above to prove that $f(\mathbf{x})$ is constant over T_1 and hence assumes its maximum at an extreme point of X. If $\mathbf{x}^0$ is on the boundary of T_1, we reapply the preceding argument and consider the intersection of the supporting hyperplane S_2 and T_1, i.e., $T_2 = S_2 \cap T_1$ at the point $\mathbf{x}^0$. This is a closed convex set and is nonempty, and the foregoing argument can be repeated. At most we need to apply the argument $n - 1$ times and, if we do, we obtain $T_n = \{\mathbf{x}^0\}$ and so, obviously, $\mathbf{x}^0$ is an extreme point of X.

CONSTRAINED OPTIMIZATION

In the preceding section we considered finding optima of functions where there were no constraints of any kind on the variables. We shall now consider the case where there are various kinds of constraints on the variables $\mathbf{x}$. The simplest kinds of constraints are of the form

$$a_j \le x_j \le b_j, \qquad j \in J \tag{56}$$

where J is the set of some or all of the indices of the components of $\mathbf{x}$. Often the lower bounds are zero for problems where the variables have physical significance, i.e., $x_j \ge 0$.

The most general types of constraints are of the form

$$g(\mathbf{x}) \{\le, =, \ge\} b \tag{57}$$

where it is understood in Eq. (57) that only one of the relations in brackets holds. Hence the general constrained optimization problem can be written as

$$\text{Max } f(\mathbf{x})$$

$$\text{subject to:} \quad g_i(\mathbf{x}) \le b_i, \qquad i = 1, 2, \ldots, m \tag{58}$$

It should be noted that the problem statement given in (58) is completely general. Constraints of the form of Eq. (56) are included since the $g_i(\cdot)$ can be any functions. Furthermore, all inequalities can be converted to the form given in (58) since $g_k(\mathbf{x}) \ge b_k$ can be converted to $-g_k(\mathbf{x}) \le -b_k$. Similarly, $g_k(\mathbf{x}) = b_k$ can be written as the pair of inequalities $g_k(\mathbf{x}) \le b_k$ and $g_k(\mathbf{x}) \ge b_k$.

Let us consider first how we might extend the methods discussed for unconstrained optimization to solve the problem:

$$\text{Max } z = f(\mathbf{x}), \qquad \mathbf{a} \le \mathbf{x} \le \mathbf{b} \tag{59}$$

$\mathbf{x}$, $\mathbf{a}$, and $\mathbf{b}$ are n-component vectors. We shall see that the procedure to solve this problem is much more complicated than for the unconstrained case. Since the previously described necessary and sufficient conditions will find only relative (local) maxima, we must also examine the possibility that a *constrained local maximum* may exist, i.e., one where one or more of the x_j are at their bound a_j or b_j. In addition, we must consider the possibility that an optimum may occur at a point on one of the bounding hyperplanes of the admissiable region, $\mathbf{a} \le \mathbf{x} \le \mathbf{b}$. We now give a general procedure for doing this. In order to do so we first define some convenient notation. Since there are 2^n combinations of the a_j and b_j (extreme points of the bounding hyperplanes of the admissible region), let us designate these $N = 2^n$ combinations as the vectors $\mathbf{y}_1, \mathbf{y}_2, \ldots, \mathbf{y}_N$. For example, if $n = 3$,

$$\mathbf{y}_1 = [\mathbf{a}_1, \mathbf{a}_2, \mathbf{a}_3], \mathbf{y}_2 = [\mathbf{b}_1, \mathbf{a}_2, \mathbf{a}_3], \mathbf{y}_3 = [\mathbf{a}_1, \mathbf{b}_2, \mathbf{a}_3], \ldots, \mathbf{y}_8 = [\mathbf{b}_1, \mathbf{b}_2, \mathbf{b}_3]$$

Also we define:

$$\begin{aligned} f_{jL}(\mathbf{x}) &= f(\mathbf{x}) \wedge (x_j = a_j), \qquad j = 1, 2, \ldots, n \\ f_{jU}(\mathbf{x}) &= f(\mathbf{x}) \wedge (x_j = b_j), \qquad j = 1, 2, \ldots, n \end{aligned} \tag{60}$$

Procedure for Finding Global Maximum of $z = f(\mathbf{x})$, $\mathbf{a} \le \mathbf{x} \le \mathbf{b}$

1. Compute $f(\mathbf{y}_k)$, $k = 1, 2, \ldots, N$.
2. Compute all solutions of $\partial f/\partial x_j = 0, j = 1, 2, \ldots, n$.

3. If any of these solutions satisfy $\mathbf{a} \le \mathbf{x} \le \mathbf{b}$, designate these $\mathbf{x}_i$, $i \in S$.
4. For these $\mathbf{x}_i$, $i \in S$, use sufficient conditions to eliminate minima and stationary points that are not optima. Designate these $\mathbf{x}_i$, $i \in S_1$.
5. Compute $z_1 = \text{Max}\,[f(x_i),\ i \in S_1;\ f(\mathbf{y}_k),\ k = 1, 2, \ldots, N]$.
6. Compute all solutions to $\partial f_{jL}/\partial x_t = 0$, $\partial f_{jU}/\partial x_t = 0, j, t = 1, 2, \ldots, n; j \neq t$.
7. If any of these solutions satisfy $\mathbf{a} \le \mathbf{x} \le \mathbf{b}$, designate these $\mathbf{x}_h$, $h \in H$.
8. For $\mathbf{x}_h$, $h \in H$, use sufficient conditions to eliminate minima and stationary points that are not optima. Designate these $\mathbf{x}_h$, $h \in H_1$.
9. Compute $z_2 = \text{Max}\,[f(\mathbf{x}_h),\ h \in H_1]$.
10. Global maximum $z^* = \text{Max}\,(z_1, z_2)$.

It is readily seen that the above procedure is a lengthy and arduous one for any function of more than a few variables. Consider the following example.

Example. Find $\mathbf{x} = (x_1, x_2)$ which maximizes

$$z = f(\mathbf{x}) = -2x_1^2 + 3x_1x_2 - x_2 + 3$$

subject to: $-4 \le x_1 \le 4$
$0 \le x_2 \le 3$

$$f(-4, 0) = -29, \qquad f(-4, 3) = -68, \qquad f(4, 0) = -29, \qquad f(4, 3) = 4$$

Next we look at the necessary conditions

$$\frac{\partial f}{\partial x_1} = -4x_1 + 3x_2 = 0$$
$$\frac{\partial f}{\partial x_2} = 3x_1 - 1 = 0$$

Therefore the only solution is $(x_1, x_2) = (\frac{1}{3}, \frac{4}{9})$. This solution satisfies the constraints $\begin{bmatrix} -4 \\ 0 \end{bmatrix} \le \mathbf{x} \le \begin{bmatrix} 4 \\ 3 \end{bmatrix}$. Therefore we check the sufficient conditions

$$\frac{\partial^2 f}{\partial x_1^2} = -4, \qquad \frac{\partial^2 f}{\partial x_2^2} = 0, \qquad \frac{\partial^2 f}{\partial x_1 \partial x_2} = 3$$

and we find that $H = \begin{bmatrix} -4 & 3 \\ 3 & 0 \end{bmatrix}$ which is indefinite. Hence $(x_1, x_2) = (\frac{1}{3}, \frac{4}{9})$ is not a maximum or minimum. We now consider $x_1 = -4$ and have $f_{1L} = -13x_2 - 29$, $\partial f_{1L}/\partial x_2 = -13$, which yields no stationary point. $f_{1U} = 11x_2 - 29$, $\partial f_{1U}/\partial x_2 = 11$, and again yields no point. We now consider $f_{2L} = -2x_3^2 + 3$ and $\partial f_{2L}/\partial x_1 = -4x_1 = 0$ which yields $x_1 = 0$ and $x_2 = 0$. Therefore $f(0, 0) = 3$ is feasible. Similarly $f_{2U} = -2x_1^2 + 9x_1$ and $\partial f_{2U}/\partial x_1 = -4x_1 + 9 = 0$ and $x_1 = \frac{9}{4}$. Therefore $f(\frac{9}{4}, 3) = 10.125$. Hence we see that z^*, the global maximum, is $z^* = \text{Max}\,[-29, -68, -29, 4, 3, 10.125] = 10.125$ and is given by $x_1 = \frac{9}{4}$, $x_2 = 3$.

We consider next the optimization of functions whose variables are subject to equality constraints, i.e.,

$$\text{Max}\, f(\mathbf{x})$$

$$\text{subject to:} \quad g_i(\mathbf{x}) = b_i, \qquad i = 1, 2, \ldots, m \tag{61}$$

An obvious thought that may occur to the reader is that the equalities in Eq. (61) could be used (if $m < n$) to eliminate m of the variables and then we would be in a position to solve an unconstrained optimization problem. In fact, as long as the conditions of the implicit function theorem (Theorem 6 in this article) are satisfied, this can be done, *in principle*. However, it may not be possible to perform such elimination in practice. Consider the following examples.

Examples.

1.

$$\text{Max } f(\mathbf{x}) = -3x_1^2 - 4x_2^3$$

subject to: $2x_1 + x_2 = 8$

This can be obviously transformed, using $x_2 = 8 - 2x_1$, to

$$\text{Max } f(\mathbf{x}) = -3x_1^2 - 4(8 - 2x_1)^3 = h(x_1)$$

and is now an unconstrained problem.

2.

$$\text{Min } f(\mathbf{x}) = x_1^2 + 5x_1x_2$$

subject to: $x_1 \cos x_2 + x_2 \sin x_1 = 4$

Here we cannot solve for either x_1 or x_2 to eliminate a variable.

The preceding examples point up the necessity to have a method in which constraints are taken into account explicitly rather than implicitly by substitution. Such a method exists and is known as the *Lagrange multiplier method*. We shall first develop the theory for a function of two variables subject to a single constraint and then generalize it to a function of n variables subject to m constraints $(m < n)$.

Suppose we wish to maximize $f(x_1, x_2)$ subject to the constraint $g(x_1, x_2) = b$. We assume that $f(\cdot)$ and $g(\cdot)$ are continuous and differentiable. If it was possible to obtain an explicit representation of x_2 from $g(x_1, x_2) = b$, then we would have $x_2 = h(x_1)$. If such a function as $h(\cdot)$ did exist, we know from the implicit function theorem that $h(\cdot)$ is differentiable and has a continuous first derivative. Hence we could obtain the following unconstrained optimization problem by substitution for x_2:

$$\text{Max } f[x_1, h(x_1)] \tag{62}$$

The problem stated in (62) is an unconstrained optimization problem and we know from Theorem 7 that a necessary condition for $f(\cdot)$ to have a maximum (or minimum) at some point $\mathbf{x}^0 = (x_1^0, x_2^0)$ is that the first derivative be zero at $\mathbf{x}^0$. We may write this as

$$\frac{d}{dx_1} f(x_1, x_2) = \frac{\partial f(x_1, x_2)}{\partial x_1} + \frac{\partial f(x_1, x_2)}{\partial x_2}\frac{dx_2}{dx_1} \tag{63}$$

However, since $x_2 = h(x_1)$, we can substitute this into Eq. (63) and evaluate the derivatives at $\mathbf{x}^0$ to obtain

$$\frac{d}{dx_1} f(x_1^0, x_2^0) = \frac{\partial f(x_1^0, x_2^0)}{\partial x_1} + \frac{\partial f(x_1^0, x_2^0)}{\partial x_2}\frac{dh(x_1)}{dx_1} \tag{64}$$

and by Theorem 7 we know the expression in (64) must be zero if $\mathbf{x}^0$ is to be a stationary point. Recall that we are trying to avoid having to find $h(x_1)$ explicitly. Hence we need a way to eliminate dh/dx_1 from Eq. (64). We shall use the constraint $g(x_1, x_2) = b$ to do this. If we differentiate the constraint, we have

$$\frac{dg(x_1, x_2)}{dx_1} = \frac{\partial g(x_1, x_2)}{\partial x_1} + \frac{\partial g(x_1, x_2)}{\partial x_2}\frac{dh(x_1)}{dx_1} = 0 \tag{65}$$

if $x_2 = h(x_1)$ existed. We can rearrange Eq. (65) to obtain

$$\frac{dg(x_1)}{dx_1} = -\frac{\partial g(x_1, x_2)}{\partial x_1}\Big/\frac{\partial g(x_1, x_2)}{\partial x_2} \tag{66}$$

We now substitute from Eq. (66), evaluated at $\mathbf{x}^0$, into Eq. (64) and set the result equal to zero to obtain

$$\frac{\partial f(x_1{}^0, x_2{}^0)}{\partial x_1} - \left(\frac{\partial f(x_1{}^0, x_2{}^0)}{\partial x_2}\right)\left(\frac{\partial g(x_1{}^0, x_2{}^0)}{\partial x_1}\Big/\frac{\partial g(x_1{}^0, x_2{}^0)}{\partial x_2}\right) = 0 \tag{67}$$

We now define a variable λ as

$$\lambda = \frac{\partial f(x_1{}^0, x_2{}^0)}{\partial x_2}\Big/\frac{\partial g(x_1{}^0, x_2{}^0)}{\partial x_2} \tag{68}$$

Substitution from Eq. (68) into (67) yields

$$\frac{\partial f(x_1{}^0, x_2{}^0)}{\partial x_1} - \lambda\frac{\partial g(x_1{}^0, x_2{}^0)}{\partial x_1} = 0 \tag{69}$$

The definition of λ from Eq. (68) yields

$$\frac{\partial f(x_1{}^0, x_2{}^0)}{\partial x_2} - \lambda\frac{\partial g(x_1{}^0, x_2{}^0)}{\partial x_2} = 0 \tag{70}$$

and the original constraint must also be satisfied, i.e.,

$$g(x_1{}^0, x_2{}^0) = b \tag{71}$$

Equations (69) through (71) constitute a set of necessary conditions for the existence of a solution to our original problem, Max $f(x_1, x_2)$ subject to $g(x_1, x_2) = b$. We need to solve three equations, (69) through (71), in three variables, x_1, x_2, and λ, in order to find the stationary point $\mathbf{x}^0$. In the process we also determine λ. The value of this approach is quite clear, viz., that we do not have to determine $h(x_1)$, although we assumed its existence when we made use of the implicit function theorem.

Formally, the way we arrive at the foregoing necessary conditions for any specific problem is to construct what is known as the *Lagrangian function*, which is defined as

$$F(x_1, x_2, \lambda) = f(x_1, x_2) + \lambda[b - g(x_1, x_2)] \tag{72}$$

We treat $F(\cdot)$ as an unconstrained function, differentiate it with respect to x_1, x_2, and λ, and equate the resulting expressions to zero to obtain

$$\begin{aligned}
\frac{\partial F(x_1, x_2, \lambda)}{\partial x_1} &= \frac{\partial f(x_1, x_2)}{\partial x_1} - \lambda\frac{\partial g(x_1, x_2)}{\partial x_1} = 0 \\
\frac{\partial F(x_1, x_2, \lambda)}{\partial x_2} &= \frac{\partial f(x_1, x_2)}{\partial x_2} - \lambda\frac{\partial g(x_1, x_2)}{\partial x_2} = 0 \\
\frac{\partial F(x_1, x_2, \lambda)}{\partial \lambda} &= b - g(x_1, x_2) = 0
\end{aligned} \tag{73}$$

It can be seen that Eqs. (69) through (71) are identical with the set given by Eqs. (73).

Example. Suppose we wish to solve

$$\text{Min } f(\mathbf{x}) = 4x_1{}^2 + 5x_2{}^2$$

subject to: $3x_1 - 2x_2 = 14$

The Lagrangian function is

$$\begin{aligned}
F(x_1, x_2, \lambda) &= 4x_1{}^2 + 5x_2{}^2 + \lambda[14 - 3x_1 + 2x_2] \\
\frac{\partial F(\mathbf{x}, \lambda)}{\partial x_1} &= 8x_1 - 3\lambda = 0 \\
\frac{\partial F(\mathbf{x}, \lambda)}{\partial x_2} &= 10x_2 + 2\lambda = 0 \\
\frac{\partial F(\mathbf{x}, \lambda)}{\partial \lambda} &= 14 - 3x_1 + 2x_2 = 0
\end{aligned} \tag{74}$$

Solving the three equations of (74) we obtain

$$x_1 = \frac{210}{61}, \qquad x_2 = -\frac{112}{61}, \qquad \lambda = \frac{560}{61}$$

It can be verified that this is a minimum. However, we will defer a discussion of sufficient conditions until we generalize the above results to any number of variables and constraints.

We now consider the following general problem with the purpose of extending the Lagrange multiplier approach.

$$\text{Max } f(\mathbf{x})$$

subject to: $g_i(\mathbf{x}) = b_i, \qquad i = 1, 2, \ldots, m < n$ (75)

where $\mathbf{x} = (x_1, x_2, \ldots, x_n)$. We know that at a relative maximum $\mathbf{x}^0$, a necessary condition is that $\mathbf{x}^0$ be a stationary point of $f(\cdot)$ and hence that the first partial derivatives of $f(\mathbf{x})$ be zero. This can be expressed in the form of a total differential:

$$df(\mathbf{x}) = \sum_{j=1}^{n} \frac{\partial f(\mathbf{x}^0)}{\partial x_j}\, dx_j = 0 \tag{76}$$

Obviously, Eq. (76) does not suffice for the problem given by (75) because a relative maximum could occur on the boundary of one of the constraints. At such a point it is not necessarily true that all the first partial derivatives of $f(\mathbf{x})$ vanish. Let us then consider the total differential of the constraints $g_i(\mathbf{x}) = b_i$, $i = 1, 2, \ldots, m$. This is given by

$$dg_i(\mathbf{x}) = \sum_{j=1}^{n} \frac{\partial g_i(\mathbf{x})}{\partial x_j}\, dx_j = 0, \qquad i = 1, 2, \ldots, m \tag{77}$$

Let us now multiply each of the functions in Eq. (77) by a *Lagrange multiplier*, λ_i, and subtract the result from the total differential $df(\mathbf{x})$ given by Eq. (76). We then obtain

$$df(\mathbf{x}) - \sum_{i=1}^{m} \lambda_i\, dg_i(\mathbf{x}) = \sum_{j=1}^{n} \left[\frac{\partial f(\mathbf{x})}{\partial x_j} - \sum_{i=1}^{m} \lambda_i \frac{\partial g_i(\mathbf{x})}{\partial x_j}\right] dx_j = 0 \tag{78}$$

Without loss of generality, let us assume that we can eliminate the first m variables $x_1, x_2, \ldots, x_m$. If this is so, then we can express these variables in terms of the remaining $n - m$ variables, which can be regarded as independent variables. Hence the dx_j, $j = m + 1, m + 2, \ldots, n$, can be considered independent variables in Eq. (78). If this is the case, then it is clear from Eq. (78) that

$$\frac{\partial f(\mathbf{x})}{\partial x_j} - \sum_{i=1}^{m} \frac{\partial g_i(\mathbf{x})}{\partial x_j} = 0, \qquad j = m + 1, m + 2, \ldots, n \tag{79}$$

Since the terms in Eq. (79) are zero, if we subtract them from Eq. (78) we obtain

$$\sum_{j=1}^{m} \left[\frac{\partial f(\mathbf{x})}{\partial x_j} - \sum_{i=1}^{m} \lambda_i \frac{\partial g_i(\mathbf{x})}{\partial x_j}\right] dx_j = 0 \tag{80}$$

We know that the dx_j, $j = 1, 2, \ldots, m$, are not independent but are uniquely determined by the values of the dx_j, $j = m + 1, m + 2, \ldots, n$. Hence, in order to satisfy Eq. (80), the coefficients of the dx_j, $j = 1, 2, \ldots, m$, must also be zero. Therefore

$$\frac{\partial f(\mathbf{x})}{\partial x_j} - \sum_{i=1}^{m} \lambda_i \frac{\partial g_i(\mathbf{x})}{\partial x_j} = 0, \qquad j = 1, 2, \ldots, m \tag{81}$$

Combining (79) and (81) and including the original constraints, we see that at a relative maximum (or minimum) $\mathbf{x}^0$, the following must be satisfied:

$$\begin{aligned} \frac{\partial f(\mathbf{x}^0)}{\partial x_j} - \sum_{i=1}^{m} \lambda_i \frac{\partial g_i(\mathbf{x}^0)}{\partial x_j} &= 0, \qquad j = 1, 2, \ldots, n \\ g_i(\mathbf{x}^0) - b_i &= 0, \qquad i = 1, 2, \ldots, m \end{aligned} \tag{82}$$

Equations (82) are $m + n$ equations in $m + n$ variables.

The necessary conditions given by Eqs. (82) can be found from a Lagrangian function. This function, for the problem given by (75), is

$$F(\mathbf{x}, \boldsymbol{\lambda}) = f(\mathbf{x}) + \sum_{i=1}^{m} \lambda_i[b_i - g_i(\mathbf{x})] \tag{83}$$

We can now prove, very simply, the following theorem.

Theorem 16. A necessary condition that $f(\mathbf{x})$ have a relative maximum or minimum at $\mathbf{x}^0$, subject to the constraints that $g_i(\mathbf{x}) = b_i$, $i = 1, 2, \ldots, m$, is that the first partial derivatives of the Lagrangian function, with respect to each of the arguments of $F(\cdot)$, vanish.

Proof: If we take the first partial derivatives of $F(\mathbf{x}, \boldsymbol{\lambda})$ with respect to each of the x_j, $j = 1, 2, \ldots, n$, and each of λ_i, $i = 1, 2, \ldots, m$, and equate the result to zero we obtain

$$\begin{aligned} \frac{\partial F(\mathbf{x}, \boldsymbol{\lambda})}{\partial x_j} &= \frac{\partial f(\mathbf{x})}{\partial x_j} - \sum_{i=1}^{m} \lambda_i \frac{\partial g_i(\mathbf{x})}{\partial x_j} = 0, \qquad && j = 1, 2, \ldots, n \\ \frac{\partial F(\mathbf{x}, \boldsymbol{\lambda})}{\partial \lambda_i} &= b_i - g_i(\mathbf{x}) = 0, && i = 1, 2, \ldots, m \end{aligned} \tag{84}$$

Equations (84) are identical to Eqs. (82).

It should be noted that in the derivation of the necessary conditions we assumed that the first m variables (more generally, that there was at least one set of m variables out of n) could be expressed in terms of the remaining $n - m$ variables. This is equivalent to the assumption that the rank of the $m \times n$ matrix $\| \partial g_i / \partial x_j \|$ is equal to m at $\mathbf{x}^0$. If this is not true, we have a much more complex situation which will be discussed subsequently.

Example. $\text{Min } f(\mathbf{x}) = x_1^2 + 2x_2^2 + 3x_3^2$

subject to:

$$\begin{aligned} x_1 + 2x_2 + 4x_3 &= 12 \\ 2x_1 + x_2 + 3x_3 &= 10 \end{aligned}$$

$$F(\mathbf{x}, \boldsymbol{\lambda}) = x_1^2 + 2x_2^2 + 3x_3^2 + \lambda_1[12 - x_1 - 2x_2 - 4x_3] + \lambda_2[10 - 2x_1 - x_2 - 3x_3]$$

$$\begin{aligned} \frac{\partial F}{\partial x_1} &= 2x_1 - \lambda_1 - 2\lambda_2 = 0 \\ \frac{\partial F}{\partial x_2} &= 4x_2 - 2\lambda_1 - \lambda_2 = 0 \\ \frac{\partial F}{\partial x_3} &= 6x_2 - 4\lambda_1 - 3\lambda_2 = 0 \\ \frac{\partial F}{\partial \lambda_1} &= 12 - x_1 - 2x_2 - 4x_3 = 0 \\ \frac{\partial F}{\partial \lambda_2} &= 10 - 2x_1 - x_2 - 3x_3 = 0 \end{aligned} \tag{85}$$

The solution to Eqs. (85) is as follows:

$$x_1 = \frac{12}{11}, \quad x_2 = \frac{8}{11}, \quad x_3 = \frac{26}{11}, \quad \lambda_1 = \frac{56}{11}, \quad \lambda_2 = \frac{-16}{11}$$

After we develop sufficient conditions for Lagrange multiplier solutions, we will show that this is indeed a minimum point.

Consider now our original problem:

$$\text{Max } f(\mathbf{x})$$

$$\text{subject to:} \quad g_i(\mathbf{x}) = b_i, \quad i = 1, 2, \ldots, m \tag{86}$$

If $\mathbf{x}^0$ is to be a solution to this problem, then $g_i(\mathbf{x}^0) = b_i$, $i = 1, 2, \ldots, m$, and therefore $b_i - g_i(\mathbf{x}^0) = 0$, $i = 1, 2, \ldots, m$. Hence we have for such a solution point that

$$F(\mathbf{x}^0, \boldsymbol{\lambda}^0) = f(\mathbf{x}^0) + \sum_{i=1}^{m} [b_i - g_i(\mathbf{x}^0)] = f(\mathbf{x}^0) \tag{87}$$

We also know that if $\mathbf{x}^0$ is a local maximum of $f(\cdot)$, then $f(\mathbf{x}^0 + \mathbf{h}) - f(\mathbf{x}^0) < 0$ in a neighborhood of $\mathbf{x}^0$. However, unlike the unconstrained case, these are additional constraints on the components of $\mathbf{h}$ because of the constraints on $\mathbf{x}$, i.e., the constraints evaluated at $\mathbf{x}^0 + \mathbf{h}$ must also be satisfied. Hence

$$g_i(\mathbf{x}^0 + \mathbf{h}) = b_i, \quad i = 1, 2, \ldots, m$$

must also be satisfied. We may then write that

$$F(\mathbf{x}^0 + \mathbf{h}, \boldsymbol{\lambda}^0) - F(\mathbf{x}^0, \boldsymbol{\lambda}^0) = f(\mathbf{x}^0 + \mathbf{h}) - f(\mathbf{x}^0) \tag{88}$$

If we expand $F(\cdot)$ in a Taylor series about $\mathbf{x}^0$, from Theorem 5 we have

$$F(\mathbf{x}^0 + \mathbf{h}, \boldsymbol{\lambda}^0) - F(\mathbf{x}^0, \boldsymbol{\lambda}^0) = \sum_{j=1}^{n} \frac{\partial F(\mathbf{x}^0, \boldsymbol{\lambda}^0)}{\partial x_j} h_j + \frac{1}{2} \sum_{j=1}^{n} \sum_{k=1}^{n} h_j h_k \frac{\partial^2 F(\mathbf{x}^0, \boldsymbol{\lambda}^0)}{\partial x_j \partial x_k} + R_2(\mathbf{x}^0 + \theta\mathbf{h}, \boldsymbol{\lambda}^0) \tag{89}$$

From the necessary conditions we know that the first partial derivatives of F are zero. Hence Eq. (89) simplifies to:

$$F(\mathbf{x}^0 + \mathbf{h}, \boldsymbol{\lambda}^0) - F(\mathbf{x}^0, \boldsymbol{\lambda}^0) = \frac{1}{2} \sum_{j=1}^{n} \sum_{k=1}^{n} h_j h_k \frac{\partial^2 F(\mathbf{x}^0, \boldsymbol{\lambda}^0)}{\partial x_j \partial x_k} + R_2(\mathbf{x}^0 + \theta\mathbf{h}, \boldsymbol{\lambda}^0) \tag{90}$$

We now have the information to establish the following theorem.

Theorem 17. A sufficient condition for $f(\mathbf{x})$ to have a relative maximum at $\mathbf{x}^0$ subject to the constraints $g_i(\mathbf{x}) = b_i$, $i = 1, 2, \ldots, m$, is that the quadratic form

$$Q = \sum_{j=1}^{n} \sum_{k=1}^{n} h_j h_k \frac{\partial^2 F(\mathbf{x}^0, \boldsymbol{\lambda}^0)}{\partial x_j \partial x_k}$$

be negative definite for all values of $\mathbf{h}$ for which the constraints $g_i(\mathbf{x}) = b_i$, $i = 1, 2, \ldots, m$, are also satisfied.

Proof: Since a local maximum implies that $f(\mathbf{x}^0 + \mathbf{h}) - f(\mathbf{x}^0) < 0$, and we saw from Eq. (88) that $F(\mathbf{x}^0 + \mathbf{h}, \boldsymbol{\lambda}^0) - F(\mathbf{x}^0, \boldsymbol{\lambda}^0) = f(\mathbf{x}^0 + \mathbf{h}) - f(\mathbf{x}^0)$, then we see from Eq. (90) that if we can show that the quadratic form Q is negative definite, we have indeed established a sufficient condition. Suppose Q is not negative definite. In this case $F(\mathbf{x}^0 + \mathbf{h}, \boldsymbol{\lambda}^0) - F(\mathbf{x}^0, \boldsymbol{\lambda}^0)$ will not be negative for all admissible $\mathbf{h}$ and therefore, by definition, $\mathbf{x}^0$ is not a relative maximum. It is also clear that this needs to be true only for the values of $\mathbf{h}$ which satisfy the constraints because the necessary conditions specify (see Theorem 16) that the constraints be satisfied. Hence the theorem is proved.

We may similarly show that a sufficient condition for a relative minimum is that Q be positive definite.

It can be shown (see Ref. 5) that a necessary condition for the quadratic form Q to be negative definite (or positive definite) for all allowable values of $\mathbf{h}$ is that all the roots μ_j of a certain polynomial defined below be negative (or positive). We wish to find the roots of $P(\mu) = 0$, i.e.,

$$P(\mu) = \begin{vmatrix} (F_{11}{}^0 - \mu) & F_{12}{}^0 & \cdots & F_{1n}{}^0 & g_{11}{}^0 & g_{21}{}^0 & \cdots & g_{m1}{}^0 \\ F_{21}{}^0 & (F_{22}{}^0 - \mu) & \cdots & F_{2n}{}^0 & g_{12}{}^0 & g_{22}{}^0 & \cdots & g_{m2}{}^0 \\ \vdots & \vdots & & \vdots & \vdots & \vdots & & \vdots \\ F_{n1}{}^0 & F_{n2}{}^0 & \cdots & (F_{nn}{}^0 - \mu) & g_{1n}{}^0 & g_{2n}{}^0 & \cdots & g_{mn}{}^0 \\ g_{11}{}^0 & g_{12}{}^0 & \cdots & g_{1n}{}^0 & 0 & 0 & \cdots & 0 \\ g_{21}{}^0 & g_{22}{}^0 & \cdots & g_{2n}{}^0 & 0 & 0 & \cdots & 0 \\ \vdots & \vdots & & \vdots & \vdots & \vdots & & \vdots \\ g_{m1}{}^0 & g_{m2}{}^0 & \cdots & g_{mn}{}^0 & 0 & 0 & \cdots & 0 \end{vmatrix} = 0 \quad (91)$$

where $F_{jk}{}^0 = \partial^2 F(\mathbf{x}^0, \boldsymbol{\lambda}^0)/\partial x_j \partial x_k$ and $g_{ij}{}^0 = \partial g_i(\mathbf{x}^0)/\partial x_j$.

If each root μ_j of $P(\mu) = 0$ is negative, then $\mathbf{x}^0$ is a relative maximum. If each root is positive, then $\mathbf{x}^0$ is a relative minimum. If some roots are positive while others are negative, then $\mathbf{x}^0$ is not an extremal point.

Example. Let us examine the problem we solved in the previous example to see if the solution was, in fact, a minimum as we claimed.

$$\begin{array}{llll} F_{11} = 2, & F_{22} = 4, & F_{33} = 6 & F_{12} = F_{21} = F_{13} = F_{31} = F_{23} = F_{32} = 0 \\ g_{11} = 1, & g_{12} = 2, & g_{13} = 4 & \\ g_{21} = 2, & g_{22} = 1, & g_{23} = 3 & \end{array}$$

Therefore

$$P(\mu) = \begin{vmatrix} (2 - \mu) & 0 & 0 & 1 & 2 \\ 0 & (4 - \mu) & 0 & 2 & 1 \\ 0 & 0 & (6 - \mu) & 4 & 3 \\ 1 & 2 & 4 & 0 & 0 \\ 2 & 1 & 3 & 0 & 0 \end{vmatrix} = 0 \quad (92)$$

Simplifying Eq. (92) yields the polynomial

$$P(\mu) = 12\mu - 38 = 0$$

Therefore $\mu = \frac{38}{12} > 0$. Therefore $\mathbf{x}^0 = (\frac{12}{11}, \frac{8}{11}, \frac{26}{11})$ is a relative minimum.

We shall now return to one last matter regarding the use of Lagrange multipliers. That is the assumption, which we mentioned previously, that there was at least one mth order Jacobian, $\| \partial g_i(\mathbf{x})/\partial x_j \|$, which was nonsingular, i.e., whose determinant was not zero. Unless this is true, the previously stated necessary conditions do not hold. Before we treat this situation, let us consider an example taken from Ref. 6 which illustrates the problem.

Example. $\operatorname{Min} f(x_1, x_2) = x_1^2 + x_2^2$

subject to: $g(x_1, x_2) = (x_1 - 1)^3 - x^2 = 0$

The unconstrained minimum of f is at $(x_1, x_2) = (0, 0)$. However, this does not satisfy the constraint. It is clear, however, that $x_1 = 1$, $x_2 = 0$ will satisfy the constraint, and so $\mathbf{x}^0 = (1, 0)$. Now let us attempt to solve this problem using a Lagrange multiplier.

$$F = x_1^2 + x_2^2 + \lambda[-(x_1 - 1)^3 + x_2^2]$$

$$\frac{\partial F}{\partial x_1} = 2x_1 - 3\lambda(x_1 - 1)^2 = 0 \tag{93a}$$

$$\frac{\partial F}{\partial x_2} = 2x_2 - 2\lambda x_2 = 0 \tag{93b}$$

$$\frac{\partial F}{\partial \lambda} = -(x_1 - 1)^3 + x_2^2 = 0 \tag{93c}$$

It can be seen that $\mathbf{x} = (1, 0)$ satisfies Eqs. (93b) and (93c) but $\partial F/\partial x_1$ at (1, 0) equals 2. The point (1, 0) is a minimum but fails to satisfy the necessary conditions. The reason is that

$$\frac{\partial g}{\partial x_1} = 3(x_1 - 1)^2 = 0 \qquad \text{at } x_1 = 1$$

$$\frac{\partial g}{\partial x_2} = -2x_2 = 0 \qquad \text{at } x_2 = 0$$

Hence both Jacobians are singular and the condition under which we derived the necessary conditions has not been met.

A method for allowing for the situation when the Jacobian matrix is singular was developed originally by Caratheodory [7]. A discussion of Caratheodory's method can also be found in Hadley [8]. The basic idea is quite simple. Since the Lagrange multiplier method, as the foregoing example exhibits, generates only points where at least one Jacobian does not vanish but *not* those points where all the Jacobians vanish, we need to find a way to enlarge the set of points considered by the usual Lagrange multiplier method. This can be accomplished by modifying the definition of the Lagrangian as follows. We define $F(\mathbf{x}, \boldsymbol{\lambda})$ to be

$$F(\mathbf{x}, \boldsymbol{\lambda}) = \lambda_0 f(\mathbf{x}) + \sum_{i=1}^{m} \lambda_i[b_i - g_i(\mathbf{x})] \tag{94}$$

where λ_0 is a multiplier that may be either 0 or 1. When $\lambda_0 = 1$, we have the usual case, i.e., the necessary conditions under the assumption that at least one Jacobian does not vanish. When $\lambda_0 = 0$, we have the necessary conditions for the case where all Jacobians are zero.

The modified Lagrangian procedure then is quite simple. Define the Lagrangian function as in Eq. (94) and solve the usual equations

$$\frac{\partial F(\mathbf{x}, \boldsymbol{\lambda})}{\partial x_j} = 0, \qquad j = 1, 2, \ldots, n \tag{95}$$
$$\frac{\partial F(\mathbf{x}, \boldsymbol{\lambda})}{\partial \lambda_i} = 0, \qquad i = 1, 2, \ldots, m$$

under two conditions, both for $\lambda_0 = 1$ and for $\lambda_0 = 0$. Then we will generate all possible points to be considered as extrema.

Example. Let us reconsider our previous example:

$$\text{Min } f(x_1, x_2) = x_1{}^2 + x_2{}^2$$

subject to: $g(x_1, x_2) = (x - 1)^3 - x_2{}^2 = 0$

We define the Lagrangian as:

$$\begin{aligned} F(\mathbf{x}, \boldsymbol{\lambda}) &= \lambda_0(x_1{}^2 + x_2{}^2) + \lambda_1[-(x_1 - 1)^3 + x_2{}^2] \\ \frac{\partial F(\mathbf{x}, \boldsymbol{\lambda})}{\partial x_1} &= 2\lambda_0 x_1 - 3\lambda_1(x_1 - 1)^2 = 0 \\ \frac{\partial F(\mathbf{x}, \boldsymbol{\lambda})}{\partial x_2} &= 2\lambda_0 x_2 + 2\lambda_1 x_2 = 0 \\ \frac{\partial F(\mathbf{x}, \boldsymbol{\lambda})}{\partial \lambda_1} &= -(x_1 - 1)^3 + x_2{}^2 = 0 \end{aligned} \tag{96}$$

If we set $\lambda_0 = 1$ in Eqs. (96), we get the same problem we solved in the previous example, which failed to find the solution. Now, in addition, we set $\lambda_0 = 0$ to obtain

$$\begin{aligned} 3\lambda_1(x_1 - 1)^2 &= 0 \\ 2\lambda_1 x_2 &= 0 \\ (x_1 - 1)^3 - x_2{}^2 &= 0 \end{aligned} \tag{97}$$

Equations (97) have one nontrivial solution, $x_1 = 1$, $x_2 = 0$, which is the minimum we seek.

Our last consideration in constrained optimization by "classical" techniques will be treatment of inequality constraints on the variables. First we shall consider a simple extension of the Lagrangian methods we have discussed previously. Then we shall describe a more extensive revision known as the Kuhn-Tucker theory.

We consider the following problem:

$$\text{Max } z = f(\mathbf{x}) \tag{98}$$

subject to: $g_i(\mathbf{x}) \leq b_i, \qquad i = 1, 2, \ldots, m$

We can convert the inequalities in (98) to equalities as follows:

$$\begin{aligned} g_i(\mathbf{x}) + y_i &= b_i, \qquad i = 1, 2, \ldots, m \\ y_i &\geq 0, \qquad i = 1, 2, \ldots, m \end{aligned} \tag{99}$$

Using the equalities of (99) and the function $f(\cdot)$, we can form the Lagrangian function

$$F(\mathbf{x}, \mathbf{y}, \boldsymbol{\lambda}) = f(\mathbf{x}) + \sum_{i=1}^{m} \lambda_i[b_i - y_i - g_i(\mathbf{x})] \tag{100}$$

Let us consider the first partial derivatives of F with respect to y_i. We have

$$\frac{\partial F}{\partial y_i} = \lambda_i = 0, \qquad i = 1, 2, \ldots, m \tag{101}$$

Equation (101) will hold at a stationary point. Consider the implications of this equation. If $y_i > 0$, then $g_i(\mathbf{x}) < b_i$ and hence the constraint is not active, i.e., it need not be considered. However, if $y_i = 0$, the constraint is active, i.e., $g_i(\mathbf{x}) = b_i$ and it may be treated as an equality and λ_i may *not* be zero.

Another general observation that we can use in treating the problem given by (98) is the following. If we solve the unconstrained problem Max $z = f(\mathbf{x})$ and find that the optimal solution $\mathbf{x}^*$ also satisfies $g_i(\mathbf{x} \leq b_i$, $i = 1, 2, \ldots, m$, then we have also solved (98). This follows from the fact that if $z^* = f(\mathbf{x}^*)$, the addition of one or more constraints can only cause z to be less than or equal to z^*. It cannot increase z. If it did, it would contradict the fact that z^* was optimal for the unconstrained problem.

We will now use the two foregoing observations to state an algorithm for finding the global maximum of inequality constrained optimization problems of the form given by (98).

1. Solve the problem Max $z = f(\mathbf{x})$, ignoring the inequality constraints. Designate this solution $\mathbf{x}^0$.
2. If $\mathbf{x}^0$ satisfies $g_i(\mathbf{x}) \leq b_i$, $i = 1, 2, \ldots, m$, we are done. $\mathbf{x}^0$ is the global maximum.
3. If one or more of the constraints are not satisfied, select any constraint, say the first, and allow it to become active, i.e., let $y_i = 0$. Now solve the problem

   $$\text{Max } z = f(\mathbf{x})$$

   subject to: $\quad g_1(\mathbf{x}) = b_1$

   Call the optimal solution $\mathbf{x}^1$. If $\mathbf{x}^1$ satisfies all the constraints, we are done. If not, this procedure is repeated, allowing one constraint at a time to become active.
4. If Step 3 did not yield an optimum, we now allow all combinations of two constraints at a time to be active and again solve the resulting problems. If a solution is found that satisfies all constraints, we are done.

5. If Step 4 did not yield an optimum, we allow three constraints at a time to become active and so forth.

The above procedure is very time consuming for more than just a few variables, even if we assume that the computation can be carried out successfully.

Let us now consider the development of a set of necessary and sufficient conditions for inequality constrained optimization problems. This theory is often called the Kuhn-Tucker theory because it was first presented in Ref. 9 by Kuhn and Tucker.

Consider the problem

$$\text{Max } f(\mathbf{x})$$

$$\text{subject to:} \quad g_i(\mathbf{x}) \leq b_i, \qquad i = 1, 2, \ldots, m \tag{102}$$

Let us convert the inequalities to equalities by adding nonnegative slack variables y_i^2 (y_i^2 cannot be negative) to obtain

$$g_i(\mathbf{x}) + y_i^2 = b_i, \qquad i = 1, 2, \ldots, m \tag{103}$$

We now construct the Lagrangian function

$$F(\mathbf{x}, \mathbf{y}, \boldsymbol{\lambda}) = f(\mathbf{x}) + \sum_{i=1}^{m} \lambda_i [b_i - g_i(\mathbf{x}) - y_i^2] \tag{104}$$

and write down the necessary conditions for obtaining the stationary points of the Lagrangian function. These are

$$\frac{\partial F(\mathbf{x}, \mathbf{y}, \boldsymbol{\lambda})}{\partial x_j} = \frac{\partial f(\mathbf{x})}{\partial x_j} - \sum_{i=1}^{m} \lambda_i \frac{\partial g_i(\mathbf{x})}{\partial x_j} = 0, \qquad j = 1, 2, \ldots, n$$

$$\frac{\partial F(\mathbf{x}, \mathbf{y}, \boldsymbol{\lambda})}{\partial y_i} = -2\lambda_i y_i = 0, \qquad i = 1, 2, \ldots, m \tag{106}$$

$$\frac{\partial F(\mathbf{x}, \mathbf{y}, \boldsymbol{\lambda})}{\partial \lambda_i} = b_i - g_i(\mathbf{x}) - y_i^2 = 0, \qquad i = 1, 2, \ldots, m \tag{107}$$

From the last two sets of equations we can obtain that

$$\lambda_i [b_i - g_i(\mathbf{x})] = 0, \qquad i = 1, 2, \ldots, m \tag{108}$$

Equations (105) and (108) together with the constraints represent part of the necessary conditions for a maximum of $f(\mathbf{x})$. The final point that needs to be established is that $\lambda_i \leq 0$, $i = 1, 2, \ldots, m$. From (106) we know that either $\lambda_i = 0$ or $y_i = 0$ or both. Let us consider what must obtain at a relative maximum. We suppose the above conditions are satisfied at some point ($\mathbf{x}^0$, $\mathbf{y}^0$, $\boldsymbol{\lambda}^0$) and suppose further that $\mathbf{x}^0$ is a relative maximum. We define $z_0 = f(\mathbf{x}^0)$ and note that $g_i(\mathbf{x}^0, y_i^0) = b_i$, $i = 1, 2, \ldots, m$. In order to show that $\lambda_i \leq 0$, we first show that

$\partial z^0/\partial b_i = -\lambda_i^0$. We see this as follows. We know from calculus that

$$\frac{\partial z^0}{\partial b_i} = \sum_{j=1}^{n} \frac{\partial f(\mathbf{x}^0)}{\partial x_j} \frac{\partial x_j}{\partial b_i} \tag{109}$$

$$\frac{\partial g_k(\mathbf{x}^0, y_k^0)}{\partial b_i} = \sum_{j=1}^{n} \frac{\partial g_k(\mathbf{x}^0, y_k^0)}{\partial x_j} \frac{\partial x_j}{\partial b_i} = \delta_{ik} \tag{110}$$

where δ_{ik} is the Kronecker delta and is given by

$$\delta_{ik} = \begin{cases} 1, & i = k \\ 0, & i \neq k \end{cases}$$

If we multiply δ_{ik} in (110) by λ_k^0 and sum over all values of k, we have

$$\sum_{k=1}^{m} \lambda_k^0 \delta_{ik} = \sum_{k=1}^{m} \lambda_k^0 \sum_{j=1}^{n} \frac{\partial g_k(\mathbf{x}^0, y_k^0)}{\partial x_j} \frac{\partial x_j}{\partial b_i} \tag{111}$$

It is clear that $\Sigma_{k=1}^{m} \lambda_k^0 \delta_{ik} = \lambda_i^0$ since $\delta_{ik} = 0$ unless $i = k$. Therefore,

$$\lambda_i^0 = \sum_{k=1}^{m} \lambda_k^0 \sum_{j=1}^{n} \frac{\partial g_k(\mathbf{x}^0, y_k^0)}{\partial x_j} \frac{\partial x_j}{\partial b_i} \tag{112}$$

If we add Eq. (112) to Eq. (109) we obtain

$$\frac{\partial z^0}{\partial b_i} + \lambda_i^0 = \sum_{j=1}^{n} \left[\frac{\partial f(\mathbf{x}^0)}{\partial x_j} + \sum_{k=1}^{m} \frac{\partial g_k(\mathbf{x}^0, y_k^0)}{\partial x_j} \right] \frac{\partial x_j}{\partial b_i} \tag{113}$$

However, the necessary conditions (Eq. 105) state that the term in brackets in Eq. (113) is zero. Therefore, we have shown that

$$\frac{\partial z^0}{\partial b_i} = -\lambda_i \tag{114}$$

We note again from (106) that either $\lambda_i^0 = 0$ or $y_i^0 = 0$ or both. Suppose $y_i^0 \neq 0$. This implies that $g_i(\mathbf{x}^0) = b_i$. If we increased b_i we would not constrain $\mathbf{x}^0$ further. Hence $\partial z^0/\partial b_i = -\lambda_i^0 = 0$. Now suppose that $\lambda_i^0 \neq 0$. From (106) this means that $y_i^0 = 0$. Therefore $g_i(\mathbf{x}^0) = b_i$. If $\lambda_i^0 > 0$, then $\partial z^0/\partial b_i < 0$, which means that as b_i is increased, z would decrease. This is not possible, since as b_i increases we increase the feasible region and so z cannot decrease. Hence, at an optimal solution $\lambda_i \leq 0$.

Let us now restate all of the previous in a simple theorem, which we have already proven.

Theorem 18. In order to maximize $f(\mathbf{x})$ subject to $g_i(\mathbf{x}) \leq b_i$, $i = 1, 2, \ldots, m$, the necessary conditions for the existence of a relative maximum at $\mathbf{x}^0$ are

$$\begin{aligned}
&(1)\ \frac{\partial F(\mathbf{x}^0, \mathbf{y}^0, \boldsymbol{\lambda}^0)}{\partial x_j} = 0, && j = 1, 2, \ldots, n \\
&(2)\ \lambda_i^0[b_i - g_i(\mathbf{x}^0)] = 0, && i = 1, 2, \ldots, m \\
&(3)\ \lambda_i^0 \leq 0, && i = 1, 2, \ldots, m \\
&(4)\ g_i(\mathbf{x}^0) \leq b_i, && i = 1, 2, \ldots, m
\end{aligned} \tag{115}$$

Theorem 19. In order to minimize $f(\mathbf{x})$ subject to $g_i(\mathbf{x}) \leq b_i$, $i = 1, 2, \ldots, m$, the necessary conditions for the existence of a relative minimum at $\mathbf{x}^0$ are

$$\begin{aligned}
&(1)\ \frac{\partial F(\mathbf{x}^0, \mathbf{y}^0, \boldsymbol{\lambda}^0)}{\partial x_j} &&= 0, \quad j = 1, 2, \ldots, n \\
&(2)\ \lambda_i^0[b_i - g_i(\mathbf{x}^0)] &&= 0, \quad i = 1, 2, \ldots, m \\
&(3)\ \lambda_i^0 &&\geq 0, \quad i = 1, 2, \ldots, m \\
&(4)\ g_i(\mathbf{x}^0) &&\leq b_i, \quad i = 1, 2, \ldots, m
\end{aligned} \tag{116}$$

Proof: Similar to that of Theorem 18.

It should be noted that even though $\mathbf{y}$ is included as an argument of F in (115) and (116), it is no longer necessary to explicitly use it as long as the constraints are written as $g_i(\mathbf{x}) \leq b_i$, $i = 1, 2, \ldots, m$. It was merely a convenient device for deriving the Kuhn-Tucker conditions, which are listed as (1) through (4) in Theorems 18 and 19.

Example.

$$\text{Max } f(x_1 x_2) = 4x_1 + 2x_1x_2 + 3x_2$$

subject to: $2x_1^2 + 3x_2 \leq 10$

Applying the Kuhn-Tucker conditions (114) to the Lagrangian

$$F = 4x_1 + 2x_1x_2 + 3x_2 + \lambda[10 - 2x_1^2 - 3x_2]$$

We then have

$$\begin{aligned}
\frac{\partial F}{\partial X_1} &= 4 + 2x_2 - 4\lambda x_1 = 0 \\
\frac{\partial F}{\partial x_2} &= 2x_1 + 3 - 3\lambda = 0 \\
\lambda(10 - 2x_1^2 - 3x_2) &= 0 \\
\lambda &\leq 0 \\
2x_1^2 + 3x_2 &\leq 10
\end{aligned}$$

If we solve the first three equations of (117), we obtain a solution satisfying the condition that $\lambda \leq 0$:

$$x_1^0 = -2.208, \qquad x_2^0 = 0.0843, \qquad \lambda^0 = -0.472$$

yielding $f(x_1^0, x_2^0) = -8.951$.

We now present by the following theorem, a result which states under what conditions the necessary Kuhn-Tucker conditions are also sufficient.

Theorem 20. If $f(\cdot)$ is a strictly concave function and $g_i(\cdot)$, $i = 1, 2, \ldots, m$, are convex functions which are continuous and differentiable, then the Kuhn-Tucker conditions (114) are sufficient as well as necessary for an absolute maximum.

Proof: If $f(\cdot)$ and $g_i(\cdot)$, $i = 1, 2, \ldots, m$, satisfy our hypothesis, then we may write the Lagrangian function as we did for Theorem 18 as

$$F(\mathbf{x}, \mathbf{y}, \boldsymbol{\lambda}) = f(\mathbf{x}) + \sum_{i=1}^{m} \lambda_i[b_i - g_i(\mathbf{x}) - y_i^2] \tag{118}$$

If $\lambda_i \leq 0$, then $\lambda g_i(\mathbf{x})$ is concave if $g_i(\mathbf{x})$ is convex. Therefore, since the sum of concave functions is concave,

$$f(\mathbf{x}) - \sum_{i=1}^{m} g_i(\mathbf{x}) \tag{119}$$

is a strictly concave function. Furthermore, since $\lambda_i y_i = 0$ and $\lambda_i b_i$ is a constant, if (119) is concave then so is $F(\mathbf{x}, \mathbf{y}, \boldsymbol{\lambda})$. We have previously shown that a necessary condition for $f(\mathbf{x})$ to have a maximum at $\mathbf{x}^0$ is that $F(\mathbf{x}, \mathbf{y}, \boldsymbol{\lambda})$ have a stationary point at $\mathbf{x}^0$ (Theorem 18). However, since $F(\mathbf{x}, \mathbf{y}, \boldsymbol{\lambda})$ is strictly concave, its derivatives can vanish at one point only. Hence this point must be the relative maximum $\mathbf{x}^0$. Therefore the Kuhn-Tucker conditions (114) are sufficient as well as necessary for an absolute maximum of $f(\cdot)$ at $\mathbf{x}^0$ under the assumed hypothesis.

Kuhn and Tucker's derivation of their conditions [9] is along quite different lines. They assume a constraint qualification and proceed to use Farkas' lemma to arrive at their result. We have assumed a perhaps stronger condition, in that we have implicitly assumed, just as we did in the previous proofs about the Lagrange multiplier method, that at least one of the Jacobian matrices $\| \partial g_i/\partial x_j \|$ was nonsingular.

The Kuhn-Tucker conditions can be written in a number of equivalent forms if one wants to specify explicitly equalities, inequalities of opposite sign, nonnegativity restrictions, etc. However, they can all be included as special cases under (102). For example, $x_t \geq 0$ can be written as

$$g_t(\mathbf{x}) = -x_t \leq 0$$

Some of these alternative formulations can be found in Ref. 8.

REFERENCES

1. T. M. Apostol, *Mathematical Analysis,* Addison-Wesley, Reading, Massachusetts, 1957.
2. L. Cooper and D. I. Steinberg, *Introduction to Methods of Optimization,* Saunders, Philadelphia, 1970.
3. A. Ralston, *A First Course in Numerical Analysis,* McGraw-Hill, New York, 1965.
4. F. B. Hildebrand, *Introduction To Numerical Analysis,* McGraw-Hill, New York, 1956.

5. H. Hancock, *Theory of Maxima and Minima,* Dover, New York, 1950.
6. R. Courant, *Differential and Integral Calculus,* Vol. 2, Wiley-Interscience, New York, 1963.
7. C. Caratheodory, *Calculus of Variations and Partial Differential Equations of the First Order,* Vol. 2, Holden-Day, San Francisco, 1967.
8. G. Hadley, *Nonlinear and Dynamic Programming,* Addison-Wesley, Reading, Massachusetts, 1964.
9. H. W. Kuhn and A. W. Tucker, *Nonlinear Programming, Proceedings of the Second Berkeley Symposium on Mathematical Statistics and Probability* (J. Neyman, ed.), University of California Press, Berkeley, 1951.

11
NONLINEAR PROGRAMMING

MORDECAI AVRIEL

Faculty of Industrial Engineering
and Management
Technion-Israel Institute of Technology
Haifa, Israel

INTRODUCTION

Mathematical programming is the finite dimensional branch of optimization theory in which a single-valued objective function f on n real variables $x_1, \ldots, x_n$ is minimized (or maximized), possibly subject to a finite number of constraints, which are written as inequalities or equations. Generally we define a mathematical program of, say, minimization as

$$\text{(MP)} \quad \min f(x) \tag{1}$$

subject to

$$g_i(x) \geq 0, \quad i = 1, \ldots, m \tag{2}$$

$$h_j(x) = 0, \quad j = 1, \ldots, p \tag{3}$$

where x denotes the column vector whose components are $x_1, \ldots, x_n$. In other words, (MP) is the problem of finding a vector x^* that satisfies (2) and (3) and such that $f(x^*)$ has a minimal—that is, optimal value. If one or more of the functions appearing in (MP) are nonlinear in x, we call it a nonlinear program, in contrast to a linear program, where all these functions must be linear.

Nonlinear programming problems arise in such various disciplines as engineering, economics, business administration, physical sciences, and mathematics, or in any other area where decisions (in a broad sense) must be taken in some complex (or conflicting) situation that can be represented by a mathematical model. In order to illustrate some types of nonlinear programs, a few examples are presented below.

Example 1. Nonlinear Curve Fitting. Suppose that in some scientific research, say in physics, a certain phenomenon F is measured in the laboratory as a function of time. Also suppose that we are given a mathematical model of the phenomenon, and from the model we know that the value of F is assumed to vary with time t as

$$F(t) = x_1 + x_2 \exp(-x_3 t) \tag{4}$$

The purpose of the laboratory experiments is to find the unknown parameters x_1, x_2, and x_3 by measuring values of F at times $t^1, t^2, \ldots, t^M$. The decision-making process involves assigning values to the parameters that are optimal in some sense. For example, we can seek values of the parameters in the least-squares sense—that is, those values for which the sum of squares of the experimental deviations from the theoretical curve is minimized. We then have the nonlinear program without constraints

$$\min f(x_1, x_2, x_3) = \sum_{i=1}^{M} [F(t^i) - x_1 - x_2 \exp(-x_3 t^i)]^2 \tag{5}$$

Note that this unconstrained program, if solved, may yield unacceptable values of the parameters. To avoid such a situation, we can impose restrictions in the form of constraints. For example, the parameter x_3 can be restricted to have a nonnegative value—that is,

$$g_1(x_1, x_2, x_3) = x_3 \geq 0 \tag{6}$$

Also suppose that, for the particular phenomenon under consideration, the mathematical model proposed can be acceptable only if the parameters are so chosen that at $t = 0$ we have $F(0) = 1$. Hence we must add a constraint

$$h_1(x_1, x_2, x_3) = x_1 + x_2 = 1 \tag{7}$$

Solving (5), subject to (6) and (7), is then a constrained nonlinear programming problem having a nonlinear objective function with linear inequality and equality constraints.

Example 2. Chemical Process Design. Consider the problem of manufacturing a given quantity F_B gram moles per hour of chemical product B from a feed consisting of a solution of reactant A in a reactor. The chemical reaction is

$$2A \rightarrow B \tag{8}$$

with an empirical rate equation,

$$-\frac{dC_A}{dt} = 8.4(C_A)^2 = 8.4[C_A^0(1 - x_A)]^2 \quad \left(\frac{\text{g-mole}}{\text{liter-hr}}\right) \tag{9}$$

where C_A = concentration of A in the reactor (g-mole/liter)

C_A^0 = concentration of A in the feed (g-mole/liter)

t = time (hours)

x_A = conversion, fraction of reactant converted into product

Suppose that the feed solution is available at a continuous range of concentrations of A and that its unit cost p_A is given by the relation

$$p_A = 4(C_A^0)^{1.4} \quad (\$/\text{liter}) \tag{10}$$

The operating cost of the reactor is given by

$$p_R = 0.75(V)^{0.6} \quad (\$/\text{hr}) \tag{11}$$

where V (liter) is the volume of the reactor. Assume that the product B can be sold at a price of 10 \$/g-mole. Our problem is to determine the rate of feed solution F_A^0 (liter/hr), its concentration C_A^0, the volume of the reactor V, and the conversion x_A for optimum operation—that is, for maximizing total profit per hour—given by

$$p_T = 10F_B - p_A F_A^0 - p_R \quad (\$/\text{hr}) \tag{12}$$

Material balance around the reactor yields

$$F_A^0 C_A^0 = F_A^0 C_A^0(1 - x_A) - \left(\frac{dC_A}{dt}\right)V \tag{13}$$

From (8) we get

$$\tfrac{1}{2}F_A^0 C_A^0 x_A = F_B \tag{14}$$

Our design problem then becomes

$$\max p_T = 5F_A^0 C_A^0 x_A - 4(C_A^0)^{1.4} F_A^0 - 0.75(V)^{0.6} \tag{15}$$

subject to the constraint obtained from (9) and (13)

$$8.4C_A^0(1 - x_A)^2 V - F_A^0 x_A = 0 \tag{16}$$

This is a nonlinear program in variables C_A^0, F_A^0, V, and x_A in which both the objective and the constraint functions are nonlinear. Note that here the objective function is maximized.

Our survey of nonlinear programming consists of two main sections. In the first section, Nonlinear Programming Analysis, we derive necessary and sufficient optimality conditions, treat convex functions and programming, and duality theory. The second section, Methods of Nonlinear Programming, deals with algorithms for the numerical solution of unconstrained and constrained nonlinear programs. The survey closely follows the subjects discussed in detail in Avriel [1], where proofs of most of the results cited below can also be found.

NONLINEAR PROGRAMMING ANALYSIS

Nonlinear programming as we know it today is a relatively new discipline, although nonlinear optimization problems were already formulated and solved by the ancient Greeks. Optimization problems have also influenced the thoughts of some of the founding fathers of modern mathematics, such as Fermat in the seventeenth century, and Euler and Lagrange in the eighteenth century. Most of the activity, however, of early mathematicians in the area of optimization was centered around infinite-dimensional problems, known as the calculus of variations.

A comprehensive treatment of necessary and sufficient conditions for optimality in finite-dimensional problems, representing the state of the art at the beginning of the twentieth century, is given by Hancock [2]. The development of complex technological and economic systems with the obvious need to optimize them in some sense on the one hand, and the spectacular advances in computer technology resulting in the fast execution of complicated calculations on the other, have jointly contributed toward an accelerated development of mathematical (linear and nonlinear) programming during the second half of this century. In fact, most of the results presented here have been derived after 1950.

Optimality Conditions

We start our discussion of optimality conditions in nonlinear programming with a few definitions. Consider a real-valued function f with domain D in R^n and let $X \subset D$ denote the set of points $x \in R^n$ defined by a finite number of equations and/or inequalities. The set X is called the feasible set and members of the feasible set are called feasible points. Then f is said to have a local minimum over X at a point $x^* \in X$ if there exists a number $\delta > 0$ such that

$$f(x) \geq f(x^*) \tag{17}$$

for all $x \in X$ satisfying $\|x - x^*\| < \delta$. We define a local maximum in a similar way but with the sense of the inequality in (17) reversed. If the inequality (17) is replaced by a strict inequality

$$f(x) > f(x^*) \tag{18}$$

for all $x \in X$, $x \neq x^*$, $\|x - x^*\| < \delta$, we have a strict local minimum; and if the sense of the inequality in (18) is reversed, we have a strict local maximum. The function f has a global minimum at $x^* \in X$ if (17) holds for all $x \in X$. A similar definition holds for a global maximum. An extremum is either a minimum or a maximum. If X = D in the above definitions, then x^* is called an unconstrained extremum.

We shall focus our attention on differentiable functions. Note that differentiability of a real function implies the existence of first partial derivatives. We define the gradient of f at x as the vector $\nabla f(x)$, given by

$$\nabla f(x) = \left(\frac{\partial f(x)}{\partial x_1}, \ldots, \frac{\partial f(x)}{\partial x_n}\right)^T \qquad (19)$$

Note that all vectors are assumed to be column vectors. The superscript T is used to indicate the transpose of a (column or row) vector. If f is twice differentiable at x, we define the Hessian matrix of f at x as the n×n symmetric matrix $\nabla^2 f(x)$, given by

$$\nabla^2 f(x) = \left(\frac{\partial^2 f(x)}{\partial x_i \partial x_j}\right), \quad i, j = 1, \ldots, n \qquad (20)$$

Unconstrained Extrema

First we present necessary and sufficient conditions for extrema of functions without constraints.

Theorem 1. Let x* be an interior point of D at which f has a local minimum or local maximum. If f is differentiable at x*, then $\nabla f(x^*) = 0$.

Theorem 2. Let x* be an interior point of D at which f is twice continuously differentiable. If

$$\nabla f(x^*) = 0 \qquad (21)$$

and

$$z^T \nabla^2 f(x^*) z > 0 \qquad (22)$$

for all nonzero vectors z, then f has a local minimum at x*. If the sense of the inequality in (22) is reversed, then f has a local maximum at x*. Moreover, the extrema are strict local extrema.

In both theorems we are utilizing the behavior of the function at x*, the extremum. If, however, we can investigate the behavior of the function in some neighborhood $N_\delta(x^*)$ of the extremum in question, where

$$N_\delta(x^*) = \{x: x \in R^n, \ \|x - x^*\| < \delta\} \qquad (23)$$

we have a result that provides additional conditions for a local extremum.

Theorem 3. Let x* be an interior point of D and assume that f is twice continuously differentiable on D. It is necessary for a local minimum of f at x* that

$$\nabla f(x^*) = 0 \qquad (24)$$

and

$$z^T \nabla^2 f(x^*) z \geq 0 \qquad (25)$$

for all z. Sufficient conditions for a local minimum are that (24) holds and that for every x in some neighborhood $N_\delta(x^*)$ and for every $z \in R^n$, we have

$$z^T \nabla^2 f(x) z \geq 0 \tag{26}$$

If the sense of the inequalities in (25) and (26) is reversed, the theorem applies to a local maximum.

Example 3. Let $f(x) = (x)^{2p}$, where p is a positive integer, and let D be the whole real line. The gradient ∇f is given by $\nabla f(x) = 2p(x)^{2p-1}$. At $x = 0$, the gradient vanishes; that is, the origin satisfies the necessary conditions for a minimum or a maximum as stated in Theorem 1.

The Hessian, $\nabla^2 f$, is given by

$$\nabla^2 f(x) = (2p - 1)(2p)(x)^{2p-2} \tag{27}$$

For $p = 1$, $\nabla^2 f(0) = 2$; that is, the sufficient conditions for a strict local minimum (Theorem 2) are satisfied.

If, however, we take $p > 1$, then $\nabla^2 f(0) = 0$ and the sufficient conditions of Theorem 2 are not satisfied; yet f has a minimum at the origin, as can be seen in Fig. 1. On the other hand, by taking any neighborhood of the origin, the reader can verify that all the conditions for a local minimum (both the necessary and the sufficient) of Theorem 3 are satisfied. In fact, the origin is actually a strict global minimum of f.

Equality Constrained Extrema

Consider now the problem of finding the minimum (or maximum) of a real-valued function f with domain $D \subset R^n$, subject to the constraints

$$h_j(x) = 0, \quad j = 1, \ldots, p \tag{28}$$

where $p < n$ and each of the h_j is a real-valued function defined on D. The assumption that the number of constraint equations is less than the number of variables will simplify subsequent discussions. The first and most intuitive method of solution of such a problem involves the elimination of p variables from the problem by using the equations in (28). The conditions for such an elimination can be found in the Implicit Function Theorem (see, e.g., Ref. 3). This theorem assumes the

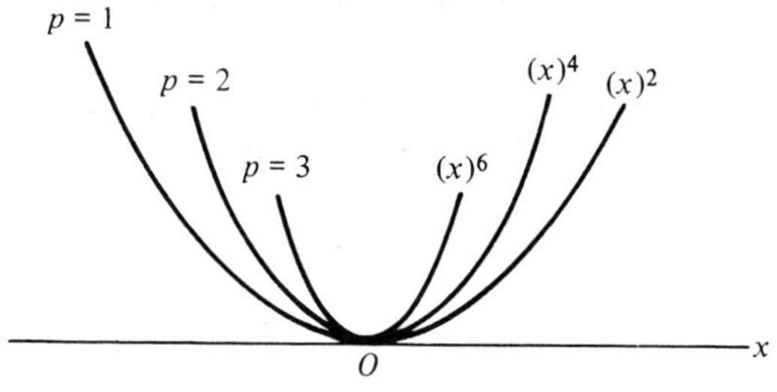

FIG. 1. The functions x^{2p} in the neighborhood of the origin.

differentiability of the functions h_j and that the n×p Jacobian matrix $[\partial h_j/\partial x_k]$ has rank p. Unfortunately, the actual solution of the constraint equations for p variables in terms of the remaining n - p can often prove a difficult if not impossible task.

Another method, also based on the idea of transforming a constrained problem into an unconstrained one, was proposed by Lagrange. Lagrange's method consists of transforming an equality constrained extremum problem into a problem of finding a stationary point of the Lagrangian $L(x, \mu)$, given by

$$L(x, \mu) = f(x) - \sum_{j=1}^{p} \mu_j h_j(x) \tag{29}$$

where the μ_j are called Lagrange multipliers. This can be seen by the following.

Theorem 4. Let f and h_j, j = 1, ..., p, be real functions on $D \subset R^n$ and continuously differentiable on a neighborhood $N_\delta(x^*) \subset D$. Suppose that x^* is a local minimum or local maximum of f over the set of points that satisfy

$$h_j(x) = 0, \quad j = 1, \ldots, p \tag{30}$$

Also assume that the Jacobian matrix of the $h_j(x^*)$ has rank p. Then there exists a vector of multipliers $\mu^* \in R^p$ such that

$$\nabla L(x^*, \mu^*) = 0 \tag{31}$$

Substituting (29) into (31) indicates that the gradient of the objective function ∇f and the gradients of the constraints $\nabla h_1, \ldots, \nabla h_p$ are linearly dependent at the optimum.

Let us turn now to sufficient conditions of optimality in equality constrained problems.

Theorem 5. Let f, h_j, j = 1, ..., p be twice continuously differentiable real functions on R^n. If there exist vectors $x^* \in R^n$, $\mu^* \in R^p$ such that

$$\nabla L(x^*, \mu^*) = 0 \tag{32}$$

and for every nonzero vector $z \in R^n$ satisfying

$$z^T \nabla h_j(x^*) = 0, \quad j = 1, \ldots, p \tag{33}$$

it follows that

$$z^T \nabla_x^2 L(x^*, \mu^*) z > 0 \tag{34}$$

then f has a strict local minimum at x^*, subject to $h_j(x) = 0$, j = 1, ..., p. If the sense of the inequality in (34) is reversed, then f has a strict local maximum at x^*.

The notation $\nabla_x^2 L(x^*, \mu^*)$ is used for the n×n matrix whose elements are the second partial derivatives of L with respect to x, evaluated at the point x^*, $\mu^*) \in R^{n+p}$.

Example 4. Consider the problem

$$\max f(x_1, x_2) = x_1x_2 \tag{35}$$

subject to the constraint

$$h(x_1, x_2) = x_1 + x_2 - 2 = 0 \tag{36}$$

First we form the Lagrangian:

$$L(x, \mu) = x_1x_2 - \mu(x_1 + x_2 - 2) \tag{37}$$

Next we look for x^*, μ^* satisfying $\nabla L(x^*, \mu^*) = 0$:

$$\frac{\partial L(x^*, \mu^*)}{\partial x_1} = x_2^* - \mu^* = 0 \tag{38}$$

$$\frac{\partial L(x^*, \mu^*)}{\partial x_2} = x_1^* - \mu^* = 0 \tag{39}$$

$$\frac{\partial L(x^*, \mu^*)}{\partial \mu} = -x_1^* - x_2^* + 2 = 0 \tag{40}$$

Solution of the last three equations yields

$$x_1 = x_2 = \mu^* = 1 \tag{41}$$

The point $(x^*, \mu^*) = (1, 1, 1)$ therefore satisfies the necessary conditions for a maximum as stated in Theorem 4.

The linear dependence between ∇f and ∇h at the optimum, as asserted by Theorem 4, is clearly illustrated in Fig. 2, where $\nabla f(x^*)$ is actually equal to $\nabla h(x^*)$.

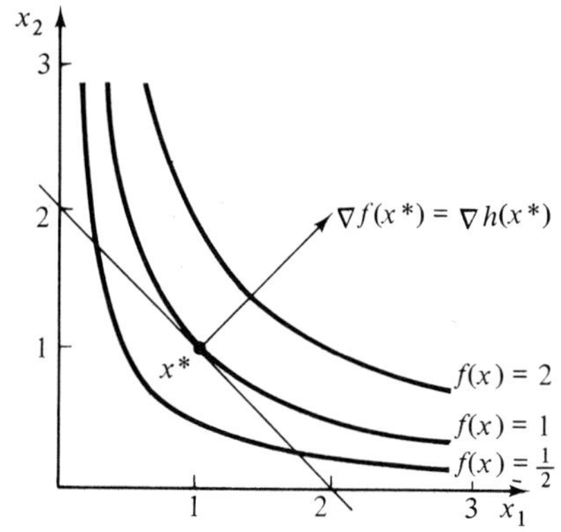

FIG. 2. Constrained maximum.

Turning to the sufficient conditions, we compute $\nabla_x^2 L(x^*, \mu^*)$:

$$\frac{\partial^2 L(x^*, \mu^*)}{\partial x_1 \partial x_1} = 0, \quad \frac{\partial^2 L(x^*, \mu^*)}{\partial x_1 \partial x_2} = 1, \quad \frac{\partial^2 L(x^*, \mu^*)}{\partial x_2 \partial x_2} = 0 \tag{42}$$

Hence

$$z^T \nabla_x^2 L(x^*, \mu^*) z = (z_1, z_2) \begin{bmatrix} 0 & 1 \\ 1 & 0 \end{bmatrix} \begin{pmatrix} z_1 \\ z_2 \end{pmatrix} = 2z_1 z_2 \tag{43}$$

and, by Theorem 5, we must determine the sign of $2z_1z_2$ for all $z \neq 0$ that satisfy $z^T \nabla h(x^*) = 0$.

Since

$$\frac{\partial h(x^*)}{\partial x_1} = 1 \quad \text{and} \quad \frac{\partial h(x^*)}{\partial x_2} = 1 \tag{44}$$

the last condition is equivalent to $z_1 + z_2 = 0$. Substituting into (43), we get

$$z^T \nabla_x^2 L(x^*, \mu^*) z = -2(z_1)^2 < 0$$

Thus (1, 1) is a strict local maximum.

Second-order necessary conditions, similar to those of Theorem 3, were derived [4, 18] for more general constrained extremum problems. In cases of problems with equality constraints, these conditions are an almost straightforward generalization of the second-order necessary and sufficient conditions for an unconstrained problem.

In Theorem 5 we assumed that the Jacobian matrix $[\partial h_j(x^*)/\partial x_k]$ has rank p ($<n$), equal to the number of constraint equations. We state now a slight generalization of Theorem 5 that does not require conditions on the rank of the Jacobian.

Theorem 6. Let f and h_j, j = 1, ..., p, be continuously differentiable real functions with domain $D \subset R^n$. If x* is a local minimum or local maximum of f for all points x in a neighborhood of x* satisfying

$$h_j(x) = 0, \quad j = 1, \ldots, p \tag{46}$$

then there exist p + 1 numbers $\mu_0^*, \mu_1^*, \ldots, \mu_p^*$, not all zero, such that

$$\mu_0^* \nabla f(x^*) - \sum_{j=1}^{p} \mu_j^* \nabla h_j(x^*) = 0 \tag{47}$$

Inequality Constrained Extrema

We consider now mathematical programming problems with inequality as well as equality constraints. In our opinion, the introduction of inequalities to optimization problems marks the end of the "classical" era of optimization and the beginning of the "modern" theory of mathematical programming. Inequality constraints are seldom strict inequalities; they can be satisfied either as equations or as strict inequalities. This feature of inequalities causes some complications in the analytic treatment of optimality conditions, if very general results are sought. We shall first sacrifice some generality and state necessary conditions for optimality using a somewhat restricting assumption on linear independence between constraint gradients or, equivalently, on the rank of a Jacobian matrix. Subsequently we shall present more general optimality conditions.

Consider the mathematical program

$$\text{(P)} \qquad \min f(x) \tag{48}$$

subject to the constraints

$$g_i(x) \geq 0, \quad i = 1, \ldots, m \tag{49}$$

$$h_j(x) = 0, \quad j = 1, \ldots, p \tag{50}$$

where the functions f, $g_1, \ldots, g_m, h_1, \ldots, h_p$ are assumed to be differentiable on some open set $D \subset R^n$ containing the feasible set X; that is, the set of all $x \in D$ satisfying (49) and (50).

For this type of problem we define the Lagrangian as

$$L(x, \lambda, \mu) = f(x) - \sum_{i=1}^{m} \lambda_i g_i(x) - \sum_{j=1}^{p} \mu_j h_j(x) \tag{51}$$

and will state optimality conditions in terms of this function. Let

$$I(x^0) = \{i: g_i(x^0) = 0\} \tag{52}$$

be the index set of active inequality constraints at x^0. We then have a generalization of Theorem 4 as follows.

Theorem 7. Let f, $g_1, \ldots, g_m, h_1, \ldots, h_p$ be real functions on an open set $D \subset R^n$ and continuously differentiable on a neighborhood $N_\delta(x^*) \subset D$. Suppose that x^* is a local minimum of f over the feasible set X given by (49) and (50). Also assume that the gradients $\nabla g_i(x^*)$, $i \in I(x^*)$, $\nabla h_j(x^*)$, $j = 1, \ldots, p$ are linearly independent. Then there exist vectors of multipliers $\lambda^* \in R^m$, $\mu^* \in R^p$ such that

$$\nabla_x L(x^*, \lambda^*, \mu^*) = \nabla f(x^*) - \sum_{i=1}^{m} \lambda_i^* \nabla g_i(x^*) - \sum_{j=1}^{p} \mu_j^* \nabla h_j(x^*) = 0 \tag{53}$$

$$\lambda_i^* g_i(x^*) = 0, \quad i = 1, \ldots, m \tag{54}$$

$$\lambda^* \geq 0 \tag{55}$$

The necessary conditions for optimality in inequality constrained nonlinear programs, a special case of which is given in the above theorem, are commonly

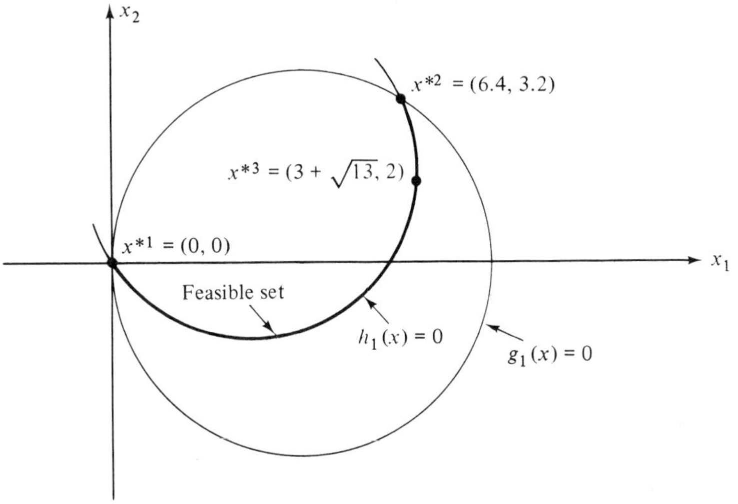

FIG. 3. The feasible set in Example 5 and the points satisfying the Kuhn-Tucker conditions.

called the Kuhn-Tucker conditions, stated in 1951 [5]. Actually, these conditions were derived much earlier by Karush [6] whose work was little known before the appearance of a historical review by Kuhn [7]. Readers interested in the theory of optimality conditions should consult Abadie [8], Dubovitskii and Milyutin [9], Gould and Tolle [10, 11], and Neustadt [12].

Example 5. Consider the following nonlinear programming problem:

$$\min f(x) = x_1 \tag{56}$$

subject to

$$g_1(x) = 16 - (x_1 - 4)^2 - (x_2)^2 \geq 0 \tag{57}$$

$$h_1(x) = (x_1 - 3)^2 + (x_2 - 2)^2 - 13 = 0 \tag{58}$$

It can be seen from Fig. 3 that f(x) has local minima at $x^{*1} = (0, 0)$ and at $x^{*2} = (6.4, 3.2)$. At both points, $I(x^{*1}) = I(x^{*2}) = \{1\}$. At the first point,

$$\nabla g_1(x^{*1}) = \begin{pmatrix} 8 \\ 0 \end{pmatrix}, \quad \nabla h_1(x^{*1}) = \begin{pmatrix} -6 \\ -4 \end{pmatrix} \tag{59}$$

and these vectors are linearly independent. Consequently, the Kuhn-Tucker conditions given in Theorem 7 hold. The reader can verify that the corresponding Kuhn-Tucker multipliers are $\lambda_1^* = 1/8$, $\mu_1^* = 0$. At the second point

$$\nabla g_1(x^{*2}) = \begin{pmatrix} -4.8 \\ -6.4 \end{pmatrix}, \quad \nabla h_1(x^{*2}) = \begin{pmatrix} 6.8 \\ 2.4 \end{pmatrix} \tag{60}$$

and, again, these vectors are linearly independent. The required multipliers have values $\lambda_1^* = 3/40$, $\mu_1^* = 8/40$. To illustrate the fact that the Kuhn-Tucker conditions are necessary but, in general, not sufficient, consider the feasible point $x^{*3} = (3 + \sqrt{13}, 2)$. It can be easily shown that the Kuhn-Tucker conditions hold also at this point with $\lambda_1^* = 0$, $\mu_1^* = \sqrt{13}/26$. Inspection of Fig. 3, however, reveals that x^{*3} is not a solution of our problem!

Turning now to the more general necessary conditions, we introduce a few definitions: Given problem (P) above, define the linearizing cone of the feasible set X at a point $x^0 \in X$ as

$$Z(X, x^0) = \{z: z^T \nabla g_i(x^0) \geq 0, \quad i \in I(x^0), \quad z^T \nabla h_j(x^0) = 0, \quad j = 1, \ldots, p\} \tag{61}$$

Note that a set $K \subset R^n$ is called a cone if $x \in K$ implies $\alpha x \in K$ for every nonnegative number α. Let $S(X, x^0)$ be the set of all vectors z for which there exists a sequence of vectors $\{x^k\} \subset X$ converging to x^0 and a sequence of nonnegative numbers $\{\alpha^k\}$ such that the sequence $\{\alpha^k(x^k - x^0)\}$ converges to z. $S(X, x^0)$ is called the closed cone of tangents of X at x^0. Finally, given a set $A \subset R^n$, the normal cone of A, denoted A', consists of all vectors $x \in R^n$ such that $x^T y \geq 0$ for all $y \in A$. Using these definitions we can state the following.

Theorem 8. Let $f, g_1, \ldots, g_m, h_1, \ldots, h_p$ be real functions on an open set $D \subset R^n$ and continuously differentiable on a neighborhood $N_\delta(x^*) \subset D$. Suppose that x^* is a local minimum of f over the feasible set X given by (49) and (50). Also assume that

$$(Z(X, x^*))' = (S(X, x^*))' \tag{62}$$

Then there exist vectors of multipliers $\lambda^* \in R^m$, $\mu^* \in R^p$ such that (53) to (55) hold.

This theorem represents the most general form of the Kuhn-Tucker necessary conditions for optimality involving only first derivatives that we present here. The condition (62) is often called a constraint qualification since it imposes some regularity conditions on the constraints of problem (P). For more on constraint qualifications see Refs. 13 and 14. The work of Kuhn and Tucker on optimality conditions [5] was also preceded by weaker conditions, due to John [15]. For the John conditions, that were subsequently extended by Mangasarian and Fromovitz [16], no constraint qualification is needed as we shall now see.

Theorem 9. Suppose that $f, g_1, \ldots, g_m, h_1, \ldots, h_p$ are continuously differentiable on an open set containing X. If x^* is a solution of problem (P), then there exist vectors $\lambda^* = (\lambda_0^*, \lambda_1^*, \ldots, \lambda_m^*)^T$ and $\mu^* = (\mu_1^*, \ldots, \mu_p^*)^T$ such that

$$\lambda_0^* \nabla f(x^*) - \sum_{i=1}^{m} \lambda_i^* \nabla g_i(x^*) - \sum_{j=1}^{p} \mu_j^* \nabla h_j(x^*) = 0 \tag{63}$$

$$\lambda_i^* g_i(x^*) = 0, \qquad i = 1, \ldots, m \tag{64}$$

$$(\lambda^*, \mu^*) \neq 0, \qquad \lambda^* \geq 0 \tag{65}$$

Comparing (63) to (65) with (53) to (55) reveals the main weakness of the John conditions: If $\lambda_0^* > 0$, we can divide (63) by it and (53) to (55) will also hold. If, however, $\lambda_0^* = 0$, then (63) to (65) are, in fact, satisfied to any differentiable objective function f whether it has a local minimum at x* or not.

Necessary conditions involving second derivatives that generalize those given in Theorem 5 for equality-constrained problems to inequalities can be found in Refs. 17-19.

We shall now present some sufficient conditions for optimality in problem (P), but first we need the following definition: Denote by $\hat{I}(x^*)$ the set of indices i for which $g_i(x^*) = 0$ and (53) to (55) are satisfied by a positive λ_i^*. Thus $\hat{I}(x^*)$ is a subset of $I(x^*)$. Let

$$\hat{Z}(X, x^*) = \{z: z^T \nabla g_i(x^*) = 0, \quad i \in \hat{I}(x^*), \quad z^T \nabla g_i(x^*) \geq 0, \quad i \in I(x^*), \quad z^T \nabla h_j(x^*) = 0, \quad j = 1, \ldots, p\} \tag{66}$$

Observe that $\hat{Z}(X, x^*) \subset Z(X, x^*)$. We have, then, the following sufficient conditions, due to McCormick [18]; see also Fiacco [20] and Hestenes [21].

Theorem 10. Suppose that $f, g_1, \ldots, g_m, h_1, \ldots, h_p$ are twice continuously differentiable on a neighborhood $N_\delta(x^*) \subset D$ and let x* be feasible for problem (P). If there exist vectors λ^*, μ^* satisfying

$$\nabla_x L(x^*, \lambda^*, \mu^*) = \nabla f(x^*) - \sum_{i=1}^{m} \lambda_i^* \nabla g_i(x^*) - \sum_{j=1}^{p} \mu_j^* \nabla h_j(x^*) = 0 \tag{67}$$

$$\lambda_i^* g_i(x^*) = 0, \qquad i = 1, \ldots, m \tag{68}$$

$$\lambda^* \geq 0 \tag{69}$$

and for every $z \neq 0$ such that $z \in \hat{Z}(X, x^*)$ it follows that

$$z^T[\nabla^2 f(x^*) - \sum_{i=1}^{m} \lambda_i^* \nabla^2 g_i(x^*) - \sum_{j=1}^{p} \mu_j^* \nabla^2 h_j(x^*)]z > 0 \tag{70}$$

then x* is a strict local minimum of problem (P).

Example 6. Consider again the problem presented in Example 5. We have shown that there are (at least) three points satisfying the necessary conditions for optimality. Let us check the sufficient conditions. At x^{*1} we have $\hat{Z}(X, x^{*1}) = \{0\}$, and there are no vectors $z \neq 0$ such that $z \in \hat{Z}(X, x^{*1})$; hence the sufficient conditions of Theorem 10 are trivially satisfied. The reader can verify that these conditions also hold at x^{*2}. At x^{*3}, however,

$$\hat{Z}(X, x^{*3}) = \{(z_1, z_2): z_1 = 0\} \tag{71}$$

and the quadratic form appearing in (70) is $(-1\sqrt{13})z^Tz$, which is negative for all $z \neq 0$. Thus x^{*3} does not satisfy the sufficient conditions.

Saddlepoints of the Lagrangian

Still another type of optimality conditions is related to the Lagrangian and expressed in terms of saddlepoints of the latter function. Here we state some of these conditions that do not require special assumptions on the nature of the functions appearing in the nonlinear programming problem (P).

Let Φ be a real function of the two vector variables $x \in D \subset R^n$ and $y \in E \subset R^m$. Thus the domain of Φ is $D \times E$. A point $(\bar{x}, \bar{y})$ with $\bar{x} \in D$ and $\bar{y} \in E$ is said to be a saddlepoint of Φ if

$$\Phi(\bar{x}, y) \leq \Phi(\bar{x}, \bar{y}) \leq (x, \bar{y}) \tag{72}$$

for every $x \in D$ and $y \in E$.

Associated with the nonlinear program (P) there is a saddlepoint problem that can be stated as follows:

(S) Find a point $\bar{x} \in R^n$, $\bar{\lambda} \in R^m$, $\bar{\lambda} \geq 0$, $\bar{\mu} \in R^p$ such that $(\bar{x}, \bar{\lambda}, \bar{\mu})$ is a saddlepoint of the Lagrangian

$$L(x, \lambda, \mu) = f(x) - \sum_{i=1}^{m} \lambda_i g_i(x) - \sum_{j=1}^{p} \mu_j h_j(x) \tag{73}$$

That is,

$$L(\bar{x}, \lambda, \mu) \leq L(\bar{x}, \bar{\lambda}, \bar{\mu}) \leq L(x, \bar{\lambda}, \bar{\mu}) \tag{74}$$

for every $x \in R^n$, $\lambda \in R^m$, $\lambda \geq 0$, and $\mu \in R^p$.

A one-sided relation between a saddlepoint of a Lagrangian and a solution of problem (P) is given in the next result.

Theorem 11. If $(\bar{x}, \bar{\lambda}, \bar{\mu})$ is a solution of problem (S), then $\bar{x}$ is a solution of problem (P).

Note that the preceding sufficient condition for optimality in problem (P) holds whether the Lagrangian is differentiable or not. If, however, the functions f, g_i, h_j are indeed differentiable, we have the following result.

Theorem 12. Suppose that f, $g_1, \ldots, g_m, h_1, \ldots, h_p$ are differentiable functions and $(\bar{x}, \bar{\lambda}, \bar{\mu})$ is a solution of problem (S). Then

$$\nabla_x L(\bar{x}, \bar{\lambda}, \bar{\mu}) = \nabla f(\bar{x}) - \sum_{i=1}^{m} \bar{\lambda}_i \nabla g_i(\bar{x}) - \sum_{j=1}^{p} \bar{\mu}_j \nabla h_j(\bar{x}) = 0 \tag{75}$$

$$\bar{\lambda}_i g_i(\bar{x}) = 0, \quad i = 1, \ldots, m \tag{76}$$

$$\bar{\lambda} \geq 0 \tag{77}$$

Note that although a saddlepoint of the Lagrangian implies that (75) to (77) hold without additional regularity conditions, an optimal solution x^* of (P) does not generally imply the existence of a pair (λ^*, μ^*) satisfying (75) to (77) unless a condition such as (62) is imposed on (P).

In general, very little can be said about the behavior of the Lagrangian function $L(x, \lambda, \mu)$ at a point (x^*, λ^*, μ^*) satisfying the Kuhn-Tucker conditions. In particular, the converse of Theorem 11 generally does not hold as can be seen in the following example.

Example 7. Suppose that we have the following program:

$$\min f(x) = x \tag{78}$$

subject to

$$-(x)^2 \geq 0 \tag{79}$$

The optimal solution is $x^* = 0$. The corresponding saddlepoint problem of the Lagrangian is to find a $\lambda^* \geq 0$ such that

$$x^* + \lambda(x^*)^2 \leq x^* + \lambda^*(x^*)^2 \leq x + \lambda^*(x)^2 \tag{80}$$

for every $x \in R$, or, equivalently,

$$0 \leq x + \lambda^*(x)^2 \tag{81}$$

Clearly, λ^* cannot vanish. But for any $\lambda^* > 0$ we can choose $x > -1/\lambda^*$ and (81) will not hold. Thus there exists no λ^* such that (x^*, λ^*) will be a saddlepoint.

This concludes our discussion of optimality conditions in a general nonlinear program. It should be apparent to the reader by now that such general programs have certain theoretical and practical disadvantages. Necessary conditions of optimality do not coincide with sufficient ones, thus the identification of optimal solutions in a problem is not an easy task. Moreover, there may be local optima that are not global, and one cannot be sure that if a solution is found that satisfies some optimality conditions, then it is the true optimum sought. By imposing some convexity requirements on the functions appearing in a nonlinear program, we find a remedy to all these difficulties. Convex programs and their properties are the subjects of the next section.

Convexity in Nonlinear Programming

We begin our discussion of convexity by defining convex sets and functions and presenting some of their properties that are relevant to mathematical programming. Readers interested in a detailed exposition of convex sets and functions from an optimization-oriented point of view should consult the lecture notes of Fenchel [22] or the later books of Roberts and Varberg [23], Rockafellar [24], and Stoer and Witzgall [25].

Convex Sets

A subset C of R^n is called a convex set if for any two points $x^1 \in C$, $x^2 \in C$, the line segment joining them is also in C. Let q denote an ordered pair of numbers (q_1, q_2), called weights, with the following properties:

$$q_1 \geq 0, \quad q_2 \geq 0, \quad q_1 + q_2 = 1 \tag{82}$$

Then a set C is convex if $x^1 \in C$, $x^2 \in C$, and q, as defined above, also imply that $(q_1x^1 + q_2x^2) \in C$.

Example 8. Examples of convex sets are given below.

(a) The empty set, the set consisting of a single point $x \in R^n$, and the whole space R^n are trivial examples of convex sets.

(b) The set

$$H = \{x: x \in R^n, \ c^Tx = b\} \tag{83}$$

where $c \neq 0$ is a given vector and b is a given number, is called a hyperplane. It is a convex set.

(c) Using the foregoing notation, we define the closed half space

$$H^c = \{x: x \in R^n, \ c^Tx \geq b\} \tag{84}$$

and the open half space

$$H^o = \{x: x \in R^n, \ c^Tx > b\} \tag{85}$$

as being generated by the hyperplane H. Note that each hyperplane also generates an opposite closed (open) half space, obtained by reversing the sense of the inequalities in (84) and (85). All these half spaces are convex sets.

(d) The hypersphere

$$S_\alpha(x^0) = \{x: x \in R^n, \ \|x - x^0\| \leq \alpha\} \tag{86}$$

where $x^0 \in R^n$ is a given vector and α is a given positive number, is a convex set.

We now present some simple algebraic and geometric properties of convex sets.

1. Given vectors $x^1, x^2, \ldots, x^p$ in R^n and numbers $\alpha_1, \ldots, \alpha_p$ satisfying

$$\alpha_1 \geq 0, \ \ldots, \ \alpha_p \geq 0, \quad \sum_{i=1}^{p} \alpha_i = 1 \tag{87}$$

the convex combination of the vectors $x^1, \ldots, x^p$ is a vector x given by

$$x = \alpha_1 x^1 + \cdots + \alpha_p x^p \tag{88}$$

Then the set $C \subset R^n$ is convex if and only if every convex combination x of the points $x^1 \in C, \ldots, x^p \in C$ is contained in C.

2. The intersection of an arbitrary collection of convex sets is also a convex set. The intersection of all the convex sets containing an arbitrary set $A \subset R^n$ is called the convex hull of A and is denoted by Co(A). It is clear that the set Co(A) is a convex set; it is actually the "smallest" convex set in R^n containing A.
3. By the sum of two sets $X \subset R^n$, $Y \subset R^n$, we mean the set

$$X + Y = \{x + y: x \in X,\ y \in Y\} \tag{89}$$

Similarly, the scalar multiple of a set $X \subset R^n$ is defined as

$$\lambda X = \{\lambda x: x \in X\}, \qquad \lambda \in R \tag{90}$$

If C and D are convex subsets of R^n, then the sets C + D and λC are also convex.

Many important results in mathematical programming can be proved by using so-called separation theorems of convex sets. These theorems deal with two non-empty convex subsets C_1 and C_2 in R^n for which there exists a hyperplane H such that C_1 lies on one side of H and C_2 on the opposite side. Such a hyperplane is said to separate C_1 and C_2. These theorems, together with applications, can be found in Refs. 22-25 mentioned above and in Eggleston [26], Mangasarian [27], and Valentine [28].

Convex Functions

The fundamentals of the theory of convex functions were laid down by Hölder [29] and Jensen [30] around the turn of the twentieth century. It is intuitive and, in most cases, also sufficient to look at convex functions from a geometrical point of view. A real function f is said to be convex provided that the chord connecting any two points of the curve f(x) lies on or above the curve. The convex functions shown in Fig. 4 illustrate the "chord above the curve" principle. A look at these drawings also reveals that convex functions are not necessarily continuous or differentiable everywhere on their domain of definition. An important observation, which will serve as a starting point in our formal discussions on convex functions, is the close relation between convex sets and convex functions. The shaded areas above the curves in Fig. 4 are convex sets that, in turn, can be used to define convex functions. For reasons that will become clear later, it will be convenient to study convex functions that may have infinite values.

Let f be a function defined on a subset D of R^n with values in the extended reals; that is, f(x) is either a real number or it is $\pm\infty$. The subset of R^{n+1}

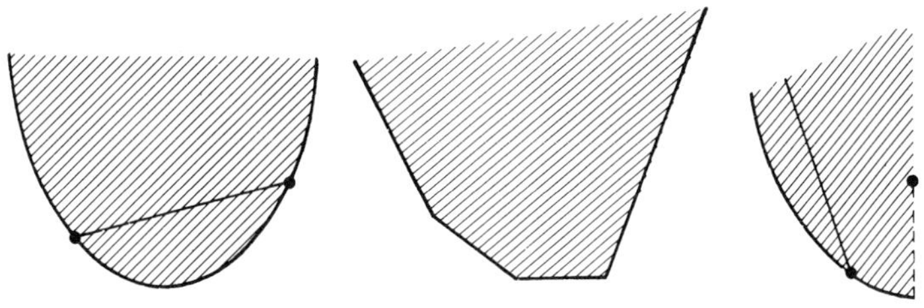

FIG. 4. Examples of convex functions.

$$P(f) = \{(x, \alpha): x \in D \subset R^n, \ \alpha \in R, \ f(x) \leq \alpha\} \tag{91}$$

is called the epigraph of f. We define f to be a convex function if P(f) is a convex set. In Fig. 4 the epigraphs are, of course, the shaded areas. Some straightforward examples of convex functions follow.

Example 9. The function $f = +\infty$ on R^n is a convex function, since $P(f) = \emptyset$ and the empty set is convex. Similarly, $f = -\infty$ on R^n is also convex, since $P(f) = R^{n+1}$.

Consider a convex function f defined on a proper subset D of R^n. Let

$$f_1(x) = \begin{cases} f(x), & \text{if } x \in D \\ +\infty, & \text{if } x \notin D \end{cases} \tag{92}$$

The epigraph P(f) of the function f defined on D is identical to $P(f_1)$, where f_1 is a function defined on all R^n. In this way, we can always construct convex functions defined throughout R^n. In particular, let $a \in R$, $b \in R^n$ be given. Then

$$f_1(x) = \begin{cases} a, & \text{if } x = b \\ +\infty, & \text{if } x \neq b \end{cases} \tag{93}$$

is a convex function defined on R^n.

As a result of Example 9, we shall assume that, unless mentioned explicitly, a convex function is defined on all R^n. The set

$$ED(f) = \{x: x \in R^n, \ f(x) < +\infty\} \tag{94}$$

is called the effective domain or domain of finiteness of f. It is actually the projection of P(f) on R^n. If f is a convex function, then ED(f) is a convex set in R^n. The converse statement generally does not hold. That is, if ED(f) is a convex set, f will not necessarily be a convex function.

A function defined on $D \subset R^n$ is concave if its negative is convex. In other words, g is a concave function if and only if -g is convex. The set

$$Q(g) = \{(x, \alpha): x \in D \subset R^n, \ \alpha \in R, \ g(x) \geq \alpha\} \tag{95}$$

is called the hypograph of g, and g is said to be concave if Q(g) is a convex subset of R^{n+1}. The reader should be able to modify results concerning properties of convex functions to the analogous properties for concave functions.

Allowing convex functions to take on infinite values requires some caution in arithmetic operations, such as the undefined operation $+\infty + (-\infty)$. Convex functions that have values $+\infty$ and $-\infty$ do not arise frequently in applications. We shall, therefore, be concerned mainly with proper convex functions defined as convex functions that nowhere have the value $-\infty$ and are not identically equal to $+\infty$. Convex functions that are not proper are called improper. Formally, f is a proper convex function if f is convex, $f(x) > -\infty$ for all $x \in R^n$, and $ED(f) \neq \emptyset$. For proper convex functions such that ED(f) contains more than one point, we can now derive a more familiar result, which can be taken as an alternative definition. Given $x^1 \in ED(f)$, $x^2 \in ED(f)$, if $(x^1, \alpha^1) \in P(f)$, $(x^2, \alpha^2) \in P(f)$, then for every vector of weights q we have $(q_1x^1 + q_2x^2, q_1\alpha^1 + q_2\alpha^2) \in P(f)$. That is,

$$f(q_1x^1 + q_2x^2) \leq q_1\alpha^1 + q_2\alpha^2 \tag{96}$$

Since the inequality in (96) must hold for every α^1, α^2 such that $f(x^1) \leq \alpha^1$, $f(x^2) \leq \alpha^2$, it can be modified to

$$f(q_1x^1 + q_2x^2) \leq q_1f(x^1) + q_2f(x^2) \tag{97}$$

The inequality in (97) is the "classical" definition of real convex functions, expressing the "chord above the curve" property. Generally (97) remains valid even if we let $f(x) = +\infty$ or $f(x) = -\infty$ for some x, provided that we adopt the arithmetic rule $0(\infty) = 0(-\infty) = 0$ and avoid the undefined operation $+\infty + (-\infty)$.

A real-valued function f defined on a convex subset C of R^n is called strictly convex if for any $x^1 \in C$, $x^2 \in C$, $x^1 \neq x^2$, and $1 > q_1 > 0$, $q_1 + q_2 = 1$, inequality (97) holds as a strict inequality. Most of the results dealing with convex functions can be also strengthened to strictly convex functions.

Some basic properties of convex functions are given below.

1. Let f be a proper convex function on R^n. Let $x^1, \ldots, x^s$ be points in R^n and $q_1, \ldots, q_s$ nonnegative members satisfying $q_1 + \cdots + q_s = 1$. Then

$$f(q_1x^1 + \cdots + q_sx^s) \leq q_1f(x^1) + \cdots + q_sf(x^s) \tag{98}$$

2. Let f be a convex function and let λ be a nonnegative number. Then λf is also a convex function. Let f and g be convex functions. Then f + g is also convex, provided that the undefined operation $+\infty + (-\infty)$ is avoided. In particular, every linear combination $\lambda_1f_1 + \cdots + \lambda_kf_k$ of convex functions with $\lambda_1 \geq 0, \ldots, \lambda_k \geq 0$ is also a convex function.
3. Let f be a real convex function on R^n and let Ψ be a nondecreasing proper convex function on R. Then $\Psi(f(x))$ is convex on R^n.
4. Let $\{f_i\}$, $i \in I$, be a finite or infinite collection of convex functions on R^n. For every $x \in R^n$, define the pointwise supremum of this collection as

$$f(x) = \sup_{i \in I} f_i(x) \tag{99}$$

Then f is a convex function.
5. The function f is convex on R^n if and only if for every $x^1 \in R^n$, $x^2 \in R^n$ the function ϕ, defined by

$$\phi(\lambda) = f(\lambda x^1 + (1 - \lambda)x^2) \tag{100}$$

is convex for all $0 \leq \lambda \leq 1$.

The reader may have noticed that we have said nothing about the continuity of convex functions. It can easily be demonstrated that convex functions can be discontinuous. Take, for example, the function f given by

$$f(x) = \begin{cases} (x)^2 - 1, & x < 1 \\ 2, & x = 1 \\ +\infty, & x > 1 \end{cases} \tag{101}$$

This is a convex function, discontinuous on the boundary of its effective domain. Roughly speaking, discontinuities in convex functions can occur only at some boundary points of their effective domain. Some of these discontinuities can be eliminated by the closure operation for convex functions. Consider a convex function f defined on R^n and, for a given point $x^0 \in R^n$, the collection of all linear affine functions h of the form $h(x) = a^T x - b$ such that $h(x^0) \leq f(x^0)$. Let us define the <u>support set</u>

$$L(f) = \{(a, b): a \in R^n, \; b \in R, \; a^T x - b \leq f(x) \text{ for every } x \in R^n\} \tag{102}$$

Then we define the <u>closure of a convex function</u>, denoted cl f, as

$$\text{cl } f(x) = \sup_{(a,b) \in L(f)} \{a^T x - b\} \tag{103}$$

Clearly, $\text{cl } f(x) \leq f(x)$ for every $x \in R^n$. A convex function f is said to be <u>closed</u> if $f = \text{cl } f$. In particular, it can be shown that a proper convex function f is closed if and only if the convex level set $\{x: x \in R^n, \; f(x) \leq \alpha\}$ is closed for every real number α.

Equivalently, the closure operation for proper convex functions is directly related to the closure operation for sets (the notation $\bar{C}$ is used for the closure of the set C). The epigraph of cl f is the closure of the epigraph of f; that is,

$$P(\text{cl } f) = \overline{P(f)} \tag{104}$$

The last relation can, of course, also be used as the definition of the closure operation for proper convex functions. For improper convex functions, the closure operation has a very simple meaning. If $f(x) = +\infty$ for every $x \in R^n$, then $\text{cl } f = f$. If, however, $f(x) = -\infty$ for some $x \in R^n$, it follows that $L(f) = \emptyset$; and since, by convention, the supremum over the empty set is taken to be $-\infty$, we get $\text{cl } f(x) = -\infty$ for <u>every</u> $x \in R^n$. Consequently, the only closed improper convex functions are those that are identically $+\infty$ or $-\infty$.

<u>Example 10</u>. Let us apply the closure operation to the convex function defined by (101). Since this is a proper convex function, we simply "close" the epigraph of f as in Fig. 5. Thus we obtain the closure of f as

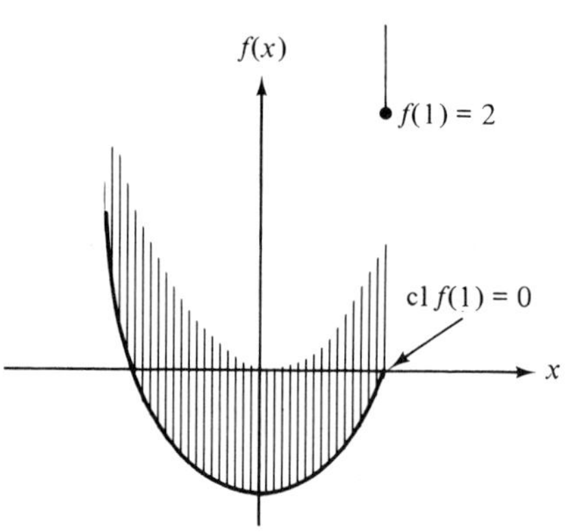

FIG. 5. Closure operation for a convex function.

$$\text{cl } f = \begin{cases} (x)^2 - 1, & x \leq 1 \\ +\infty, & x > 1 \end{cases} \tag{105}$$

and

$$\text{cl } f(1) = 0 \tag{106}$$

In other words, the closure operation in this example lowered the value of f at the boundary of its effective domain until the function became continuous over ED(f).

As we just saw in a simple example, the closure operation for convex functions eliminates certain discontinuities at boundary points of the effective domain. The function cl f agrees with f at every point in the interior of ED(f). The reason is that convex functions are continuous on the interior of their effective domains. Actually, a slightly more general result holds, but before stating it we introduce the notion of relative interior of a nonempty convex set.

Consider a subset B of R^n with the following property. For every two points $x^1 \in B$, $x^2 \in B$ and for every value of the real number α, also $\alpha x^1 + (1-\alpha)x^2 \in B$. A set satisfying this property is called an <u>affine set</u>. Single points, lines, and hyperplanes are examples of affine sets in R^n. Given a convex set $C \subset R^n$, the intersection of all affine sets containing C is called the <u>affine hull</u> of C. The <u>relative interior</u> of C, denoted by ri C, is defined as the interior of C, viewed as a subset of its affine hull.

<u>Example 11</u>. Consider the convex set $C \subset R^2$ consisting of the line segment between the points (a, 0) and (b, 0), a < b; that is,

$$C = \{(x_1, 0): a \leq x_1 \leq b\} \tag{107}$$

This set has no interior if we view C as a subset of R^2 (since one cannot find an open hypersphere in R^2 contained in C). The affine hull of C is, of course, the whole x_1 (one-dimensional) axis, and the relative interior of C is given by

$$\text{ri } C = \{(x_1, 0): a < x_1 < b\} \tag{108}$$

which is the interior of C viewed as a subset of the x_1 axis.

Note that if the convex set $C \subset R^n$ is n dimensional, such as a square in R^2 or a cube in R^3, then the affine hull of C is R^n itself, so the relative interior of C coincides with the interior of C.

We can now state the following theorem.

<u>Theorem 13</u>. A convex function is continuous on the relative interior of its effective domain. In particular, a real-valued convex function on R^n is continuous everywhere.

Differential Properties of Convex Functions

We begin by recalling a few notions from the differential calculus of real functions and applying them to convex functions. Let us define a direction as any vector

$y \in R^n$. Given a real function f defined on a subset S of R^n and a point x^0 in the interior of S, the derivative of f at x^0 in the direction of y or the directional derivative of f at x^0 in the direction of y is defined as

$$Df(x^0; y) = \lim_{t \to 0} \frac{f(x^0 + ty) - f(x^0)}{t} \tag{109}$$

if the limit exists.

A straightforward extension of this definition is as follows: Let f be any function defined on R^n with values in the extended reals. Let x^0 be a point where f(x) has a finite value and let y be a direction. The right-sided derivative of f at x^0 in the direction of y is defined as

$$D^+f(x^0; y) = \lim_{t \to 0^+} \frac{f(x^0 + ty) - f(x^0)}{t} \tag{110}$$

if the limit, which can be $\pm\infty$, exists. Here $t \to 0^+$ means that t approaches 0 through positive numbers. Similarly, the left-sided derivative of f at x^0 in the direction of y is defined as

$$D^-f(x^0; y) = \lim_{t \to 0^-} \frac{f(x^0 + ty) - f(x^0)}{t} \tag{111}$$

For y = 0, both $D^+f(x^0; 0)$ and $D^-f(x^0; 0)$ are defined to be zero.

The reader can easily verify that

$$-D^+f(x^0; -y) = D^-f(x^0; y) \tag{112}$$

If for some x^0 and y

$$D^+f(x^0; y) = D^-f(x^0; y) \tag{113}$$

then we have a directional derivative in the sense of (109). Letting $y = (1, 0, \ldots, 0)^T \in R^n$, the directional derivative of f at x^0 in the direction of y is just $\partial f(x^0)/\partial x_1$, the partial derivative of f with respect to x_1, and we can similarly get the partial derivatives of f with respect to $x_2, \ldots, x_n$. If a function is differentiable at a point x^0, then the directional derivatives of f at x^0 in all directions y are finite and are given by

$$Df(x^0; y) = y^T \nabla f(x^0) \tag{114}$$

where, as before, $\nabla f(x)$ is the gradient of f, evaluated at x. Before stating the first result on directional derivatives of convex functions, the reader might wish to recall that a function f on R^n is said to be positively homogeneous (of degree one) if for every $x \in R^n$ and positive numbers t

$$f(tx) = tf(x) \tag{115}$$

We have the following theorem.

Theorem 14. Let f be a convex function and let $x \in R^n$ be a point such that f(x) is finite. Then the right- and left-sided derivatives of f at x exist in every direction y. Moreover, D^+f and D^-f are positively homogeneous convex functions of y and

$$D^{+}f(x; y) \geq D^{-}f(x; y) \tag{116}$$

The next concept to be introduced is the subgradient, which is related to the ordinary gradient in the case of differentiable convex functions and to the directional derivatives in the more general case. A subgradient of a convex function f at a point $x \in R^n$ is a vector $\xi \in R^n$ satisfying

$$f(y) \geq f(x) + \xi^T(y - x) \tag{117}$$

for every $y \in R^n$. For a convex function f it is possible that, at some point x, (i) no vector satisfying (117) exists, or (ii) there is a unique ξ satisfying (117), or (iii) there exist more than one such ξ. We denote by $\partial f(x)$ the set of all subgradients of a convex function f at x. Some basic properties of subgradients can be summarized as follows: The set $\partial f(x)$, also called the subdifferential of f, is a closed convex set. It contains a single vector $\xi \in R^n$ if and only if the convex function f is differentiable (in the ordinary sense) at x and then $\xi = \nabla f(x)$. That is,

$$\xi_j = \frac{\partial f(x)}{\partial x_j}, \quad j = 1, \ldots, n \tag{118}$$

Subgradients can be characterized by the directional derivatives, as can be seen in the next theorem.

Theorem 15. A vector $\xi \in R^n$ is a subgradient of a convex function f at a point x where f(x) is finite if and only if

$$D^{+}f(x; y) \geq \xi^T y \tag{119}$$

for every direction y.

From the last two theorems we have the following.

Theorem 16. Let f be a convex function on R^n and suppose that f(x) is finite. Then

$$f(y) \geq f(x) + D^{+}f(x; y - x) \tag{120}$$

for every $y \in R^n$. In particular, if f is differentiable at x, then

$$f(y) \geq f(x) + (y - x)^T \nabla f(x) \tag{121}$$

Suppose now that f(x) and f(y) are finite. Then

$$D^{+}f(y; y - x) \geq D^{+}f(x; y - x) \tag{122}$$

and

$$D^{-}f(y; y - x) \geq D^{-}f(x; y - x) \tag{123}$$

If particular, if f is differentiable at x and y, then

$$(y - x)^T[\nabla f(y) - f(x)] \geq 0 \tag{124}$$

Example 12. To illustrate the notions and results obtained so far, consider the convex function f defined on R by

$$f(x) = \begin{cases} +\infty, & x < -1 \\ 2, & x = -1 \\ (x)^2, & -1 < x \leq 0 \\ x, & 0 \leq x \leq 1 \\ +\infty, & 1 < x \end{cases} \tag{125}$$

The right- and left-sided derivatives in the direction y = 1 are given by

$$D^+f(x; 1) = \begin{cases} \text{undefined}, & x < -1 \\ -\infty, & x = -1 \\ 2x, & -1 < x < 0 \\ 1, & 0 \leq x < 1 \\ +\infty, & x = 1 \\ \text{undefined}, & 1 < x \end{cases} \tag{126}$$

and

$$D^-f(x; 1) = \begin{cases} \text{undefined}, & x < -1 \\ -\infty, & x = -1 \\ 2x, & -1 < x \leq 0 \\ 1, & 0 < x \leq 1 \\ \text{undefined}, & 1 < x \end{cases} \tag{127}$$

We can see that $D^+f(x; 1) = D^-f(x; 1)$ for $-1 \leq x < 0$ and $0 < x < 1$, and $D^+f(x; 1) > D^-f(x; 1)$ for $x = 0$ and $x = 1$. By Theorem 15, the number ξ is a subgradient of f if and only if

$$D^+f(x; y) \geq \xi y \tag{128}$$

for every $y \in R$. Since one-sided derivatives are positively homogeneous, we get

$$D^+f(x, y) = \begin{cases} yD^+f(x; 1), & y > 0 \\ 0, & y = 0 \\ -yD^+f(x; -1), & y < 0 \end{cases} \tag{129}$$

From the last two relations we conclude that $\xi \in \partial f(x)$ if and only if

$$D^+f(x; 1) \geq \xi \geq D^-f(x; 1) \tag{130}$$

Consequently,

$$\partial f(x) = \begin{cases} \emptyset, & x \leq -1 \\ 2x, & -1 < x < 0 \\ \{\xi : 0 \leq \xi \leq 1\}, & x = 0 \\ 1, & 0 < x < 1 \\ \{\xi : \xi \geq 1\} & x = 1 \\ \emptyset, & 1 < x \end{cases} \tag{131}$$

In the next few results we restrict our attention to real-valued differentiable convex functions, although most of the following results can also be extended to convex functions defined in the wider sense at the beginning of this section by the general concepts of one-sided (right and left) derivatives and subgradients. The differential results established in Theorem 16 were necessary conditions for convex functions. The next theorem shows that, in the special case of differentiable convex functions, these conditions are also sufficient.

Theorem 17. Let f be a real-valued differentiable function on R^n. If

$$f(y) \geq f(x) + (y - x)^T \nabla f(x) \tag{132}$$

or

$$(y - x)^T [\nabla f(y) - \nabla f(x)] \geq 0 \tag{133}$$

for every two points $x \in R^n$, $y \in R^n$, then f is convex on R^n.

We conclude the discussion of differential properties with some results involving twice differentiable convex functions.

Theorem 18. Let f be a real-valued function on an open convex set $C \subset R^n$ with continuous second partial derivatives. Then f is convex on C if and only if the Hessian of f evaluated at every $x \in C$ is positive semidefinite. That is, for each $x \in C$ we have

$$y^T \nabla^2 f(x) y \geq 0 \tag{134}$$

for every $y \in R^n$.

A word of caution is in order here. Theorem 18 cannot be fully sharpened in the case of strictly convex functions by replacing the words "positive semidefinite" in the statement of the theorem by "positive definite." In fact, the reader can easily find examples of strictly convex functions whose Hessians are not positive definite. The situation is, however, somewhat better in the converse case; that is, under the hypotheses of the theorem a positive definite Hessian matrix does imply strict convexity. For a thorough treatment of these points, the reader is referred to Bernstein and Toupin [31].

Extrema of Convex Functions

As indicated earlier, convex functions play a central role in nonlinear programming. The main importance of convex functions lies in some basic properties that are summarized below.

Theorem 19. Let f be a proper convex function on R^n. Then every local minimum of f is a global minimum of f on R^n.

Suppose that in problem (P), defined by (48) to (50), the functions $g_1, \ldots, g_m$ are all concave functions and the $h_1, \ldots, h_p$ are all linear functions. Then the feasible set X is a convex set. If the objective function f to be minimized is a proper convex function, we can define a new objective function $\hat{f}$, given by

$$\hat{f}(x) = \begin{cases} f(x), & \text{if } x \in X \\ +\infty, & \text{if } x \notin X \end{cases} \tag{135}$$

and $\hat{f}$ will be a proper convex function on R^n, coinciding with f on X. From the previous theorem we conclude that every local minimum of $\hat{f}$ is also a global minimum, or if X is nonempty, every local minimum of f at some point $x^* \in X$ is also a global minimum of f on all X. Formally, we have the following.

Theorem 20. Let f be a proper convex function on R^n and let $X \subset R^n$ be a convex set. Then every local minimum of f at $x^* \in X$ is a global minimum of f over all X.

Note that generally the minimal value of a convex function can be attained at more than one point. The next theorem states that the set of minimizing points of a proper convex function is a convex set.

Theorem 21. Let f be a convex function on R^n and let α be a real number. Then the level sets of f, given by

$$S(f, \alpha) = \{x: x \in R^n,\ f(x) \le \alpha\} \tag{136}$$

are convex sets for any α. In particular, the set of points at which f attains its minimum is convex. If f is a strictly convex function defined on a convex set $X \subset R^n$ and attains its minimum on X, it is attained at a unique point of X.

In many applications, when the minimum of a differentiable function is sought, one looks for the stationary points of the function—that is, points at which the gradient vanishes. This situation can be justified in the case of convex functions by the following.

Theorem 22. Let f be a convex function on R^n. Then $0 \in \partial f(x^*)$ if and only if f attains its minimum at x^*. In particular, let f be a differentiable convex function on an open convex set $X \subset R^n$ such that $x^* \in X$. Then

$$\nabla f(x^*) = 0 \qquad (137)$$

if and only if f attains its minimum at x^*.

This theorem also indicates that in seeking an unconstrained minimum of a convex function, no second-order conditions need be checked at points where the gradient vanishes.

Optimality Conditions for Convex Programs

Here we consider convex programs, a special case of the general nonlinear programming problem (P), introduced earlier. The optimality conditions derived there become simpler for convex programs. Consider, then, the following nonlinear program, called a convex program:

$$\text{(CP)} \qquad \min f(x) \qquad (138)$$

subject to the constraints

$$g_i(x) \geq 0, \quad i = 1, \ldots, m \qquad (139)$$

$$h_j(x) = 0, \quad j = 1, \ldots, p \qquad (140)$$

where f is a proper convex function on R^n, the g_i are proper concave functions, and the h_j are linear (affine) functions of the form

$$h_j(x) = \sum_{k=1}^{n} a_{jk} x_k - b_j \qquad (141)$$

Such a program is called convex because the objective function is a convex function, and the set of all $x \in R^n$ satisfying the inequalities in (139) and the equations in (140) is a convex set. Note that generally the set of $x \in R^n$ satisfying an equation $h(x) = 0$, where h is a nonlinear convex or concave function, is not a convex set.

We shall now show that with an appropriate assumption of differentiability, the Kuhn-Tucker necessary conditions for optimality, as stated in Theorems 7 and 8, are also sufficient when applied to a convex program.

Theorem 23. Suppose that the functions f and $g_1, \ldots, g_m$ are real-valued, differentiable, convex and concave functions on R^n, respectively, and let $h_1, \ldots, h_p$ be linear. If there exist vectors x^*, λ^*, μ^*, with x^* satisfying (139) and (140), and

$$\nabla f(x^*) - \sum_{i=1}^{m} \lambda_i^* \nabla g_i(x^*) - \sum_{j=1}^{p} \mu_j^* \nabla h_j(x^*) = 0 \qquad (142)$$

$$\lambda_i^* g_i(x^*) = 0, \qquad i = 1, \ldots, m \qquad (143)$$

$$\lambda^* \geq 0 \qquad (144)$$

then x^* is a global optimum of (CP).

It is clear from this result that under the same hypotheses the John conditions, as stated in Theorem 9, are also sufficient, provided that $\lambda_0^* > 0$.

The regularity condition (62) assumed in the necessary conditions for optimality in the general case can be replaced in convex programs by a simpler and easier computable, although stronger condition, due to Slater [32]. A convex program (CP) is said to be <u>strongly consistent</u> if there exists a point $x^0 \in R^n$ satisfying

$$g_i(x^0) > 0, \quad i = 1, \ldots, m \tag{145}$$

$$h_j(x^0) = 0, \quad j = 1, \ldots, p \tag{146}$$

and the vectors of coefficients $a^j = (a_{j1}, \ldots, a_{jn})^T$ in the linear functions $h_j(x) = (a^j)^T x - b_j$ are linearly independent. Such a program is also said to satisfy Slater's condition.

We conclude the discussion of optimality conditions in convex programs with a converse of Theorem 11 for the case of convex programs. We shall see that every solution of a convex program is a saddlepoint of the corresponding Lagrangian.

<u>Theorem 24</u>. Let the functions f, $g_1, \ldots, g_m, h_1, \ldots, h_p$ satisfy the hypotheses of Theorem 23, possibly without the differentiability assumption. Suppose that (CP) is strongly consistent. If x^* is a solution of (CP), then there exist vectors $\lambda^* = (\lambda_1^*, \ldots, \lambda_m^*)$ and $\mu^* = (\mu_1^*, \ldots, \mu_p^*)$ such that (x^*, λ^*, μ^*) is a solution of problem (S), that is,

$$L(x^*, \lambda, \mu) \leq L(x^*, \lambda^*, \mu^*) \leq L(x, \lambda^*, \mu^*) \tag{147}$$

for every $\lambda \in R^m$, $\lambda \geq 0$, and $\mu \in R^p$. Furthermore,

$$\lambda_i^* g_i(x^*) = 0, \quad i = 1, \ldots, m \tag{148}$$

<u>Duality in Convex Programming</u>

In discussing duality for nonlinear programming problems, it is necessary to start with linear programming where the first and most complete duality results in mathematical programming were obtained. According to Dantzig [33], the notion of duality was first introduced into linear programming by Von Neumann in 1947 and was subsequently formulated in a precise form by Gale, Kuhn, and Tucker [34]. The idea of duality is to associate with each linear program, called the <u>primal program</u>, another linear program, called the <u>dual program</u>. Dual linear programs have some interesting properties that, in addition to being elegant from a theoretical point of view, are also significant for computational purposes and economic interpretations.

The most important properties of primal-dual linear programs are as follows:

1. If the primal program is a minimization of a linear function over a set of linear constraints, then the dual is a maximization of another linear function over another set of linear constraints. Given a (primal) linear program, its dual can easily be formulated from the data of the primal alone. Primal variables do not appear in the dual program and vice versa. Relations between primal

and dual programs are "involutions"; that is, the dual of the dual is again the primal.

2. For all primal and dual feasible solutions, the primal objective function (to be minimized) has a value greater than or equal to the value of the dual objective function (to be maximized). If the primal has an optimal solution, then the dual also has an optimal solution and the respective objective functions have the same value at optimum. By knowing the optimal solution of either the primal or the dual, it is possible to obtain an optimal solution of the other program without solving it. It is possible, therefore, to solve a linear program via its dual if the dual program is computationally more attractive.
3. Linear programming plays an important role in economic theory. An optimal dual solution can be considered the vector of Lagrange multipliers or "shadow prices" of the primal program and vice versa. Many economic problems can be stated as linear programs. The dual programs in these cases turn out to represent related economic problems.

The elegant duality theorems in linear programming, as well as the important paper of Kuhn and Tucker [5], have encouraged many researchers to extend the notion of duality to nonlinear programming problems. Early results in this direction were not encouraging. It quickly became clear that in order to obtain significant results, only convex primal programs could be considered. Unfortunately, the dual of such a program, as formulated in the earlier works [35-38], was usually a non-convex program, and there was no "involution" between primal and dual programs, as in the linear case, where the dual of the dual is the primal.

During the late 1960s a more complete duality theory for convex programs emerged, the mean features of which are similar to those of the primal-dual linear programs listed above. This theory is based on the concept of conjugate functions, introduced by Fenchel [39] and developed to its full strength by Rockafellar [24].

According to the modern approach, nonlinear convex programs will be formulated together with certain kinds of perturbations in the data of the problem. Such perturbations usually represent small changes in the parameters of the program and are important from both a theoretical and a practical point of view. This approach will allow us to study questions on the sensitivity of optimal solutions to small perturbations in the parameters of the program, in addition to the more standard questions on existence and characterization of feasible and optimal solutions.

Conjugate Functions

We defined the closure of a convex function f on R^n as the pointwise supremum of all linear (affine) functions h with values $h(x) \leq f(x)$. Suppose that $h(x) = \xi^T x - b$. We can ask the following question: Given a vector of coefficients $\xi \in R^n$, for what values of b will h(x) be less than or equal to f(x) for every $x \in R^n$? Clearly, b must satisfy

$$b \geq \xi^T x - f(x) \tag{149}$$

for every $x \in R^n$. Hence

$$b \geq \sup_x \{\xi^T x - f(x)\} \tag{150}$$

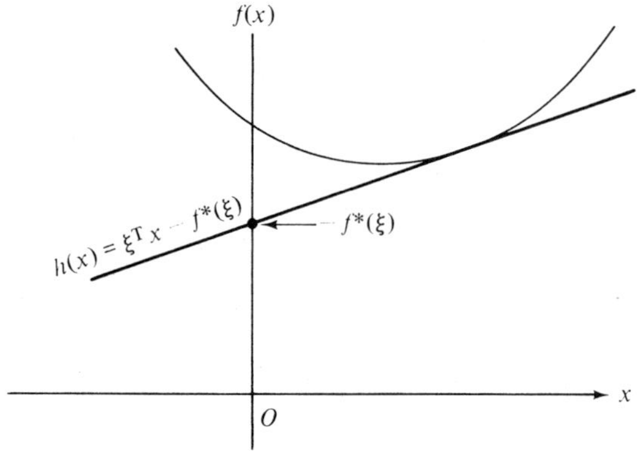

FIG. 6. Geometric interpretation of conjugate functions.

where, unless explicitly mentioned, the supremum is understood to be taken on all R^n. Generally, the greatest lower bound of b satisfying (150) is a function of ξ, called the conjugate function of f and denoted f^*. Thus

$$f^*(\xi) = \sup_x \{\xi^T x - f(x)\} \tag{151}$$

The geometric interpretation of conjugate functions can be seen in Fig. 6. The intercept with the f axis of the highest linear function having a vector of coefficients ξ and lying below the function f is equal to $-f^*(\xi)$. Changing ξ will accordingly raise or lower the value of $-f^*(\xi)$ and shift the point x where the linear function h is tangent to f (if there is such a point). Although conjugate functions may look like a pure mathematical concept, they have sound economic interpretations. Suppose, for example, that the cost of manufacturing quantities $x_1, x_2, \ldots, x_n$ of n goods is given by f(x). The goods can be sold at prices $\xi_1, \xi_2, \ldots, \xi_n$, respectively. The manufacturer's problem is to choose values of $x_1, \ldots, x_n$ such that his profit from selling the goods is maximized. Clearly, his profit is given by the difference between his revenues $\Sigma_j \xi_j x_j$ and the production cost f(x). The highest possible value of his profit, as a function of the prices, is then given by $f^*(\xi)$, as can be seen from (151). Conjugate functions can also be regarded as an extension of the classical Legendre transformation [40, 41].

Some properties of conjugate functions are summarized in the following.

Theorem 25. Let f be a convex function on R^n. The conjugate function f^* is also a convex function and $f^{**} = \text{cl } f$. Moreover, f^* is a closed convex function. It is proper if and only if f is proper.

The following elementary properties of conjugate functions can be easily shown. Let f be a convex function on R^n, α a given real number, and z a given vector in R^n. Then,

(i) If $\phi(x) = f(x) + \alpha$, then $\phi^*(\xi) = f^*(\xi) - \alpha$ (152)

(ii) If $\phi(x) = f(x + z)$, then $\phi^*(\xi) = f^*(\xi) - \xi^T z$ (153)

(iii) If $\phi(x) = f(\alpha x)$, $\alpha \neq 0$, then $\phi^*(\xi) = f^*\left(\frac{\xi}{\alpha}\right)$ (154)

(iv) If $\phi(x) = \alpha f(x)$, $\alpha > 0$, then $\phi^*(\xi) = \alpha f^*\left(\frac{\xi}{\alpha}\right)$ (155)

Example 13. Consider a few simple pairs of conjugate functions.

(a) Let $\phi(x) = a^T x - b$, $x \in R^n$. Compute first the conjugate function of $f(x) = a^T x$:

$$f^*(\xi) = \sup_x \{\xi^T x - a^T x\} = \begin{cases} 0, & \text{if } \xi = a \\ +\infty, & \text{if } \xi \neq a \end{cases} \tag{156}$$

Hence, by (152), $\phi^*(\xi) = b$ if $\xi = a$ and $\phi^*(\xi) = +\infty$ otherwise.

(b) Let $\phi(x) = \frac{1}{2}x^T x$, $x \in R^n$, with $f(x) = x^T x$. Then

$$f^*(\xi) = \sup_x \{\xi^T x - x^T x\} \tag{157}$$

We can see by simple differentiation that the supremum is achieved for $x = \frac{1}{2}\xi$. Thus $f^*(\xi) = \frac{1}{4}\xi^T\xi$ and, by (155), $\phi^*(\xi) = \frac{1}{2}\xi^T\xi$.

(c) Let $x \in R$ and let $\phi(x) = -\log x$ if $x > 0$ and $\phi(x) = +\infty$ otherwise. Then $\phi^*(\xi) = \sup_{x>0}\{\xi x + \log x\}$. For $\xi \geq 0$ we have $\phi^*(\xi) = +\infty$, since $\xi x + \log x$ can be made arbitrarily large by choosing $x \to +\infty$. For $\xi < 0$ we can find the supremum by differentiation and obtain $\phi^*(\xi) = -1 - \log(-\xi)$. Thus

$$\phi^*(\xi) = \begin{cases} -1 - \log(-\xi), & \text{if } \xi < 0 \\ +\infty, & \text{if } \xi \geq 0 \end{cases} \tag{158}$$

There is a close relationship between subgradients and conjugate functions, as can be seen in the following results.

Theorem 26. Let f be a convex function on R^n. Then $\xi \in \partial f(x)$ if and only if $f^*(\xi) = \xi^T x - f(x)$. If $\xi \in \partial f(x)$, then $x \in \partial f^*(\xi)$. If f is also closed at x and $x \in \partial f^*(\xi)$, then $\xi \in \partial f(x)$.

Let us briefly discuss the conjugates of concave functions. Generally speaking, every result in conjugate function theory for convex functions can be similarly developed for concave functions, remembering that a function g is concave if and only if $f = -g$ is convex. In particular, the conjugate function g^* of a concave function g is defined as

$$g^*(\xi) = \inf_x \{\xi^T x - g(x)\} \tag{159}$$

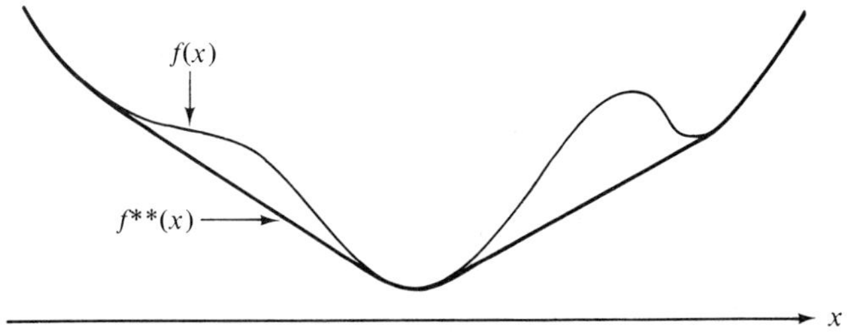

FIG. 7. The conjugate of the convex conjugate of a nonconvex function.

Here g^* is a closed concave function and its conjugate is, in turn, cl g, with the closure operation defined in analogy to the convex case. This g^* is, however, not the same as $-f^*$, where $f = -g$. In fact, $g^*(\xi) = -f^*(-\xi)$.

We may also note that conjugate functions can be defined as above for functions that are neither convex nor concave. Suppose that f is an arbitrary function defined on R^n with values in the extended reals. Then its <u>convex conjugate</u> f^* is defined by (151) and it is a closed convex function. It is the conjugate of the closure of the greatest convex function f_c satisfying $f_c \le f$. In other words, f^* is the conjugate of the closure of the convex function that is the lower envelope of the convex hull of P(f), see Fig. 7. For such an arbitrary function, we can define a concave conjugate in an obvious analogy.

Dual Convex Programs

As noted, we shall develop duality relations for nonlinear convex programming problems in a somewhat broad sense that will give us insight into the sensitivity of the optima to certain perturbations in the parameters of the programs. We deal in this section with the following version of (CP), introduced earlier:

$$\text{(SP)} \qquad \min_x f(x) \tag{160}$$

subject to

$$g_i(x) \ge 0, \quad i = 1, \ldots, m \tag{161}$$

where f is a proper convex function on R^n and the g_i are proper concave functions on R^n. We call program (SP) the <u>standard primal</u> convex programming problem. This program is called the "canonical" primal problem by Geoffrion [42] when f and g_i are real valued and the "ordinary" convex program by Rockafellar [24]. It differs from (CP) in assuming that no linear equality constraints are present. The perturbations mentioned earlier can be introduced, for example, at the right-hand side of the constraints. That is, the inequalities in (161) can be replaced by the more general constraints

$$g_i(x) \ge w_i, \quad i = 1, \ldots, m \tag{162}$$

where the w_i are real parameters. The optimal solution of (SP) with (162) replacing (161) can be regarded as a function of w, and consequently one can examine the behavior of the optimum for vectors w near the origin.

Denoting by C the effective domain of f, we then define a new convex function ϕ on R^n by the relation

$$\phi(x) = \begin{cases} f(x), & \text{if } x \in C \text{ and } g_i(x) \geq 0, \quad i = 1, \ldots, m \\ +\infty, & \text{otherwise} \end{cases} \tag{163}$$

and (SP) can be equivalently written as the (unconstrained) minimization problem of ϕ. If we denote some vector of perturbations by $w \in R^k$, we can generalize (163) to

$$\phi(x, w) = \begin{cases} f(x, w), & \text{if } (x, w) \in ED(f) \subset R^{n+k} \text{ and } g_i(x, w) \geq 0, \\ & \qquad i = 1, \ldots, m \\ +\infty, & \text{otherwise} \end{cases} \tag{164}$$

and by a proper choice of the form in which w is incorporated into the functions involved, (SP) can become the problem of minimizing $\phi(x, 0)$ over all $x \in R^n$. Note that in (162) we actually assumed that $w \in R^m$ and $g_i(x, w) = g_i(x) - w_i$. The general formulation given in (164) motivates the forthcoming analysis, which is based on the works of Geoffrion [42], Hamala [43], and Rockafellar [24, 44].

Let ϕ be a proper convex function of two vectors $x \in R^n$ and $w \in R^k$. The conjugate function of ϕ is defined as

$$\phi^*(\xi, \lambda) = \sup_{x, w} \{\xi^T x + \lambda^T w - \phi(x, w)\} \tag{165}$$

Hence for every $x \in R^n$, $\xi \in R^n$, $w \in R^k$, and $\lambda \in R^k$, we have

$$\phi(x, w) + \phi^*(\xi, \lambda) \geq \xi^T x + \lambda^T w \tag{166}$$

In particular, we can set $w = 0$, $\xi = 0$, and for every $x \in R^n$ and $\lambda \in R^k$ we obtain

$$\phi(x, 0) + \phi^*(0, \lambda) \geq 0 \tag{167}$$

As a result of the last relation, we define the following pair of convex programs:

$$(P_\phi) \quad \min_x \phi(x, 0) \quad \text{(primal program)} \tag{168}$$

$$(D_{\phi^*}) \quad \max_\lambda -\phi^*(0, \lambda) \quad \text{(dual program)} \tag{169}$$

It should be pointed out that in these two programs we are looking for those x and λ that give a minimal value of $\phi(x, 0)$ and a maximal value of $-\phi^*(0, \lambda)$, respectively; but in many cases no such x and λ exist, although $\phi(x, 0)$ may have a finite infimum and $-\phi^*(0, \lambda)$ a finite supremum. We shall call the infimum of $\phi(x, 0)$ and the supremum of $-\phi^*(0, \lambda)$ the <u>optimal values</u> of (P_ϕ) and (D_{ϕ^*}), respectively.

Program (P_ϕ) is said to be <u>consistent</u> if there exists an $\hat{x}$ such that $\phi(\hat{x}, 0) < +\infty$, or, equivalently, if $(\hat{x}, 0) \in ED(\phi)$. Similarly, (D_{ϕ^*}) is consistent if there exists a $\hat{\lambda}$ such that $\phi^*(0, \hat{\lambda}) < +\infty$. If for every $x \in R^n$, $(x, 0) \notin ED(\phi)$, then (P_ϕ) is said to be <u>inconsistent</u>. If $(0, \lambda) \notin ED(\phi^*)$ for every $\lambda \in R^k$, then (D_{ϕ^*}) is inconsistent. The following result is a straightforward consequence of (167).

Theorem 27 (Weak Duality Theorem). For ϕ and ϕ^* as defined above,

$$\inf_x \phi(x, 0) \geq \sup_\lambda -\phi^*(0, \lambda) \tag{170}$$

If

$$\phi(x^*, 0) = -\phi^*(0, \lambda^*) \tag{171}$$

then x^* and λ^* are optimal solutions of (P_ϕ) and (D_{ϕ^*}), respectively.

If

$$\inf_x \phi(x, 0) = -\infty \tag{172}$$

then (D_{ϕ^*}) is inconsistent. If

$$\sup_\lambda -\phi^*(0, \lambda) = +\infty \tag{173}$$

then (P_ϕ) is inconsistent.

Convexity plays a central role in the forthcoming results, but it is interesting to note here that relations (165) to (173) also hold for functions ϕ that are not necessarily convex, provided that the undefined sum $+\infty + (-\infty)$ is avoided. This condition is ensured, for example, if $\phi(x, 0)$ has finite values on R^n. Any feasible solution λ of a dual program provides a value of $-\phi^*(0, \lambda)$ that can serve as a lower bound on the optimal value of (P_ϕ). It would be desirable, however, to have a stronger result than (170), one stating that the optimal values of (P_ϕ) and (D_{ϕ^*}) are actually equal. Here is one of the main differences between convex and nonconvex programs: for a convex ϕ, under suitable regularity assumptions, (170) does hold as an equality, whereas for ϕ nonconvex one can get a duality gap, which is the difference between the optimal values of (P_ϕ) and (D_{ϕ^*}) when (170) holds as a strict inequality.

Suppose now that ϕ is a closed proper convex function on R^{n+k}. Then $\phi^{**} = \phi$, and by reformulating (D_{ϕ^*}) as a primal program

$$(P_{\phi^*}) \qquad \min_\lambda \phi^*(0, \lambda) \tag{174}$$

we get a dual program

$$(D_\phi) \qquad \max_x -\phi^{**}(x, 0) = \max_x -\phi(x, 0) \tag{175}$$

which is equivalent to the original primal program (168) in the sense that the vector x is optimal for (P_ϕ) if and only if it is optimal for (D_ϕ). Thus the duality relations for closed proper convex functions are symmetric: the dual of (D_{ϕ^*}) is (P_ϕ).

Associated with the proper convex functions ϕ and ϕ^*, we define the primal and dual perturbation functions Φ and Ψ as follows:

$$\Phi(w) = \inf_x \phi(x, w) \tag{176}$$

$$\Psi(\xi) = \inf_\lambda \phi^*(\xi, \lambda) \tag{177}$$

The perturbation functions Φ and Ψ are also convex. We need the notion of perturbation functions in order to define stability of convex programs. Stability is, in turn, a condition on (P_ϕ) that we use to establish equality for the optimal values of (P_ϕ) and (D_{ϕ^*}), the existence of an optimal solution for the dual program (D_{ϕ^*}), and a characterization of such a solution in terms of subgradients of Φ. Primal program (P_ϕ) is said to be stable if either (a) $\Phi(0)$ is finite and there is no direction y such that $D^+\Phi(0, y) = -\infty$, or (b) $\Phi(0) = -\infty$. In other words, (P_ϕ) is stable either if $\inf_x \phi(x, 0)$ is finite and the perturbation function $\Phi(w)$ does not decrease infinitely steeply for any w in the neighborhood of 0 or if the optimal value of (P_ϕ) is unbounded from below. We can also give an alternative characterization of stability: Program (P_ϕ) is stable if and only if $\partial\Phi(0)$ is a nonempty set. Note that if $\Phi(0) = -\infty$, then $\partial\Phi(0) = R^k$. Stability of (D_{ϕ^*}) is defined in a completely analogous manner.

Example 14. Let us consider some examples of stable and unstable programs.

(a) Let

$$\phi_1(x, w) = \begin{cases} x, & \text{if } x \geq w \\ +\infty, & \text{if } x < w \end{cases} \tag{178}$$

where $x \in R$, $w \in R$. It is easy to show that

$$\Phi_1(w) = \inf_x \phi_1(x, w) = w \tag{179}$$

for every $w \in R$ and $\Phi_1(0) = 0$. Hence Φ_1 is finite at 0. Indeed $\partial\Phi_1(0)$ consists of a single point

$$\partial\Phi_1(0) = \{\nabla\Phi_1(0)\} = \{1\} \tag{180}$$

and (P_{ϕ_1}) is stable.

(b) Let

$$\phi_2(x, w) = \begin{cases} -x, & \text{if } x \geq w \\ +\infty, & \text{if } x < w \end{cases} \tag{181}$$

Now we get

$$\Phi_2(w) = \inf_x \phi_2(x, w) = -\infty \tag{182}$$

for every $w \in R$. Hence $\Phi_2(0) = -\infty$ and (P_{ϕ_2}) is again stable.

(c) Next take

$$\phi_3(x, w) = \begin{cases} x, & \text{if } (x)^2 \leq w \\ +\infty, & \text{if } (x)^2 > w \end{cases} \tag{183}$$

Then

$$\Phi_3(w) = \begin{cases} -\sqrt{w}, & \text{if } w \geq 0 \\ +\infty, & \text{if } w < 0 \end{cases} \tag{184}$$

Here Φ_3 is a closed proper convex function, as can be seen in Fig. 8. At $w = 0$, we get $\Phi_3(0) = 0$, but

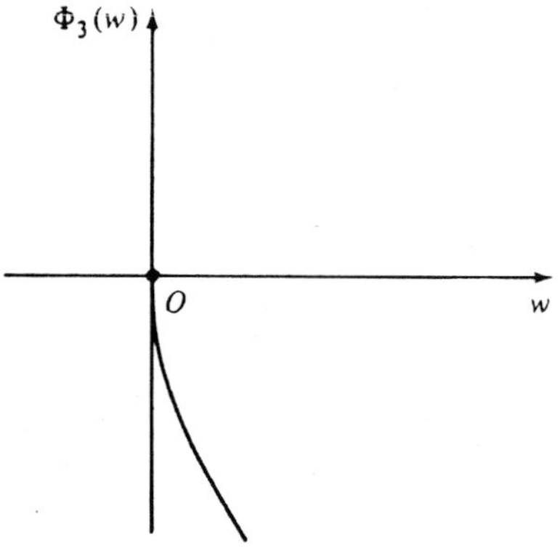

FIG. 8. Perturbation function for an unstable program.

$$D^+\Phi_3(0;\ 1) = \lim_{w\to 0^+} \frac{\Phi_3(w) - \Phi_3(0)}{w} = \lim_{w\to 0^+} \frac{-\sqrt{w}}{w} = -\infty \tag{185}$$

Thus (P_{ϕ_3}) is unstable.

We now state a strong duality theorem based on stability.

Theorem 28 (Strong Duality Theorem). Let ϕ be a closed proper convex function on R^{n+k} and let $\Phi(0)$ be finite. Then program (D_{ϕ^*}) has an optimal solution λ^*, and

$$\Phi(0) = \inf_x \phi(x,\ 0) = \max_\lambda -\phi^*(0,\ \lambda) = -\phi^*(0,\ \lambda^*) \tag{186}$$

if and only if (P_ϕ) is stable. Moreover, λ^* is a subgradient of $\Phi(0)$ if and only if (186) holds. Dually, let $\Psi(0)$ be finite. Then program (P_ϕ) has an optimal solution x^* and

$$\phi(x^*,\ 0) = \min_x \phi(x,\ 0) = \sup_\lambda -\phi^*(0,\ \lambda) \tag{187}$$

if and only if (D_{ϕ^*}) is stable. Moreover, x^* is a subgradient of $\Psi(0)$ if and only if (187) holds.

Consequently, if (P_ϕ) and (D_{ϕ^*}) are both stable, then both have optimal solutions and

$$+\infty > \min_x \phi(x,\ 0) = \max -\phi^*(0,\ \lambda) > -\infty \tag{188}$$

Example 15. Consider again the functions that appeared in the previous example.

(a) First let

$$\phi_1(x,\ w) = \begin{cases} x, & \text{if } x \geq w \\ +\infty, & \text{if } x < w \end{cases} \tag{189}$$

Then

$$\phi_1^*(\xi, \lambda) = \begin{cases} 0, & \text{if } \xi + \lambda = 1, \quad \lambda \geq 0 \\ +\infty, & \text{otherwise} \end{cases} \tag{190}$$

Hence the stable program (P_{ϕ_1}) can be written as

$$\min_x \phi_1(x, 0) = \min_{x \geq 0} x \tag{191}$$

and $(D_{\phi_1^*})$ is given by

$$\max_\lambda -\phi_1^*(0, \lambda) = \max_{\lambda=1} 0 \tag{192}$$

Clearly, $\min_{x \geq 0} x = 0$, and the optimal solutions of (P_{ϕ_1}) and $(D_{\phi_1^*})$ are $x^* = 0$, $\lambda^* = 1$, respectively. We have already seen that $\Phi_1(w) = w$ for every $w \in R$, $\Phi_1(0) = 0$, and $\partial\Phi_1(0) = \{1\}$.

Similarly,

$$\Psi_1(\xi) = \begin{cases} 0, & \text{if } \xi \leq 1 \\ +\infty, & \text{if } \xi > 1 \end{cases} \tag{193}$$

It follows that $\Psi_1(0) = 0$ and $\partial\Psi_1(0) = \{0\}$. Hence $(D_{\phi_1^*})$ is also stable. In addition, we have $x^* \in \partial\Psi_1(0)$ and $\lambda^* \in \partial\Phi_1(0)$.

(b) Next, take

$$\phi_2(x, w) = \begin{cases} -x, & \text{if } x \geq w \\ +\infty, & \text{if } x < w \end{cases} \tag{194}$$

Then

$$\phi_2^*(\xi, \lambda) = \begin{cases} 0, & \text{if } \xi + \lambda = -1, \quad \lambda \geq 0 \\ +\infty, & \text{otherwise} \end{cases} \tag{195}$$

The optimal value of (P_{ϕ_2}) approaches $-\infty$; that is,

$$\Phi_2(0) = \inf_x \phi_2(x, 0) = \inf_{x \geq 0} -x = -\infty \tag{196}$$

Turning now to the dual problem $(D_{\phi_2^*})$, we can see that $(0, \lambda) \notin ED(\phi_2^*)$ for every real λ (since $\lambda = -1$, $\lambda \geq 0$ has no solution). Thus $(D_{\phi_2^*})$ is inconsistent, and although (P_{ϕ_2}) is stable the dual has no solution since the assumption of a finite $\Phi_2(0)$ is not satisfied.

(c) Let

$$\phi_3(x, w) = \begin{cases} x, & \text{if } (x)^2 \leq w \\ +\infty, & \text{if } (x)^2 > w \end{cases} \tag{197}$$

We saw in Example 14 that program (P_{ϕ_3}) is unstable. Computing the conjugate function of ϕ_3, it can be shown that

$$\phi_3^*(\xi, \lambda) = \begin{cases} 0, & \text{if } \xi = 1, \lambda = 0 \\ -\dfrac{(\xi - 1)^2}{4\lambda}, & \text{if } \lambda < 0 \\ +\infty, & \text{otherwise} \end{cases} \tag{198}$$

The optimal solution of (P_{ϕ_3}) is $x^* = 0$ and

$$\Phi_3(0) = \min_x \phi_3(x, 0) = 0 \tag{199}$$

Turning to the dual program $(D_{\phi_3^*})$, we obtain

$$\sup_{\lambda<0} -\phi_3^*(0, \lambda) = \sup_{\lambda<0} \frac{1}{4\lambda} = 0 \tag{200}$$

but there is no real λ^* for which the supremum is attained.

We have seen that stability is an important property for duality of convex programs. Unfortunately, it is not always easy to show that a convex program is stable. There are, however, stronger conditions which imply stability and which are easier to establish. Earlier we defined strongly consistent programs that we redefine now for our more general convex program (P_ϕ) as given in (168): Program (P_ϕ) is said to be strongly consistent if there exists a positive number ϵ such that for every w satisfying $\|w\| < \epsilon$, an x^0 can be found such that $\phi(x^0, w) < +\infty$. For a convex program of type (SP), given by (160) and (161), strong consistency means that there exists an x^0 such that

$$g_i(x^0) > 0, \quad i = 1, \ldots, m \tag{201}$$

and this condition, in turn, implies that the corresponding (P_ϕ), derived from a program formulation in which the perturbation vector w is on the right-hand sides of the constraints, is strongly consistent in the sense just stated.

Lagrange Multipliers and Optimality Conditions

Consider again the convex program (P_ϕ) as defined in (168). We define the Lagrangian that corresponds to (P_ϕ) as [43, 44]

$$L(x, \lambda) = \inf_w \{\phi(x, w) - \lambda^T w\} \tag{202}$$

where, as before, λ is the vector of Lagrange multipliers.

The relation between primal and dual programs and the Lagrangian is given by the following.

Theorem 29. Suppose that ϕ is a closed proper convex function on R^{n+k}. Then

$$\phi(x, w) = \sup_\lambda \{L(x, \lambda) + \lambda^T w\} \tag{203}$$

$$-\phi^*(\xi, \lambda) = \inf_x \{L(x, \lambda) - \xi^T x\} \tag{204}$$

In particular,

$$\phi(x, 0) = \sup_{\lambda} L(x, \lambda) \tag{205}$$

and

$$-\phi^*(0, \lambda) = \inf_{x} L(x, \lambda) \tag{206}$$

The relation between saddlepoints of the Lagrangian and optimal solutions of the corresponding convex programs is given in Theorem 30.

Theorem 30. Let ϕ be a closed proper convex function on R^{n+k} and let L be the Lagrangian corresponding to (P_ϕ). If (x^*, λ^*) is a saddlepoint of L, then x^* is an optimal solution of (P_ϕ), λ^* is an optimal solution of (D_{ϕ^*}), and

$$\phi(x^*, 0) = L(x^*, \lambda^*) = -\phi^*(0, \lambda^*) \tag{207}$$

Conversely, if (P_ϕ) is stable and x^* is optimal for (P_ϕ), then there exists a λ^* such that (x^*, λ^*) is a saddlepoint of the Lagrangian L. If (D_{ϕ^*}) is stable and λ^* is optimal for (D_{ϕ^*}), then there exists an x^* such that (x^*, λ^*) is a saddlepoint of the Lagrangian L.

Let us return to (SP) as given in (160) and (161) and introduce the perturbations on the right-hand side of (161). The perturbed problem is given by

$$\min f(x) \tag{208}$$

subject to

$$g_i(x) \geq w_i, \qquad i = 1, \ldots, m \tag{209}$$

In order to get the dual program corresponding to (SP), we write

$$\phi(x, w) = \begin{cases} f(x), & \text{if } g_i(x) \geq w_i, \quad i = 1, \ldots, m \\ +\infty, & \text{otherwise} \end{cases} \tag{210}$$

Assume that ϕ is a closed convex function and compute its conjugate

$$\phi^*(\xi, \lambda) = \sup_{x, w} \{\xi^T x + \lambda^T w - \phi(x, w)\} \tag{211}$$

$$= \sup_{\substack{g_i(x) \geq w_i \\ i=1, \ldots, m}} \{\xi^T x + \lambda^T w - f(x)\} \tag{212}$$

Introducing nonnegative "slack variables" s_i, we can convert constraints (209) into

$$w_i = g_i(x) - s_i, \qquad i = 1, \ldots, m \tag{213}$$

$$s_i \geq 0 \qquad i = 1, \ldots, m \tag{214}$$

and obtain

$$\phi^*(\xi, \lambda) = \sup_x \{\xi^T x + \sum_{i=1}^m \lambda_i g_i(x) - f(x)\} + \sup_{s \geq 0} \{-\lambda^T s\} \tag{215}$$

Hence

$$\phi^*(\xi, \lambda) = \begin{cases} \sup_x \{\xi^T x + \sum_{i=1}^m \lambda_i g_i(x) - f(x)\}, & \text{if } \lambda \geq 0 \\ +\infty, & \text{otherwise} \end{cases} \tag{216}$$

The dual problem of (SP), consisting of maximizing $-\phi^*(0, \lambda)$ is given by

$$\text{(DSP)} \qquad \max -\phi^*(0, \lambda) = \max_{\lambda \geq 0} \{\inf_x [f(x) - \sum_{i=1}^m \lambda_i g_i(x)]\} \tag{217}$$

Comparing (206) with (216) and (217), we can see that the Lagrangian associated with (SP) is given by

$$L(x, \lambda) = \begin{cases} f(x) - \sum_{i=1}^m \lambda_i g_i(x), & \text{if } \lambda \geq 0 \\ -\infty, & \text{otherwise} \end{cases} \tag{218}$$

For those x and λ that yield finite values of $L(x, \lambda)$, the preceding formula corresponds to the Lagrangian associated with problem (SP), as defined earlier in this survey.

Kuhn-Tucker type optimality conditions for (SP) can also be readily derived. By the notion of stability, these conditions become stronger than similar results presented earlier.

Theorem 31. Assume that (SP) has an optimal solution x^*. Then there exists a vector $\lambda^* \in R^m$ such that

$$\lambda^* \geq 0 \tag{219}$$

$$\lambda_i^* g_i(x^*) = 0, \qquad i = 1, \ldots, m \tag{220}$$

and the Lagrangian L has a saddlepoint at (x^*, λ^*). That is,

$$L(x^*, \lambda) \leq L(x^*, \lambda^*) \leq L(x, \lambda^*) \tag{221}$$

for all $\lambda \geq 0$ and $x \in R^n$ if and only if (SP) is stable.

Further, suppose that f and $g_1, \ldots, g_m$ are real-valued differentiable convex and concave functions on R^n, respectively. Then there exists a vector $\lambda^* \in R^m$ such that

$$\nabla f(x^*) - \sum_{i=1}^m \lambda_i^* \nabla g_i(x^*) = 0 \tag{222}$$

$$\lambda_i^* g_i(x^*) = 0, \qquad i = 1, \ldots, m \tag{223}$$

$$\lambda^* \geq 0 \tag{224}$$

if and only if (SP) is stable.

Dual programs of (SP) have been investigated by many researchers and several formulations were proposed, each of which can possibly be derived from the results presented here. It is important to note that for every primal convex program we can formulate many different dual programs, one for each type of perturbation introduced.

For (SP), the type of perturbation used in this section results in a dual program (DSP) whose objective function [see (217)] is given in terms of an additional optimization problem—that is, that of

$$\inf_{x} \{f(x) - \sum_{i=1}^{m} \lambda_i g_i(x)\} \tag{225}$$

There are, however, special cases of (SP), or other types of perturbations, in which this additional optimization problem can be explicitly solved. This point will be illustrated in the following example.

Example 16. Let us derive the dual of a linear program. The primal linear program is given by

$$\text{(LP)} \qquad \min_{x} c^T x \tag{226}$$

subject to

$$Ax \geq b \tag{227}$$

$$x \geq 0 \tag{228}$$

where A is an $m \times n$ real matrix and b and c are m and n vectors, respectively. Introducing the perturbations on the right-hand sides of the constraints, we get

$$\phi(x, w, v) = \begin{cases} c^T x, & \text{if } Ax - b \geq w, \ x \geq v \\ +\infty, & \text{otherwise} \end{cases} \tag{229}$$

and

$$\phi^*(\xi, \lambda, \nu) = \sup_{\substack{x, w, v \\ Ax-b \geq w \\ x \geq v}} \{\xi^T x + \lambda^T w + \nu^T v - c^T x\} \tag{230}$$

This expression reduces to

$$\phi^*(\xi, \lambda, \nu) = \begin{cases} \sup_{x} \{(\xi - c + A^T\lambda + \nu)^T x - \lambda^T b\}, & \text{if } \lambda \geq 0, \ \nu \geq 0 \\ +\infty, & \text{otherwise} \end{cases} \tag{231}$$

Hence

$$\phi^*(\xi, \lambda, \nu) = \begin{cases} -\lambda^T b, & \text{if } \lambda \geq 0, \ \nu \geq 0, \ A^T\lambda + \nu = c - \xi \\ +\infty, & \text{otherwise} \end{cases} \tag{232}$$

The dual linear program becomes

$$\text{(DLP)} \quad \max \lambda^T b \tag{233}$$

subject to

$$A^T \lambda \leq c \tag{234}$$

$$\lambda \geq 0 \tag{235}$$

where the vector of dual variables ν has been eliminated from the problem, for it is simply a vector of slack variables.

Optimality conditions and duality theorems for linear programs can now be derived from the general results for convex programs.

Next consider the convex program

$$\text{(PQP)} \quad \min f(x) = a + c^T x + \tfrac{1}{2} x^T Q x \tag{236}$$

subject to

$$A^T x \geq b \tag{237}$$

where $a \in R$, $c \in R^n$, $b \in R^m$ are given vectors, Q is an $n \times n$ symmetric positive definite matrix, and A is an $n \times m$ matrix. A nonlinear program in which the objective function is quadratic and the constraints are linear, as in the case of (PQP), is called a quadratic program. For simplicity, assume that $m < n$ and that A has rank m.

We now derive a dual program corresponding to (PQP). First we transform (PQP) into an "unconstrained" program with perturbations:

$$\min \phi(x, w) = \begin{cases} f(x), & \text{if } A^T x - b \geq w \\ +\infty, & \text{otherwise} \end{cases} \tag{238}$$

where $x \in R^n$, $w \in R^m$. The conjugate function of ϕ is given by

$$\phi^*(\xi, \lambda) = \sup_{x,w} \{\xi^T x + \lambda^T w - \phi(x, w)\} \tag{239}$$

$$= \sup_{A^T x - b \geq w} \{\xi^T x + \lambda^T w - a - c^T x - \tfrac{1}{2} x^T Q x\} \tag{240}$$

and

$$\phi^*(\xi, \lambda) = \begin{cases} \sup_x \{(\xi + A\lambda - c)^T x - \tfrac{1}{2} x^T Q x - b^T \lambda - a\}, & \text{if } \lambda \geq 0 \\ +\infty, & \text{otherwise} \end{cases} \tag{241}$$

The supremum of the strictly concave function in (241) can be found by differentiation, and we obtain

$$\phi^*(\xi, \lambda) = \begin{cases} \tfrac{1}{2}(\xi + A\lambda - c)^T Q^{-1}(\xi + A\lambda - c) - b^T \lambda - a, & \text{if } \lambda \geq 0 \\ +\infty, & \text{otherwise} \end{cases} \tag{242}$$

The dual program of (PQP) becomes

$$\text{(DQP)} \quad \max v(\lambda, \eta) = a + b^T\lambda - \tfrac{1}{2}\eta^T Q^{-1}\eta \tag{243}$$

subject to

$$A\lambda - \eta = c \tag{244}$$

$$\lambda \geq 0 \tag{245}$$

Thus (DQP) is also a quadratic program. Note that the vector variable η was merely introduced for notational convenience.

It should be mentioned here that Cottle [45], Dennis [46], and Dorn [47, 48] were the first to state duality theorems for quadratic programs without using conjugate function theory as we have done here. In these early works the dual program is formulated in terms of both primal and dual variables. To obtain this early formulation, note that in an optimal solution of (PQP) we have from (53)

$$c + Qx^* - A\lambda^* = 0 \tag{246}$$

where λ^* is the vector of Lagrange multipliers.

Letting

$$\eta = Qx \tag{247}$$

and comparing (244) to (246) justifies the following formulation of a dual quadratic program:

$$\max v(x, \lambda) = a + b^T\lambda - \tfrac{1}{2}x^T Qx \tag{248}$$

subject to

$$A\lambda - Qx = c \tag{249}$$

$$\lambda \geq 0 \tag{250}$$

This concludes our discussion of some of the important aspects of convexity from a mathematical programming point of view. For the elegant results presented above, convexity is a sufficient condition, but it is by no means a necessary one. The notions of convexity have been considerably generalized so that most of the previously obtained results that used convexity will still hold if one replaces convex sets and functions by some of their generalizations. For the more advanced subjects in this areas, the reader is referred to Refs. 49-57.

COMPUTATIONAL METHODS OF NONLINEAR PROGRAMMING

In contrast to the unifying role of the simplex method in linear programming, no such unified approach to obtaining an optimal solution of a nonlinear program exists. Whereas the simplex method can efficiently solve a linear program in thousands of variables, the question of how to minimize an unconstrained nonlinear function in a few variables is still an important one.

We start our survey of computational methods with the simplest case of minimizing single-variable functions. Next we present nonderivative methods—that is, algorithms that find extrema without using derivatives of the function whose extremum is sought. These methods are important, for, in many practical applications, derivatives are either unavailable or difficult to compute. Significant progress has occurred in the 1970s in developing efficient algorithms for unconstrained optimization with derivatives. We present here some of the most widely used techniques that require partial derivatives of the function to be minimized.

The study of unconstrained optimization methods is important, since penalty methods, one of the most successful classes of algorithms for solving constrained nonlinear programs, are based on transforming constrained problems into unconstrained ones. Other methods handle constraints directly by modifying movement from one point in the space of variables to another, so that the constraints are taken into account. A discussion of constrained optimization methods concludes this survey.

Single Variable Optimization

The reason for discussing one-dimensional optimizations separately is that some of the iterative methods for n-dimensional problems include steps in which extrema are sought along certain directions in R^n, and these steps are essentially equivalent to one-dimensional optimizations.

Consider, therefore, the problem of minimizing a real function f of a single real variable x. We already saw that if f has a local extremum at x^* and is differentiable there, then the first derivative of f must vanish at x^*. That is, local extrema are solutions of the equation

$$f'(x) = 0 \tag{251}$$

We can try to solve this generally nonlinear equation and obtain all its solutions, among which will be local minima if such minima exist. If the second derivative of f is also available, then by evaluating f'' at the solutions of (251), we can usually determine which ones correspond to minima.

In most cases, however, no analytic solution of (251) can be obtained, so this equation must be solved by some iterative method that generates a sequence of points $\{x^k\}$ and derivative values $\{f'(x^k)\}$ such that the limit of the latter sequence converges to zero. The classic technique for solving a nonlinear equation is Newton's method which, in our case, requires the knowledge of f''(x), the value of the second derivative. The idea behind Newton's method is to use a guess x^k for the solution of (251), linearize f' around x^k, and solve for the point where the linear function vanishes. This point is the next guess x^{k+1}. Formally, let x^k be the current guess for solving $f'(x) = 0$. The root of the linear equation

$$f'(x^k) + f''(x^k)(x - x^k) = 0 \tag{252}$$

is given by

$$x^{k+1} = x^k - \frac{f'(x^k)}{f''(x^k)} \tag{253}$$

This iterative method is also illustrated in Fig. 9.

A closely related root-finding method can be obtained by approximating the second derivative f''(x) by the difference quotient

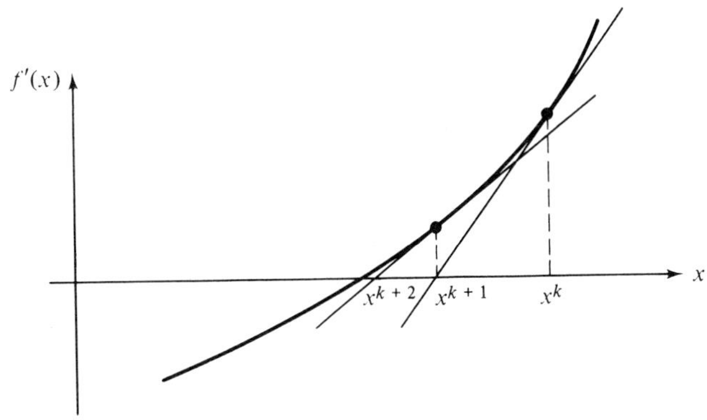

FIG. 9. Newton's method.

$$f''(x^k) \cong \frac{f'(x^k) - f'(x^{k-1})}{x^k - x^{k-1}} \tag{254}$$

When this expression is substituted into (253), we get the iteration formula of the secant method

$$x^{k+1} = x^k - \frac{f'(x^k)(x^k - x^{k-1})}{f'(x^k) - f'(x^{k-1})} \tag{255}$$

This method is shown in Fig. 10. Note that if f happened to be a convex quadratic function of a single variable, the above two methods would find the minimum of f in one iteration.

If no derivatives or only f'(x) is available, then the most popular one-dimensional optimization methods are based on polynomial (quadratic or cubic) approximations of the function to be minimized. In the quadratic method we approximate f by a quadratic function ϕ, given by

$$\phi(x) = a + bx + c(x)^2 \tag{256}$$

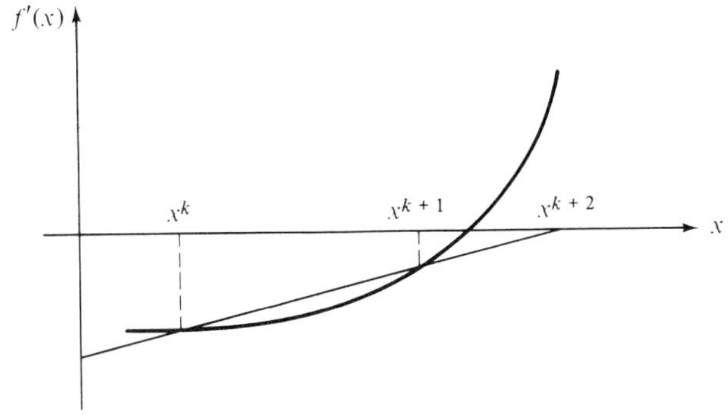

FIG. 10. The secant method.

Suppose that we evaluate f at three points, x^1, x^2, x^3, such that $x^1 < x^2 < x^3$.

Letting $\phi(x^i) = f(x^i)$ for i = 1, 2, 3, we can solve for the coefficients a, b, and c. The minimum of the quadratic function ϕ (if it has a minimum) can be found analytically by setting $\phi'(x) = 0$, and, for a first approximation of a minimum of f, we obtain the point $\hat{x}$, given by

$$\hat{x} = -b/2c \tag{257}$$

provided that $c > 0$. If $c < 0$, the quadratic function is actually a parabola with a maximum and so the point $\hat{x}$ obtained is unusable. A situation that will ensure that c is positive is

$$f(x^1) > f(x^2) \quad \text{and} \quad f(x^2) < f(x^3) \tag{258}$$

If it holds, we are also ensured that a local minimum of f has been bracketed somewhere between x^1 and x^3. The minimum of ϕ so found will also satisfy

$$f(x^1) > \phi(\hat{x}) \quad \text{and} \quad \phi(\hat{x}) < f(x^3) \tag{259}$$

and we can find a new quadratic approximation and approximate minimum by choosing a new set of three points as follows: Evaluate $f(\hat{x})$ and choose as the new x^2 one of the four points at which f has been computed and which yielded the lowest value of f (it is either $\hat{x}$ or the "old" x^2). Let the new x^1 and x^3 be the two points adjacent to the new x^2 from the left and right, respectively, and repeat the iteration. This algorithm can be terminated if either the difference between the actual and approximate values of the function at the predicted minimum is less than some tolerance $\epsilon > 0$—that is, if

$$|f(\hat{x}) - \phi(\hat{x})| < \epsilon \tag{260}$$

or if estimates of the minimum point in two or more successive iterations are closer than some predetermined distance.

The next polynomial approximation method is the cubic method, in which a given function f to be minimized is approximated by a third-order polynomial

$$\phi(x) = a + bx + c(x)^2 + d(x)^3 \tag{261}$$

The method described here was originally derived by Davidon [58]. The algorithm is based on the assumption that the first derivatives of f are available, but it can readily be modified for use without calculating derivatives. We start at an arbitrary point $x^1 \in R$ and compute $f(x^1)$, $f'(x^1)$. Assume that $f'(x^1)$ is negative. Then we find a point x^2, $x^2 > x^1$, by some iterative procedure such that either $f'(x^2)$ is nonnegative or $f(x^2) > f(x^1)$. The coefficients a, b, c, and d can now be computed by solving a system of four linear equations in four variables:

$$\begin{aligned} f(x^i) &= a + bx^i + c(x^i)^2 + d(x^i)^3, \qquad i = 1, 2 \\ f'(x^i) &= b + 2cx^i + 3d(x^i)^2, \qquad i = 1, 2 \end{aligned} \tag{262}$$

The solution of these equations can be expedited by a simple change of variables. Define a new variable z and new functions g and ψ by

$$z = x - x^1 \tag{263}$$

and

$$g(z) = f(x^1 + z), \qquad \psi(z) = \phi(x^1 + z) \tag{264}$$

Then the first derivative of the cubic function ψ with respect to z is given by

$$\psi'(z) = g'(0) - \frac{2z}{\lambda}(g'(0) + \alpha) + \frac{(z)^2}{(\lambda)^2}(g'(0) + g'(\lambda) + 2\alpha) \tag{265}$$

where $\lambda = x^2 - x^1$ and

$$\alpha = \frac{3(g(0) - g(\lambda))}{\lambda} + g'(0) + g'(\lambda) \tag{266}$$

and the point $\hat{z}$, where $\psi'(z)$ vanishes and where a minimum is predicted, is given by

$$\hat{z} = \lambda(1 - \beta) \tag{267}$$

where

$$\beta = \frac{g'(\lambda) + [(\alpha)^2 - g'(0)g'(\lambda)]^{\frac{1}{2}} - \alpha}{g'(\lambda) - g'(0) + 2[(\alpha)^2 - g'(0)g'(\lambda)]^{\frac{1}{2}}} \tag{268}$$

If $|g'(\hat{z})| < \epsilon$, where ϵ is some predetermined tolerance, the procedure is terminated; otherwise the algorithm must be restarted by using a new set of two points selected by a procedure similar to that described for the quadratic approximation method.

Among many other single variable optimization methods that exist in the literature, we mention the Fibonacci and Golden Section methods [59-61] and Brent's method [62].

Multidimensional Unconstrained Optimization without Derivatives

The more we know about a function, the better or the more efficient algorithms we can derive for seeking its extremum. For example, the availability of first and second derivatives of a function on R^n can greatly facilitate locating an extremum, but the number of arithmetic operations required for computing derivatives (especially for large values of n) can be so large that one may try to derive algorithms without using derivatives. In many practical problems such derivatives are simply unavailable, for sometimes even an analytic expression for the function to be minimized cannot be found. These considerations justify presenting a number of algorithms for unconstrained multidimensional minimization in which only values of the function are computed.

The Simplex Method

For a long time it was known that finding minima of functions of n variables by the simplest concepts—such as setting up a grid on R^n and evaluating the function at every point of this grid—was quite inefficient. Improved empirical methods emerged in the early 1960s, the first of which, called the simplex method, is described now. Parenthetically, it should be mentioned that the simplex method of unconstrained minimization should not be confused with the simplex method in linear programming, although the origin of the name is the same for both. A simplex is the convex hull of n + 1 points in R^n—for example, a line segment in R or a triangle in R^2. The simplex method of unconstrained minimization was devised by Spendley, Hext, and Himsworth [63] and later improved by Nelder and Mead [64].

Consider the minimization of the real function f(x), $x \in R^n$ and let $x^0, x^1, \ldots, x^n$ be points in R^n that form a current simplex. Let x^h and x^ℓ be defined by

$$f(x^h) = \max \{f(x^0), f(x^1), \ldots, f(x^n)\} \tag{269}$$

and

$$f(x^\ell) = \min \{f(x^0), f(x^1), \ldots, f(x^n)\} \tag{270}$$

Denote by $\bar{x}$ the centroid of all the vertices of the simplex except x^h:

$$\bar{x} = \frac{1}{n} \sum_{i=0}^{n} x^i, \quad x^i \neq x^h \tag{271}$$

The main idea of the algorithm is to replace x^h, the vertex of the current simplex that has the highest function value, by a new and better point. The replacement of this point involves three types of steps: reflection, expansion, and contraction.

In the reflection step we compute x^r by the formula

$$x^r = \bar{x} + \alpha(\bar{x} - x^h) \tag{272}$$

where α is a positive constant, called the reflection coefficient.

Let us consider three possible cases:

1. If $f(x^\ell) > f(x^r)$—that is, the reflection step has generated a new minimum—then we take an expansion step by computing

$$x^e = \bar{x} + \gamma(x^r - \bar{x}) \tag{273}$$

where $\gamma > 1$, the expansion coefficient, is a given constant. If $f(x^r) > f(x^e)$, then x^e replaces x^h and a new simplex is obtained. If, however, $f(x^e) \geq f(x^r)$, then the expansion step failed and x^h is replaced by x^r in the new simplex.

2. If

$$\max_i \{f(x^i), x^i \neq x^h\} \geq f(x^r) \geq f(x^\ell) \tag{274}$$

then x^h is replaced by x^r and the result is a new simplex.

3. If

$$f(x^r) > \max_i \{f(x^i), x^i \neq x^h\} \tag{275}$$

then replacing x^h by x^r would make x^r the new x^h. In this case, we define a point $x^{h'}$ by

$$f(x^{h'}) = \min \{f(x^h), f(x^r)\} \tag{276}$$

and take a contraction step

$$x^c = \bar{x} + \beta(x^{h'} - \bar{x}) \tag{277}$$

where $0 < \beta < 1$ is the contraction coefficient. The point x^c replaces x^h in the new simplex unless $f(x^c) > f(x^{h'})$, in which case we replace all the x^i by the new points $\hat{x}^i$, defined as

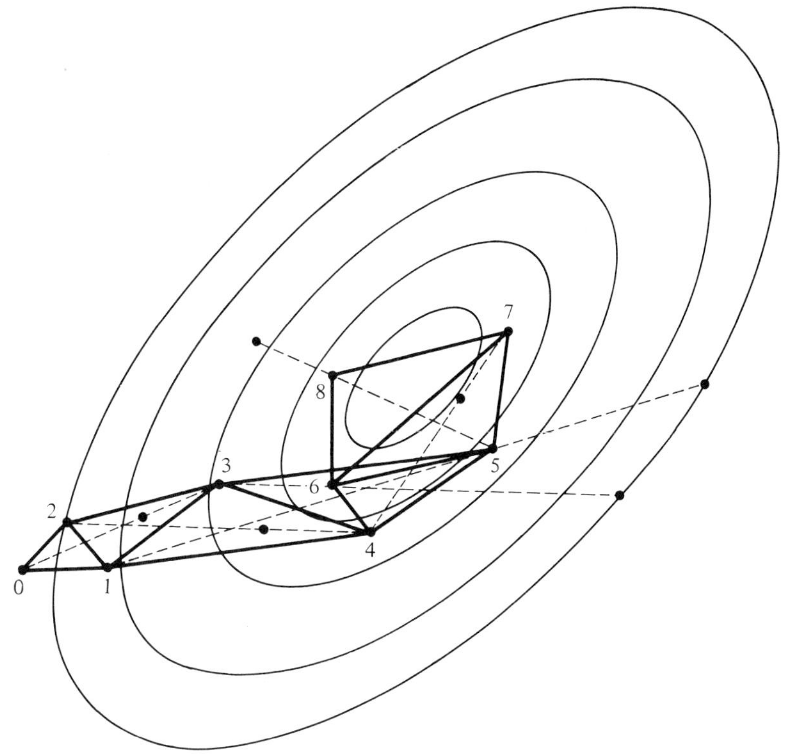

FIG. 11. The simplex method illustrated on a quadratic function.

$$\hat{x}^i = x^i + \tfrac{1}{2}(x^\ell - x^i), \qquad i = 0, \ldots, n \tag{278}$$

That is, the distances between the vertices of the old simplex and the point with the lowest function value are halved.

The evaluation of the performance of the simplex method on several test functions indicated that the choice of $\alpha = 1$, $\beta = \frac{1}{2}$, $\gamma = 2$, gave good results. A typical search for the minimum by the simplex method is shown in Fig. 11. The numbers indicate vertices of successive simplexes. Dashed lines show moves made, some of which ended in failure.

The simplex method just described has been successfully tested on many problems, and its efficiency was found to be considerably affected by the scale and orientation chosen for the first simplex [65].

Pattern Search

This simple and easily implemented algorithm is due to Hooke and Jeeves [66]. Suppose again that we are seeking the minimum of a real-valued function f(x), $x \in R^n$. Given an initial point x_B^0 (it is also marked as t_0^1 for reasons that will become clear later), we perform a sequence of exploratory moves around x_B^0 as follows: Compute $f(x_B^0)$ and $f(x_B^0 + \Delta_1)$, where Δ_j is an n vector whose j-th element is $d_j > 0$ and the rest of whose elements are zeroes; that is,

$$\Delta_1 = (d_1, 0, 0, \ldots, 0)^T \tag{279}$$

and d_1 is some prescribed positive number (step length). Hence

$$(x_B^0 + \Delta_1) = (x_{B1}^0 + d_1, x_{B2}^0, \ldots, x_{Bn}^0) \tag{280}$$

If $f(x_B^0) > f(x_B^0 + \Delta_1)$, the exploratory move was successful, we set $t_1^1 = x_B^0 + \Delta_1$, and proceed to make a move in the x_2 direction; otherwise we reverse the search direction and compute $f(x_B^0 - \Delta_1)$. If $f(x_B^0 - \Delta_1) < f(x_B^0)$, then this exploratory move is declared successful and we set $t_1^1 = x_B^0 - \Delta_1$. If the last move has also failed—that is,

$$f(x_B^0 - \Delta_1) \geq f(x_B^0) \tag{281}$$

we set $t_1^1 = x_B^0$.

Having completed the exploratory moves in the direction of the x_1 axis, we compute $f(t_1^1 + \Delta_2)$, where $\Delta_2 = (0, d_2, 0, \ldots, 0)^T$. If $f(t_1^1 + \Delta_2) < f(t_1^1)$, we set $t_2^1 = t_1^1 + \Delta_2$; otherwise we compute $f(t_1^1 - \Delta_2)$ and compare it with $f(t_1^1)$. We set $t_2^1 = t_1^1 - \Delta_2$ if $f(t_1^1 - \Delta_2) < f(t_1^1)$ and $t_2^1 = t_1^1$ otherwise. Continuing in this way, we make exploratory moves along all n axial directions and arrive at a point t_n^1. This point is called the new base point x_B^1, and we immediately move to

$$t_0^2 = x_B^1 + (x_B^1 - x_B^0) \tag{282}$$

in the "promising" direction $x_B^1 - x_B^0$. This is called a <u>pattern move</u>. If $t_0^2 \neq x_B^0$, we perform a new set of exploratory moves and establish a new base point x_B^2; otherwise the exploratory moves around t_0^1 were all failures, and we reduce the d_j for additional exploratory moves around t_0^1 with smaller step lengths. If $f(x_B^2) < f(x_B^1)$, we make our next pattern move to

$$t_0^3 = x_B^2 + (x_B^2 - x_B^1) \tag{283}$$

and the exploratory procedure is restarted around t_0^3.

We continue in this way until we arrive at a base point x_B^k such that $f(x_B^k) \geq f(x_B^{k-1})$. In this case we return to x_B^{k-1}, set $t_0^k = x_B^{k-1}$, and recommence the exploratory moves from there. If these exploratory moves are successful, the alternating pattern and exploratory moves are restarted; otherwise we reduce the d_j for additional exploratory moves around t_0^k with smaller step lengths. The algorithm terminates when the d_j become smaller than some predetermined values. This algorithm is illustrated on a function of two variables in Fig. 12. The circled numbers indicate the sequence of points visited.

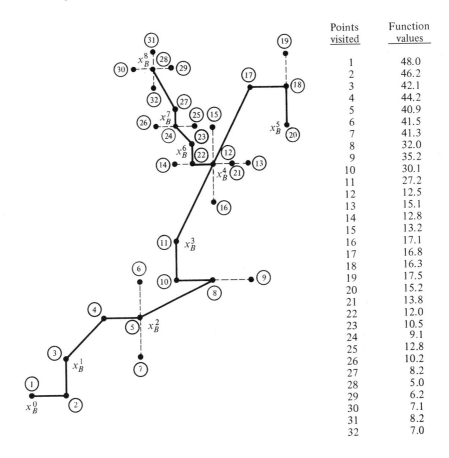

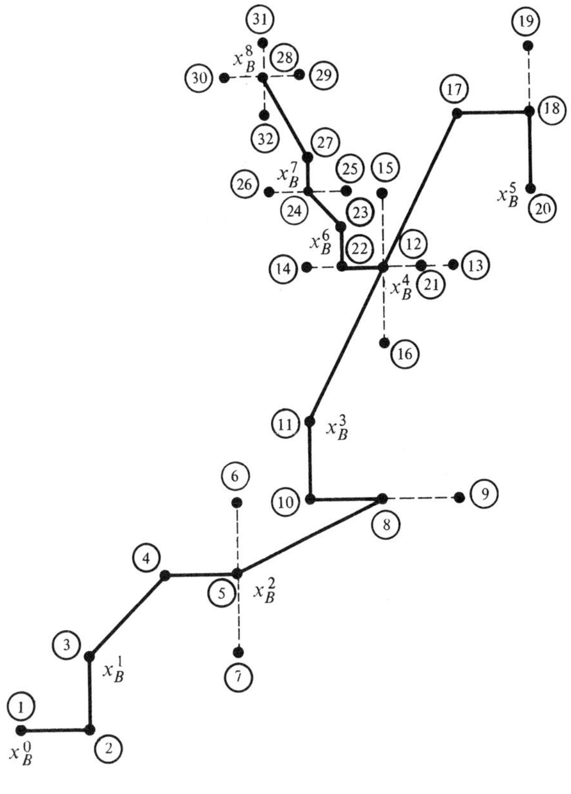

Points visited	Function values
1	48.0
2	46.2
3	42.1
4	44.2
5	40.9
6	41.5
7	41.3
8	32.0
9	35.2
10	30.1
11	27.2
12	12.5
13	15.1
14	12.8
15	13.2
16	17.1
17	16.8
18	16.3
19	17.5
20	15.2
21	13.8
22	12.0
23	10.5
24	9.1
25	12.8
26	10.2
27	8.2
28	5.0
29	6.2
30	7.1
31	8.2
32	7.0

FIG. 12. Pattern search on a function of two variables.

Although the pattern search method may require a great number of function evaluations in order to reach a point that approximates the minimum of a function, it is regarded as an easily programmed and reliable method. Its main feature consists of following "ridges" and "valleys." The pattern moves can take long steps in the assumed direction of valleys, whereas the exploratory moves find the way back to these valleys if a pattern move has climbed out of them.

Conjugate Directions

We now introduce an important concept for quadratic functions, one on which a whole class of unconstrained minimization methods is based.

Two vectors $x \in R^n$, $y \in R^n$ are said to be conjugate directions with respect to the n×n symmetric positive definite matrix A if

$$x^T Ay = 0 \tag{284}$$

Note that the notion of conjugate directions is a generalization of orthogonality—that is, the case when A is the n×n identity matrix I. It is well known that a symmetric

$n \times n$ matrix A has n orthogonal eigenvectors. This set of n vectors is also mutually conjugate, for if x^1, x^2 are such eigenvectors of A, then $Ax^2 = \lambda x^2$, where λ is the corresponding eigenvalue and

$$(x^1)^T Ax^2 = (x^1)^T \lambda x^2 = \lambda (x^1)^T (x^2) = 0 \tag{285}$$

Thus for every $n \times n$ symmetric positive definite matrix there is at least one set of n mutually conjugate directions.

We can construct n mutually conjugate directions with respect to A from a set of n linearly independent vectors $u^1, \ldots, u^n$ in R^n by a procedure similar to the Gram-Schmidt orthogonalization method [67].

Let

$$z^1 = u^1 \tag{286}$$

and

$$z^j = u^j - \sum_{k=1}^{j-1} \frac{(u^j)^T Az^k}{(z^k)^T Az^k} z^k, \qquad j = 2, \ldots, n \tag{287}$$

Then the vectors $z^1, \ldots, z^n$ are mutually conjugate with respect to A and are also linearly independent.

A geometric interpretation of conjugate vectors can be given as follows: Let f be the quadratic function

$$f(x) = a + b^T x + \tfrac{1}{2} x^T Qx, \qquad x \in R^n \tag{288}$$

and assume that Q is a symmetric positive definite matrix. Let x^* be the point minimizing f(x) for all $x \in R^n$. Then the surfaces $f(x) = c$ (constant) are generally ellipsoids with center at x^*. Let x^0 be a point satisfying $f(x^0) = c$. Construct the hyperplane tangent to the surface $f(x) = c$ at x^0. Then the vector joining x^0 and x^* is conjugate with respect to Q to every vector in the tangent hyperplane. The two-dimensional case is illustrated in Fig. 13.

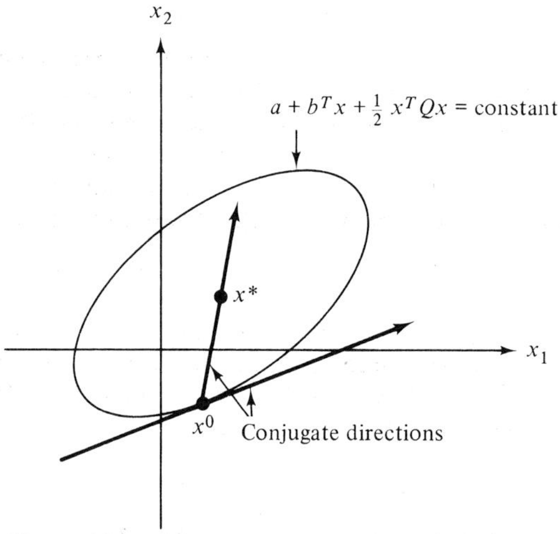

FIG. 13. Conjugate directions with respect to Q.

Suppose next that we have m nonzero vectors $z^1, \ldots, z^m$ in R^n, $m \leq n$, mutually conjugate with respect to a positive definite matrix Q. Then these vectors are linearly independent, and thus they span an m-dimensional subspace of R^n, given by the vectors satisfying

$$x = \sum_{j=1}^{m} \alpha_j z^j \tag{289}$$

where the α_j are arbitrary real numbers. For a given point $x^0 \in R^n$, the set of vectors satisfying

$$x = x^0 + \sum_{j=1}^{m} \alpha_j z^j \tag{290}$$

where z^j are m linearly independent vectors and the α_j are arbitrary numbers, is an affine set. It is said to be generated by the point x^0 and the vectors $z_1, \ldots, z^m$. For m = n, affine sets become the whole space R^n.

We can now state an important result, due to Powell [68], which relates conjugate directions to unconstrained minimization of quadratic functions.

Theorem 32. If $z^1, \ldots, z^m$ are nonzero, mutually conjugate directions with respect to the symmetric positive definite matrix Q, then the minimum of the quadratic function

$$f(x) = a + b^T x + \tfrac{1}{2} x^T Q x, \qquad x \in R^n \tag{291}$$

over the affine set generated by the point $x^0 \in R^n$ and the vectors $z^1, \ldots, z^m$ will be found by searching along each of the conjugate directions once only.

Letting m = n in Theorem 32 implies that we can find the minimum of the quadratic function (288) over R^n by performing n line searches along n nonzero directions, mutually conjugate with respect to Q.

The next theorem indicates a way to generate conjugate directions with respect to the Q matrix of the function f in (288) without the explicit use of Q. Two affine sets S and T, $S \neq T$, are said to be parallel if they are generated by the same directions, $z^1, \ldots, z^m$, but different points, $x(S) \in S$ and $x(T) \in T$. We then have the following.

Theorem 33. Let $x^*(S)$ and $x^*(T)$ be the points minimizing f(x), as given in (288), over two parallel affine sets S and T, respectively. Then the vector $x^*(T) - x^*(S)$ is conjugate with respect to Q to any direction that is contained in S and in T.

Theorems 32 and 33 are necessary in order to understand Powell's method, which will be described next.

Powell's Method

The two algorithms described above were derived for finding minima of general unconstrained functions, usually in an infinite number of iterations. On the other hand, if we had a method that could find the minimum of a quadratic function on R^n by a finite number of steps, then such a method should also be efficient for minimizing a general function having continuous second derivatives. The reason for this claim can be seen in the Taylor expansion of a general function f around its minimum $x^* \in R^n$:

$$f(x) \cong f(x^*) + (x - x^*)^T \nabla f(x^*) + \tfrac{1}{2}(x - x^*)^T \nabla^2 f(x^*)(x - x^*) \tag{292}$$

The term involving the gradient vanishes and it follows that in a sufficiently small neighborhood of x* the function f behaves like a quadratic function. In the rest of this survey we present a whole class of methods that share the common property of quadratic termination: If the method of finding the minimum of a quadratic function on R^n as given by (288) with a positive definite Q is used, it will terminate in at most n steps. The word "step" prescribed by an algorithm should be interpreted broadly as moving from one point in R^n to another, and clearly, in some cases, these moves can be carried out in an exact manner only by an infinite procedure.

Powell's method [68] can be described as follows: Each stage of the procedure consists of n + 1 successive one-dimensional line searches, first along n linearly independent directions and then along the direction connecting the best point (obtained at the end of the n one-dimensional line searches) with the starting point of that stage. After these searches, one of the first n directions is replaced by the (n + 1)-th and a new stage begins.

The k-th stage of the method is given by the following steps. Let $x_B^{k-1} = t_0^k \in R^n$ be the starting point of the k-th stage and suppose that n linearly independent directions $\Delta_1^k, \ldots, \Delta_n^k$ are given (for k = 1, the coordinate directions are usually chosen). Find numbers θ_j^* such that

$$f(t_{j-1}^k + \theta_j^* \Delta_j^k) = \min_{\theta_j} f(t_{j-1}^k + \theta_j \Delta_j^k) \tag{293}$$

for j = 1, ..., n and define

$$t_j^k = t_{j-1}^k + \theta_j^* \Delta_j^k, \qquad j = 1, \ldots, n \tag{294}$$

Let

$$\Delta_j^{k+1} = \Delta_{j+1}^k, \qquad j = 1, \ldots, n - 1 \tag{295}$$

and

$$\Delta_n^{k+1} = \Delta_{n+1}^k = t_n^k - t_0^k \tag{296}$$

Find θ_{n+1}^* such that

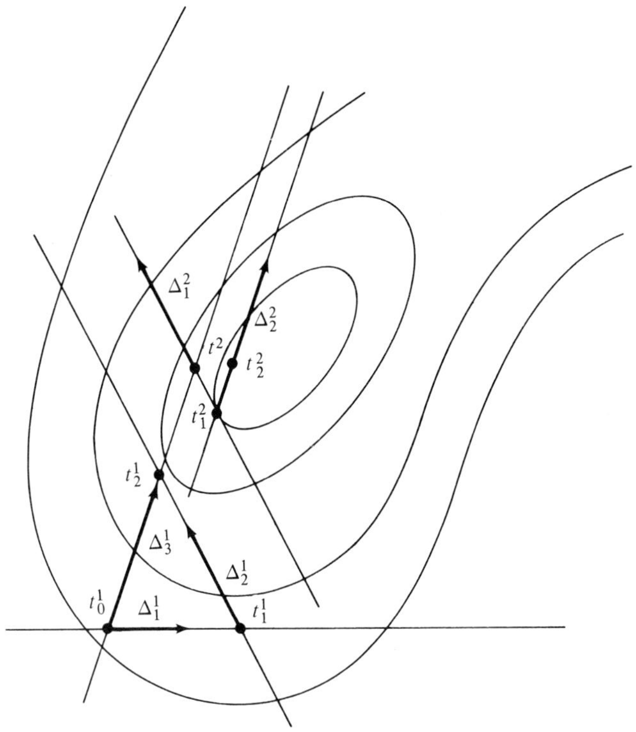

FIG. 14. Powell's method.

$$f(t_n^k + \theta_{n+1}^*(t_n^k - t_0^k)) = \min_{\theta_{n+1}} f(t_n^k + \theta_{n+1}(t_n^k - t_0^k)) \tag{297}$$

and let

$$x_B^k = t_n^k + \theta_{n+1}^*(t_n^k - t_0^k) \tag{298}$$

If $\|x_B^k - x_B^{k-1}\| < \epsilon$, where $\epsilon > 0$ is some predetermined number, stop, otherwise proceed to stage k + 1.

A few steps of this algorithm are illustrated in Fig. 14, in which a general function of two variables is minimized.

Example 17. Let us demonstrate Powell's method on minimizing the quadratic function in two variables

$$f(x) = \tfrac{3}{2}(x_1)^2 + \tfrac{1}{2}(x_2)^2 - x_1x_2 - 2x_1 \tag{299}$$

The minimum of this function is at $x^* = (1, 1)$. Suppose that we start the search at $x_B^0 = t_0^1 = (-2, 4)$ in the coordinate directions $\Delta_1^1 = (1, 0)^T$ and $\Delta_2^1 = (0, 1)^T$. The first minimization is in the Δ_1^1 direction:

$$\min_{\theta_1} f(t_0^1 + \theta_1\Delta_1^1) = \tfrac{3}{2}(-2 + \theta_1)^2 + \tfrac{1}{2}(4)^2 - 4(-2 + \theta_1) - 2(-2 + \theta_1) \tag{300}$$

and the optimum is at $\theta_1^* = 4$. Hence $t_1^1 = (2, 4)$. Now we minimize in the Δ_2^1 direction:

$$\min_{\theta_2} f(t_1^1 + \theta_2\Delta_2^1) = \tfrac{3}{2}(2)^2 + \tfrac{1}{2}(4 + \theta_2)^2 - 2(4 + \theta_2) - 4 \tag{301}$$

and obtain $\theta_2^* = -2$. Hence $t_2^1 = (2, 2)$. Consequently, $\Delta_3^1 = (4, -2)^T$ and we perform a minimization in this direction:

$$\min_{\theta_3} f(t_2^1 + \theta_3\Delta_3^1) = \tfrac{3}{2}(2 + 4\theta_3)^2 + \tfrac{1}{2}(2 - \theta_3)^2 - (2 + 4\theta_3)(2 - 2\theta_3) - 2(2 + 4\theta_2) \tag{302}$$

The optimum is at $\theta_3^* = -\frac{2}{17}$ and

$$x_B^1 = t^2 = \begin{pmatrix} 2 - \frac{8}{17} \\ \\ 2 + \frac{4}{17} \end{pmatrix} = \begin{pmatrix} \frac{26}{17} \\ \\ \frac{38}{17} \end{pmatrix} \tag{303}$$

This concludes the first iteration of the algorithm. The first two search directions of the second iteration are, by (295) and (296),

$$\Delta_1^2 = \begin{pmatrix} 0 \\ 1 \end{pmatrix}, \quad \Delta_2^2 = \begin{pmatrix} 4 \\ -2 \end{pmatrix} \tag{304}$$

We solve

$$\min_{\theta_1} f(t_0^2 + \theta_1\Delta_1^2) = \frac{3}{2}\left(\frac{26}{17}\right) + \frac{1}{2}\left(\frac{38}{17} + \theta_1\right)^2 - \frac{26}{17}\left(\frac{38}{17} + \theta_1\right) - \frac{52}{17} \tag{305}$$

and find $\theta_1^* = -\frac{12}{17}$, $t_1^2 = \left(\frac{26}{17}, \frac{26}{17}\right)$. Next we solve

$$\min_{\theta_2} f(t_1^2 + \theta_2\Delta_2^2) = \frac{3}{2}\left(\frac{26}{17} + 4\theta_2\right)^2 + \frac{1}{2}\left(\frac{26}{17} - 2\theta_2\right)^2 - \left(\frac{26}{17} + 4\theta_2\right)\left(\frac{26}{17} - 2\theta_2\right) - 2\left(\frac{26}{17} + 4\theta_2\right) \tag{306}$$

The optimum is at $\theta_2^* = -\frac{18}{289}$. Hence $t_2^2 = \left(\frac{370}{289}, \frac{478}{289}\right)$.

Now we find the third direction of the second iteration

$$\Delta_3^2 = \begin{pmatrix} \frac{370}{289} - \frac{26}{17} \\ \\ \frac{478}{289} - \frac{38}{17} \end{pmatrix} = -\begin{pmatrix} \frac{72}{289} \\ \\ \frac{168}{289} \end{pmatrix} \tag{307}$$

and solve

$$\min_{\theta_3} f(t_2^2 + \theta_3\Delta_3^2) = \frac{3}{2}\left(\frac{370}{289} - \frac{72}{289}\theta_3\right)^2 + \frac{1}{2}\left(\frac{478}{289} - \frac{168}{289}\theta_3\right)^2$$
$$- \left(\frac{370}{289} - \frac{72}{289}\theta_3\right)\left(\frac{478}{289} - \frac{168}{289}\theta_3\right) - 2\left(\frac{370}{289} - \frac{72}{289}\theta_3\right) \qquad (308)$$

The optimal solution is at $\theta_3^* = \frac{9}{8}$ and $x_B^2 = (1, 1)$. The exact minimum of the quadratic function was found in two iterations, as asserted.

Note that the search directions Δ_1^k and Δ_2^k, for k = 1, 2, were linearly independent in this example. This condition is quite important, for examples exist [69] that demonstrate that if the search directions were linearly dependent, Powell's method, as outlined above, may fail to reach the minimum of f in n iterations and, in fact, may not reach it in any number of iterations.

To see the quadratic termination property of Powell's method, suppose that we are given a quadratic function f on R^n with a positive definite Q, a starting point $x_B^0 \in R^n$, and n linearly independent directions $\Delta_1^1, \ldots, \Delta_n^1$. After performing the steps of the first stage, we have a direction $z^1 = t_n^1 - t_0^1 = \Delta_n^2$ and a new starting point $x_B^1 = t_0^2$ for the second stage. Note that if a point in R^n is optimal in n linearly independent directions, then it must be the global optimum of the quadratic function. Suppose that $t_n^1 \neq t_0^1$; that is, the direction z^1 is nonzero. By (298), the point x_B^1 is a minimum in the z^1 direction; and assuming that the directions $\Delta_1^2, \ldots, \Delta_n^2$ are linearly independent, we arrive at a point t_n^2 that is also a minimum in the z^1 direction, contained in a parallel affine set. By Theorem 33, the direction $z^2 = t_n^2 - t_0^2$ is conjugate to z^1 with respect to Q. Suppose now that k stages of the algorithm have been completed and that k nonzero directions $z^1, \ldots, z^k$, mutually conjugate with respect to Q, were generated. If the directions $\Delta_1^k, \ldots, \Delta_{n-k}^k; z^1, \ldots, z^k$ are linearly independent, then, by Theorems 32 and 33, the direction $z^{k+1} = t_n^{k+1} - t_0^{k+1}$ is mutually conjugate to $z^1, \ldots, z^k$. After completing n such stages, all the search directions are mutually conjugate with respect to Q. And so, by Theorem 33, the minimum of f over R^n has been reached.

A more recent trend in unconstrained minimization without derivatives is to modify efficient optimization methods which use partial derivatives. The modification mainly involves approximating partial derivatives by finite differences. Examples are the methods of Fiacco and McCormick [17], Gill and Murray [70], Gill, Murray, and Pitfield [71], Greenstadt [72, 73], Mifflin [74], and Stewart [75]. We should also mention a few methods derived especially for the minimization of nondifferentiable functions, such as the methods of Bertsekas and Mitter [76], Goldstein [77], Lemarechal [78], and Wolfe [79].

Multidimensional Unconstrained Optimization Using Derivatives

We start our discussion of unconstrained minimization methods that use derivatives with two classical algorithms—Newton's method and Cauchy's steepest descent method. Both methods are quite unsatisfactory for minimizing general nonlinear functions in many variables, since Newton's method, for example, may not converge to the minimum sought, and the convergence rate of the steepest descent method is very slow. They are important, however, for the understanding of their various modifications, which are considered the most successful methods in practice. Next we return to the concept of conjugate directions in connection with the method of conjugate gradients. Finally, we shall discuss variable metric methods, which are considered the best unconstrained minimization algorithms.

Newton-Type and Steepest Descent Methods

Suppose that the real function f is differentiable on R^n. We know that a necessary (and sometimes sufficient) condition for a minimum of f at some point $x^* \in R^n$ is

$$\nabla f(x^*) = 0 \tag{309}$$

If we have sufficient information about the function indicating that a solution of (309) is a minimum, then we can solve the above system of n generally nonlinear equations in n variables. Just as in the case of n = 1, the classic method here for finding a solution of (309) is Newton's method for solving systems of equations. In order to apply this method, we must assume that f is at least twice continuously differentiable. That is, in addition to the gradient vector $\nabla f(x)$, we also need the n×n Hessian matrix $\nabla^2 f(x)$ at every point $x \in R^n$. We expand (linearize) each component of ∇f around a point x^k and set the linear functions equal to zero:

$$\frac{\partial f(x^k)}{\partial x_j} = \sum_{i=1}^{n} \frac{\partial^2 f(x^k)}{\partial x_i \partial x_j}(x_i - x_i^k) = 0, \qquad j = 1, \ldots, n \tag{310}$$

or in vector notation

$$\nabla f(x^k) + \nabla^2 f(x^k)(x - x^k) = 0 \tag{311}$$

Assuming that $\nabla^2 f(x^k)$ is nonsingular, we can solve the preceding system of linear equations for x. Letting x^{k+1} be a solution, we get

$$x^{k+1} = x^k - [\nabla^2 f(x^k)]^{-1} \nabla f(x^k) \tag{312}$$

Newton's method consists of using (312) iteratively.

There is another way of looking at Newton's method. Suppose that we make a quadratic approximation of f at some point x by expanding it around x^k in a Taylor series:

$$f(x) \cong f(x^k) + (x - x^k)^T \nabla f(x^k) + \tfrac{1}{2}(x - x^k)^T \nabla^2 f(x^k)(x - x^k) \tag{313}$$

The point x^{k+1}, given by (312), is then the minimum of this quadratic approximation, provided that $\nabla^2 f(x^k)$ is positive definite. Most difficulties with Newton's method occur if $\nabla^2 f$ is not positive definite at x^k or if the point x^{k+1}, obtained by (312), is not close to x^k, so that the quadratic approximation of f around x^k is not valid at x^{k+1}. Partial remedies have been proposed. We can, for example, use a limited-step Newton method where we define a Newton direction z^{k+1} by

$$\nabla^2 f(x^k) z^{k+1} = -\nabla f(x^k) \tag{314}$$

and the iteration formula is

$$x^{k+1} = x^k + \alpha_k z^{k+1} \tag{315}$$

where α_k is chosen so that $f(x^{k+1}) < f(x^k)$, and, in some versions of the algorithm, it is chosen to minimize f along the Newton direction.

If $\nabla^2 f$ is positive definite, then a numerically stable method to compute z^{k+1} is to factorize $\nabla^2 f(x^k)$ by the method of Cholesky into the form

$$\nabla^2 f(x^k) = L^k D^k (L^k)^T \tag{316}$$

where L^k is a unit diagonal lower triangular matrix and D^k a positive definite diagonal matrix. For j = 1, ..., n, each step of the Cholesky factorization is given by

$$d_{jj} = \frac{\partial^2 f(x^k)}{\partial x_j \partial x_j} - \sum_{r=1}^{j-1} (\ell_{jr})^2 d_{rr} \tag{317}$$

$$\ell_{ij} = \left(\frac{\partial^2 f(x^k)}{\partial x_i \partial x_j} - \sum_{r=1}^{j-1} \ell_{jr} \ell_{ir} d_{rr} \right) \Big/ d_{jj}, \qquad i = j+1, \ldots, n \tag{318}$$

For more details on this method and on extensions to functions whose Hessian is not positive definite, the reader is referred to Gill and Murray [80].

Among optimization methods that use only first derivatives, the classic one is the method of steepest descent due to Cauchy [81]. It can be derived as follows: The directional derivative of a differentiable function f is defined as

$$Df(x^0; y) = y^T \nabla f(x^0) = \lim_{t \to 0} \frac{f(x^0 + ty) - f(x^0)}{t} \tag{319}$$

Consider now all the directions $y \in R^n$ such that, for a given point $x^0 \in R^n$, we have

$$y^T \nabla f(x^0) < 0 \tag{320}$$

Then it follows from (319) that, for sufficiently small positive t, we obtain

$$f(x^0 + ty) < f(x^0) \tag{321}$$

In other words, if we are seeking the minimum of f on R^n and at some point $x^0 \in R^n$ the gradient of f does not vanish, then a sufficiently small move in a direction y that satisfies (320) will result in a function decrease (the directional derivative $Df(x^0; y)$ actually measures the instantaneous increase or decrease in the value of f at x^0 along the direction y). We can, therefore, seek among all directions y having some bounded length, say $\|y\| \leq 1$, that particular direction that yields the steepest descent in the value of f at a given point x^0 for which $\nabla f(x^0) \neq 0$. We have then the nonlinear programming problem

$$\min_y y^T \nabla f(x^0) = \sum_{j=1}^{n} \frac{\partial f(x^0)}{\partial x_j} y_j \tag{322}$$

subject to

$$\|y\| = \left\{ \sum_{j=1}^{n} (y_j)^2 \right\}^{\frac{1}{2}} \leq 1 \tag{323}$$

The optimal solution of this problem is

$$y^* = \frac{-\nabla f(x^0)}{\|\nabla f(x^0)\|} \tag{324}$$

The steepest descent in the function value is thus in the direction of the negative gradient. In the method of steepest descent we proceed as follows: Given a point $x^0 \in R^n$, compute, for $k = 0, 1, \ldots,$ the sequence of points

$$x^{k+1} = x^k - \alpha_k^* \nabla f(x^k) \tag{325}$$

where $\alpha_k^* > 0$ satisfies

$$f(x^k - \alpha_k^* \nabla f(x^k)) = \min_{\alpha_k \geq 0} f(x^k - \alpha_k \nabla f(x^k)) \tag{326}$$

In Cauchy's steepest descent method, the global minimum of f is found along the negative gradient direction. Curry [82] has modified this method by choosing α_k^* to be the first stationary point of f along the direction $-\nabla f(x^k)$; that is, that stationary point for which α_k has the least positive value, and Armijo [83] considered that version of the method in which α_k^* was a value that yielded a sufficiently large decrease in the function value of f. Computational experience with all versions of the steepest descent method has generally been disappointing, and modern methods are either based on conjugate directions or some approximations of Newton's method.

Conjugate Gradient Methods

Conjugate gradient methods for minimizing a differentiable function f over R^n are first of all descent methods; that is, at every iteration k we have $f(x^k) < f(x^{k-1})$. Second, they possess the quadratic termination property by choosing search directions that are mutually conjugate with respect to the Hessian of the quadratic function, if such a function is to be minimized. The search direction at iteration k is equal to the negative gradient of f at the current point to which a scaled value of the previous direction is added. The function is usually minimized along each direction by a one-dimensional line search. Formally, the conjugate gradient method, derived by Fletcher and Reeves [84], can be described as follows: Given a starting point $x^0 \in R^n$, evaluate $\nabla f(x^0)$ and let

$$z^1 = -\nabla f(x^0) \tag{327}$$

Move to $x^1, x^2, \ldots, x^n$ by minimizing f along the directions $z^1, z^2, \ldots, z^n$ in turn, where the z^{k+1} are chosen by

$$z^{k+1} = -\nabla f(x^k) + \beta_k z^k \tag{328}$$

where

$$\beta_k = \frac{(\nabla f(x^k))^T \nabla f(x^k)}{(\nabla f(x^{k-1}))^T \nabla f(x^{k-1})} \qquad (329)$$

Restart the procedure by letting x^n and $-\nabla f(x^n)$ be the new x^0 and z^1, respectively. Terminate the algorithm if $\|\nabla f(x^k)\| \leq \epsilon$ where $\epsilon > 0$ is some predetermined small number.

Example 18. Consider the problem of minimizing the quadratic function

$$f(x) = \frac{3}{2}(x_1)^2 + \frac{1}{2}(x_2)^2 - x_1x_2 - 2x_1 \qquad (330)$$

by the conjugate gradient algorithm of Fletcher and Reeves. Suppose that we begin the search at $x^0 = (-2, 4)$. We get

$$\nabla f(x^0) = \begin{pmatrix} -12 \\ 6 \end{pmatrix} \quad \text{and} \quad z^1 = \begin{pmatrix} 12 \\ -6 \end{pmatrix} \qquad (331)$$

Minimizing $f(x^0 + \alpha_1 z^1)$ with respect to α_1, we find that $\alpha_1^* = \frac{5}{17}$. Thus

$$x^1 = \begin{pmatrix} -2 + \frac{(5)(12)}{17} \\ 4 - \frac{(5)(6)}{17} \end{pmatrix} = \begin{pmatrix} \frac{26}{17} \\ \frac{38}{17} \end{pmatrix} \quad \text{and} \quad \nabla f(x^1) = \begin{pmatrix} \frac{6}{17} \\ \frac{12}{17} \end{pmatrix} \qquad (332)$$

Now we must find z^2, the second search direction, by (328) and (329):

$$z^2 = -\nabla f(x^1) + \frac{(\nabla f(x^1))^T \nabla f(x^1)}{(\nabla f(x^0))^T \nabla f(x^0)} z^1 \qquad (333)$$

and

$$z^2 = \begin{pmatrix} -\frac{6}{17} \\ -\frac{12}{17} \end{pmatrix} + \frac{\left(\frac{6}{17}\right)^2 + \left(\frac{12}{17}\right)^2}{(-12)^2 + (6)^2} \begin{pmatrix} 12 \\ -6 \end{pmatrix} = \begin{pmatrix} -\frac{90}{289} \\ -\frac{210}{289} \end{pmatrix} \qquad (334)$$

Minimizing $f(x^1 + \alpha_2 z^2)$ with respect to α_2, we obtain $\alpha_2^* = \frac{17}{10}$. Consequently,

$$x^2 = \begin{pmatrix} \frac{26}{17} - \frac{(17)(90)}{(10)(289)} \\ \frac{38}{17} - \frac{(17)(210)}{(10)(289)} \end{pmatrix} = \begin{pmatrix} 1 \\ 1 \end{pmatrix} \qquad (335)$$

and this is the global minimum of f, as asserted. The search directions and steps of the algorithm are illustrated in Fig. 15.

There are at least three more equivalent formulas (if used for minimizing quadratic functions) of the coefficient β_k as given in (329). They are

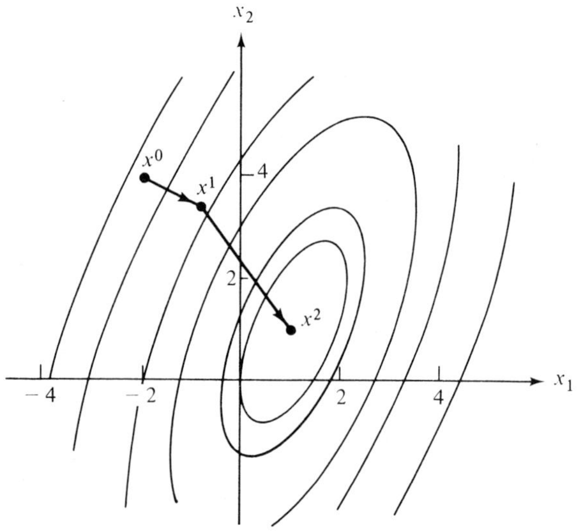

FIG. 15. Conjugate gradient method.

$$\beta_k = \frac{(z^k)^T \nabla^2 f(x^k) \nabla f(x^k)}{(z^k)^T \nabla^2 f(x^k) z^k} \tag{336}$$

$$= \frac{(\gamma^k)^T \nabla f(x^k)}{(\gamma^k)^T z^k} \tag{337}$$

and

$$\beta_k = \frac{(\gamma^k)^T \nabla f(x^k)}{(\nabla f(x^{k-1}))^T \nabla f(x^{k-1})} \tag{338}$$

where

$$\gamma^k = \nabla f(x^k) - \nabla f(x^{k-1}) \tag{339}$$

These formulas are equivalent in the sense that all yield the same search directions when used in minimizing a quadratic function that has a positive definite Q matrix with a starting direction $z^1 = -\nabla f(x^0)$. Formula (336) was suggested by Daniel [85] and (337) by Sorenson [86] and Wolfe [87], whereas (338) is used in the conjugate gradient algorithms of Polak and Ribière [88] and Polyak [89].

For nonquadratic functions the search directions generated by the various formulas of β_k are not identical. There is no clear-cut advantage for any particular β_k over the others. Among the many factors that may influence the performance of a conjugate gradient algorithm, the accuracy of the line search at each iteration occupies an important place. As long as these line searches are carried out with great accuracy, the efficiency of conjugate gradient methods is close to that of the variable metric algorithms to be presented next. If, however, the line searches are inexact, the performance of conjugate gradient algorithms generally becomes poorer. The main advantage of conjugate gradient methods is their relatively small computer storage requirement, an important factor for considering numerical methods for minimizing general functions of very many variables.

Variable Metric Algorithms

Earlier we discussed two classical optimization methods—Newton's method, in which the search direction is given by $z^{k+1} = -[\nabla^2 f(x^k)]^{-1}\nabla f(x^k)$, and the steepest descent method, in which $z^{k+1} = -\nabla f(x^k)$. It can be shown that by defining a non-Euclidean norm $x^T Ax$, where A is a symmetric positive definite matrix, the direction that minimizes the directional derivative of a differentiable function f at a point x^k is given by $-A^{-1}\nabla f(x^k)$. Such a direction can be more efficient than the steepest descent direction, as can easily be seen by an example of a quadratic function with ellipsoidal contours $f(x) = \frac{1}{2}x^T Qx$ on which the norm $x^T Qx$ is defined. Here, of course, the non-Euclidean steepest descent direction is $-Q^{-1}\nabla f(x^k)$, which is also the Newton direction, since $Q = \nabla^2 f(x^k)$. Since methods that do not use second derivatives are usually preferred, because of the extra work involved in such computations, the suggestion was made to approximate the inverse Hessian required for computing the Newton direction or, equivalently, to vary the non-Euclidean norm (metric) at each iteration. An iterative minimization algorithm that, at the current point x^k, uses a search direction z^{k+1}, given by

$$z^{k+1} = -H_k \nabla f(x^k) \tag{340}$$

where H_k is an $n \times n$ matrix, varying from iteration to iteration, is called a <u>variable metric</u> method. Note that this definition is somewhat broader than the ones that can be found in some references where H_k is restricted to be symmetric and positive definite. Indeed, as we shall see later, such symmetric positive definite matrices H_k form the basis of a distinguished subclass of variable metric algorithms. The similarity of (340) to the Newton direction also justifies calling these methods quasi-Newton algorithms, especially since we can choose the matrices H_k so that the minimization of a quadratic function $\frac{1}{2}x^T Qx$ on R^n terminates after n iterations with $H_n = Q^{-1}$ and, in the case of more general nonlinear functions, the H_k tend to $[\nabla^2 f(x^k)]^{-1}$, the inverse Hessian. The reader can find a large variety of not necessarily coinciding definitions of "variable metric" and "quasi-Newton" methods in the literature. We shall restrict the name quasi-Newton to that subclass of variable metric methods in which the matrices H_k satisfy the so-called secant relation. These methods will be discussed later.

The first and perhaps best-known variable metric algorithm was proposed by Davidon [58]. It was subsequently simplified by Fletcher and Powell [90], and we shall refer to it as the Davidon-Fletcher-Powell or DFP method. This algorithm for minimizing a differentiable function f on R^n can be described as follows: Choose an $x^0 \in R^n$ and an arbitrary symmetric $n \times n$ positive definite matrix H_0. At iteration k, we have a point x^{k-1} and a matrix H_{k-1}. Compute the search direction z^k by (340). Find the minimum of $f(x^{k-1} + \alpha_k z^k)$ with respect to α_k by one of the line search methods and let α_k^* be the value of α_k at this minimum. Set $x^k = x^{k-1} + \alpha_k^* z^k$ and compute the matrix H_k by the updating formula

$$H_k = H_{k-1} + \frac{p^k(p^k)^T}{(p^k)^T\gamma^k} - \frac{H_{k-1}\gamma^k(\gamma^k)^T H_{k-1}}{(\gamma^k)^T H_{k-1}\gamma^k} \tag{341}$$

where

$$p^k = x^k - x^{k-1} \quad \text{and} \quad \gamma^k = \nabla f(x^k) - \nabla f(x^{k-1}) \tag{342}$$

If $\|\nabla f(x^k)\| \leq \epsilon$, stop; otherwise proceed to iteration k + 1. Computational experience with the DFP algorithm has generally been satisfactory. If the function to be minimized is quadratic with a positive definite Q matrix, some important properties of the algorithm can be established: It has quadratic termination, the search directions are mutually conjugate with respect to Q, and at iteration n we have $H_n = Q^{-1}$. Thus the algorithm has features of conjugate directions as well as features of Newton-type algorithms. For some convergence properties of this method, see Powell [91]. We shall see below that the DFP method is only one choice in an infinite variety of variable metric methods possessing the same properties.

A family of variable metric methods whose members possess several common properties has been developed by Huang [92]. Members of the family are characterized by the formula for updating the H_k matrix appearing in (340). In Huang's family of methods the matrices H_k are updated according to the general formula

$$H_k = H_{k-1} + p^k(u^k)^T + H_{k-1}\gamma^k(v^k)^T \tag{343}$$

where u^k and v^k are given by

$$u^k = a_{11}^k p^k + a_{12}^k H_{k-1}^T \gamma^k \tag{344}$$

$$v^k = a_{21}^k p^k + a_{22}^k H_{k-1}^T \gamma^k \tag{345}$$

and such that

$$(u^k)^T \gamma^k = \omega \tag{346}$$

$$(v^k)^T \gamma^k = -1 \tag{347}$$

Here p^k and γ^k are as in (342) and ω is an arbitrary real number. There are five parameters in the formulas above, restricted by the two equations (346) and (347). Thus, in general, three free parameters remain. The search directions in Huang's family of methods are mutually conjugate if applied to the minimization of a quadratic function. If the line searches are carried out exactly, the methods also have the quadratic termination property. Thus in order to minimize a quadratic function with a positive definite Q matrix, the choice of the parameters in the updating formula of the H_k matrices is unimportant since, for a given H_0 and x^0, all members of Huang's family generate the same sequence of points. For a general nonlinear function, Dixon has shown [93, 94] that the sequence of points generated depends only on ω. Letting $\omega = 1$ and $a_{12}^k = a_{21}^k = 0$, the reader can verify that the H_k matrix of the DFP method, as given in (341), is a member of Huang's family.

Many variable metric algorithms have been formulated since the appearance of Davidon's method, and most use updating formulas that belong to Huang's family. Let us consider a few examples.

An important subfamily of such updating formulas can be constructed by the following considerations. At iteration k, we are given points x^{k-1}, x^k, a matrix H_{k-1}, and the next matrix is sought. We expand the gradient of f

$$\nabla f(x^{k-1}) \cong \nabla f(x^k) + \nabla^2 f(x^k)(x^{k-1} - x^k) \tag{348}$$

or

$$\gamma^k \cong \nabla^2 f(x^k) p^k \tag{349}$$

If $\nabla^2 f$ is nonsingular, we have

$$p^k \cong [\nabla^2 f(x^k)]^{-1} \gamma^k \tag{350}$$

Note that, for a quadratic function f, the Hessian $\nabla^2 f$ is a constant matrix; and if it is nonsingular, then (350) exactly holds for any step p^k. Since we are interested in methods that do not use inverse Hessian matrices directly but only approximate them, such as in the secant method discussed earlier, we can define a matrix H_k that approximates $[\nabla^2 f(x^k)]^{-1}$ and satisfies the so-called <u>secant relation</u>

$$p^k = H_k \gamma^k \tag{351}$$

Several properties of methods based on the secant relation will be discussed below. Here we only mention that if the matrix H_k is updated by a correction matrix of Huang's family so that (351) holds, then it must satisfy

$$p^k = H_{k-1}\gamma^k + p^k (u^k)^T \gamma^k + H_{k-1}\gamma^k (v^k)^T \gamma^k \tag{352}$$

and it follows from (346) and (347) that we must choose $\omega = 1$. If we further restrict the choice of parameters so that starting with a symmetric matrix H_0 all subsequent matrices H_k will be symmetric, then $a^k_{12} = a^k_{21}$ and only one free parameter, say a^k_{21}, remains.

Such a subfamily of variable metric algorithms was suggested by Broyden [95] and subsequently investigated by him [96, 97]. The updating formula is

$$H_k = H_{k-1} - \frac{H_{k-1}\gamma^k (\gamma^k)^T H_{k-1}}{(\gamma^k)^T H_{k-1}\gamma^k} + \frac{p^k (p^k)^T}{(p^k)^T \gamma^k} - a^k_{12} \frac{(p^k)^T \gamma^k}{(\gamma^k)^T H_{k-1}\gamma^k} w^k (w^k)^T \tag{353}$$

where

$$w^k = H_{k-1}\gamma^k - \frac{(\gamma^k)^T H_{k-1}\gamma^k}{(p^k)^T \gamma^k} p^k \tag{354}$$

An important aspect of the Broyden updating formula (353) is that if H_0 is chosen to be positive definite and $a^k_{12} \le 0$, then all H_k will also be positive definite. Several good reasons exist as to why positive definite matrices H_k are preferred. The first is that, in the neighborhood of a minimum, most functions can be reasonably well approximated by a second-order Taylor expansion (quadratic function) having a positive definite Hessian matrix. Because the H_k matrices are approximations of the inverse Hessian, requiring positive definiteness of the H_k seems to be a good choice. Next, since here we are discussing variable metric methods that require line searches to find α_k^* at each iteration, these searches should be as efficient as possible. The direction $z^k = -H_{k-1} \nabla f(x^{k-1})$ is a descent direction if H_{k-1} is positive definite. Thus we need only search in the positive α_k direction for the minimum of f along z^k. Another reason for using positive definite matrices is to avoid singularity of the H_k. A singular matrix H_k at iteration k may cause the breakdown of most variable metric algorithms, for if $H_k y = 0$ for some nonzero vector $y \in R^n$, then all subsequent search directions z^{k+r} will be orthogonal to y and will be restricted to an affine subset of R^n. Consequently, they do not span the whole space R^n, and the unconstrained minimum generally cannot be attained.

An updating formula that satisfies the secant relation is given by setting $a_{12}^k = a_{21}^k$, $\omega = 1$, and $a_{22}^k = 0$. In this case no free parameter is left in Huang's family of matrices, $a_{12}^k = -1/(p^k)^T \gamma^k$, and we get a special case of (353) by

$$H_k = H_{k-1} + \left[1 + \frac{(\gamma^k)^T H_{k-1} \gamma^k}{(p^k)^T \gamma^k}\right] \frac{p^k (p^k)^T}{(p^k)^T \gamma^k} - \frac{p^k (\gamma^k)^T H_{k-1}}{(p^k)^T \gamma^k} - \frac{H_{k-1} \gamma^k (p^k)^T}{(p^k)^T \gamma^k} \tag{355}$$

Algorithms based on this formula were suggested by Broyden [96, 97], Fletcher [98], Goldfarb [99], and Shanno [100]. Accordingly, a variable metric method based on (355) was named the BFGS method.

Another correction matrix can be obtained by letting $a_{11}^k = a_{22}^k = -a_{12}^k = -a_{21}^k$ and $\omega = 1$. The following formula results:

$$H_k = H_{k-1} + \frac{(p^k - H_{k-1}\gamma^k)(p^k - H_{k-1}\gamma^k)^T}{(p^k - H_{k-1}\gamma^k)^T \gamma^k} \tag{356}$$

This is the "symmetric rank one" formula, proposed by Broyden [95], Davidon [101], Murtagh and Sargent [102], and Powell [103].

Huang's family of variable metric algorithms with quadratic termination properties is based on conjugate directions and exact line searches. One can derive additional variable metric algorithms that do not require exact line searches for their successful application. For example, we can have the basic formulas

$$z^{k+1} = -H_k \nabla f(x^k) \tag{357}$$

$$x^{k+1} = x^k + \alpha_{k+1} z^{k+1} \tag{358}$$

but the H_k must be chosen so that the secant relation holds, that is

$$p^k = H_k \gamma^k \tag{359}$$

where the α_{k+1} does not necessarily minimize $f(x^k + \alpha z^{k+1})$. Such a method is called a quasi-Newton method.

Turning to the question of how to update H_k, we assume

$$H_k = H_{k-1} + \Delta H_k \tag{360}$$

Then by the secant relation,

$$\Delta H_k \gamma^k = p^k - H_{k-1}\gamma^k \tag{361}$$

If we let

$$\Delta H_k = \frac{(p^k - H_{k-1}\gamma^k)(y^k)^T}{(y^k)^T \gamma^k} \tag{362}$$

where y^k is an arbitrary n vector chosen so that $(y^k)^T \gamma^k \neq 0$, then (361), and hence (359), holds. The matrix ΔH_k in (362) has rank one. The general rank-one updating formula can, therefore, be written as [95, 104]

$$H_k = H_{k-1} + \frac{(p^k - H_{k-1}\gamma^k)(y^k)^T}{(y^k)^T \gamma^k} \tag{363}$$

subject to $(y^k)^T \gamma^k \neq 0$.

A quasi-Newton method derived by Broyden [105] uses

$$y^k = H_{k-1} p^k \tag{364}$$

There is no special reason to determine the step lengths by exact line searches, and α_{k+1} can therefore be set to one. The method does not possess the quadratic termination property, although it is quite efficient for quadratic functions as we can see in the following example.

Example 19. Let us illustrate Broyden's method on the quadratic function

$$f(x) = \tfrac{3}{2}(x_1)^2 + \tfrac{1}{2}(x_2)^2 - x_1 x_2 - 2x_1 \tag{365}$$

by performing a few iterations.

Suppose that we choose $H_0 = I$ and $\alpha_k = 1$ for all k. We start the search at $x^0 = (-2, 4)$, where $f(x^0) = 26$, $\nabla f(x^0) = (-12, 6)^T$. The next point is given by

$$x^1 = x^0 - H_0 \nabla f(x^0) = \begin{pmatrix} 10 \\ -2 \end{pmatrix} \tag{366}$$

Here $f(x^1) = 152$ and $\nabla f(x^1) = (30, -12)^T$. Consequently,

$$\gamma^1 = \begin{pmatrix} 42 \\ -18 \end{pmatrix}, \quad p^1 = \begin{pmatrix} 12 \\ -6 \end{pmatrix} \tag{367}$$

and

$$H_1 = \begin{bmatrix} 1 & 0 \\ 0 & 1 \end{bmatrix} - \frac{\left[\begin{pmatrix} 42 \\ -18 \end{pmatrix} - \begin{pmatrix} 12 \\ -6 \end{pmatrix}\right] (12, -6) \begin{bmatrix} 1 & 0 \\ 0 & 1 \end{bmatrix}}{(12, -6) \begin{bmatrix} 1 & 0 \\ 0 & 1 \end{bmatrix} \begin{pmatrix} 42 \\ -18 \end{pmatrix}} \tag{368}$$

or

$$H_1 = \begin{bmatrix} 0.41176 & 0.29412 \\ 0.23529 & 0.88235 \end{bmatrix} \tag{369}$$

The next step is given by

$$p^2 = -H_1 \nabla f(x^1) = -\begin{bmatrix} 0.41176 & 0.29412 \\ 0.23529 & 0.88235 \end{bmatrix} \begin{bmatrix} 30 \\ -12 \end{bmatrix} = \begin{bmatrix} -8.82353 \\ 3.52941 \end{bmatrix} \tag{370}$$

Hence

$$x^2 = \begin{pmatrix} 10 \\ -2 \end{pmatrix} + \begin{pmatrix} -8.82353 \\ 3.52941 \end{pmatrix} = \begin{pmatrix} 1.17647 \\ 1.52941 \end{pmatrix} \tag{371}$$

At this point, $f(x^2) = -0.90657$ and $\nabla f(x^2) = (0.00000, 0.35294)^T$. Thus

$$\gamma^2 = \begin{pmatrix} 0.00000 \\ 0.35294 \end{pmatrix} - \begin{pmatrix} 30 \\ -12 \end{pmatrix} = \begin{pmatrix} -30.00000 \\ 12.35294 \end{pmatrix} \tag{372}$$

and

$$H_2 = H_1 - \frac{(H_1\gamma^2 - p^2)(p^2)^T H_1}{(p^2)^T H_1\gamma^2} \tag{373}$$

$$= \begin{bmatrix} 0.41498 & 0.29352 \\ 0.24494 & 0.88058 \end{bmatrix} \tag{374}$$

Consequently,

$$p^3 = -H_2 \nabla f(x^2) = \begin{pmatrix} -0.10359 \\ -0.31079 \end{pmatrix} \tag{375}$$

and

$$x^3 = \begin{pmatrix} 1.17647 \\ 1.52941 \end{pmatrix} - \begin{pmatrix} 0.10359 \\ 0.31079 \end{pmatrix} = \begin{pmatrix} 1.07288 \\ 1.21862 \end{pmatrix} \tag{376}$$

At this point,

$$f(x^3) = -0.98407 \quad \text{and} \quad \nabla f(x^3) = \begin{pmatrix} 0.00002 \\ 0.14574 \end{pmatrix} \tag{377}$$

We can see that after the first step, which is taken in the steepest descent direction (since H_0 was chosen to be I), convergence is good, but there is no quadratic termination. If we compute H_3, we obtain

$$H_3 = \begin{bmatrix} 0.49560 & 0.49981 \\ 0.48683 & 1.49945 \end{bmatrix} \tag{378}$$

and this matrix is a good approximation to Q^{-1}, which is

$$Q^{-1} = \begin{bmatrix} \frac{1}{2} & \frac{1}{2} \\ \frac{1}{2} & \frac{3}{2} \end{bmatrix} \tag{379}$$

Note that although H_0 was chosen to be symmetric, the subsequent matrices may be unsymmetric, as we can clearly see in H_1, H_2, and H_3. Of course, as the matrices H_k converge to Q^{-1}, they must again approach symmetry.

This example demonstrated that the H_k matrices in Broyden's method are not necessarily symmetric, not even in the case of a symmetric H_0. Since H_k is supposed to approximate the inverse Hessian—a symmetric matrix—it is reasonable to find an updating formula such that all matrices H_k are symmetric and satisfy the secant relation.

Such an updating formula is the basis of the symmetric rank-one (SR1) algorithm, mentioned earlier. The H_k matrices are given by choosing the y^k in (363) to be

$$y^k = p^k - H_{k-1}\gamma^k \tag{380}$$

It is clear that, for any symmetric H_{k-1}, H_k will also be symmetric. Interestingly, algorithms based on this matrix updating formula possess the quadratic termination without exact line searches. A disadvantage of the SR1 algorithm is that the updating formula given by (363) and (380) does not ensure positive definiteness of the updated matrix and it may become ill-conditioned.

Theoretical considerations and numerical experience with variable metric algorithms guided Davidon [106] in deriving a more advanced variable metric algorithm in which the desirable properties of previously developed methods are preserved. In Davidon's method no line searches are made, the secant relation is maintained, and the quadratic termination property is also preserved. Moreover, positive definiteness of the updated matrices is maintained, and for a given H_k the next matrix, H_{k+1}, is optimally conditioned in the sense that the ratio of the largest to smallest eigenvalue of $(H_k)^{-1}H_{k+1}$ is minimized. The matrices H_k are factorized in the form of $J_k(J_k)^T$, where J_k is a nonsingular square matrix that is updated by a rank-one formula, resulting in less roundoff errors than a direct update of H_k. For a complete description of Davidon's method, the reader is referred to Ref. 106.

In concluding our survey of unconstrained optimization methods we mention some generalizations of conjugate gradient and variable metric algorithms. Fried [107] has extended the conjugate gradient method such that convergence in a finite number of steps is ensured for functions more general than quadratic. Similarly, Jacobson and Oksman [108] have derived a method that terminates in a finite number of steps for homogeneous functions. Numerical stability of this algorithm has been considerably improved by Kowalik and Ramakrishnan [109]. Another type of modification of variable metric algorithms is concerned with replacing movement along a straight line by one-dimensional searches along curvilinear paths. The reader can learn about these approaches from Botsaris and Jacobson [110], Vial and Zang [111], and Avriel and Dauer [112].

Penalty Function Methods

Here we begin the discussion of methods for solving constrained nonlinear programs. First we deal with the problem of how to transform a constrained program into one or more equivalent unconstrained programs that can be solved by techniques discussed earlier.

The intuitive idea behind all penalty function methods is simple. Suppose that we seek a minimum of a real function f on a proper subset X of R^n. This is, of course, a constrained optimization problem that can be transformed into an unconstrained optimization problem after some modification of the objective function, as we shall now see. Define

$$P(x) = \begin{cases} 0, & x \in X \\ +\infty, & x \notin X \end{cases} \tag{381}$$

and consider the unconstrained minimization of the augmented objective function F, given by

$$\min_{x \in R^n} F(x) = f(x) + P(x) \tag{382}$$

where f is assumed to be defined on R^n. A point x^* minimizes F if and only if it also minimizes f over X. The function P is called a penalty function, for it imposes an (infinite) penalty on points lying outside the feasible set. In practice, however, the unconstrained optimization (382) cannot be carried out (except perhaps in some trivial cases) because of the discontinuity in F on the boundary of X and the infinite values outside X. Replacing $+\infty$ by some "large" finite penalty will not simplify the problem, since the numerical difficulties would still remain, and without additional assumptions, the minimum of the augmented, everywhere-finite objective function may not coincide with the minimum of f over X.

The earliest idea for solving constrained problems by penalty functions involved a sequence of unconstrained minimizations in which a penalty parameter is adjusted from one minimization to another so that the sequence of unconstrained minima converges to a feasible point of the constrained problem that satisfies some necessary or sufficient optimality conditions. These methods have been successfully implemented in several applications and are considered one of the effective tools available at present for solving constrained problems.

Exterior Penalty Functions

Consider the nonlinear program introduced earlier:

$$(P) \qquad \min f(x) \tag{383}$$

subject to the constraints

$$g_i(x) \geq 0, \quad i = 1, \ldots, m \tag{384}$$

$$h_j(x) = 0, \quad j = 1, \ldots, p \tag{385}$$

where f, $g_1, \ldots, g_m, h_1, \ldots, h_p$ are assumed to be continuous on R^n. Let X denote the feasible set; that is,

$$X = \{x: x \in R^n,\ g_i(x) \geq 0,\ i = 1, \ldots, m;\ h_j(x) = 0,\ j = 1, \ldots, p\} \tag{386}$$

Exterior penalty function methods usually solve (P) by a sequence of unconstrained minimization problems whose optimal solutions approach the solution of (P) from outside the feasible set. In the sequence of unconstrained optimizations, a penalty is imposed on every $x \notin X$ such that this penalty is increased from problem to problem, thereby forcing the unconstrained optima toward the feasible set. The development of the algorithm presented here is due to Zangwill [113], where further details can be found.

Define real-valued continuous functions ψ and ξ of the variable $\eta \in R$ by

$$\psi(\eta) = |\min(0, \eta)|^{\alpha} \tag{387}$$

and

$$\xi(\eta) = |\eta|^{\beta} \tag{388}$$

where $\alpha \geq 1$ and $\beta \geq 1$ are given constants, usually equal to 1 or 2. Let

$$s(x) = \sum_{i=1}^{m} \psi(g_i(x)) + \sum_{j=1}^{p} \xi(h_j(x)) \tag{389}$$

or

$$= \sum_{i=1}^{m} |\min(0, g_i(x))|^{\alpha} + \sum_{j=1}^{p} |h_j(x)|^{\beta} \tag{390}$$

This continuous function is called a <u>loss function</u> for problem (P). Note that

$$s(x) = 0, \quad \text{if } x \in X \tag{391}$$

and

$$s(x) > 0, \quad \text{if } x \notin X \tag{392}$$

For any positive number p we can define the augmented objective function for problem (P) as

$$F(x, p) = f(x) + \frac{1}{p} s(x) \tag{393}$$

and observe that $F(x, p) = f(x)$ if and only if x is feasible; otherwise $F(x, p) > f(x)$. The $s(x)/p$ term approximates the discontinuous penalty function P(x) in (381) as $p \to 0$. The exterior penalty function method consists of solving a sequence of unconstrained optimizations for $k = 0, 1, 2, \ldots$, given by

$$(EP^k) \qquad \min_{x \in R^n} F(x, p^k) = f(x) = \frac{1}{p^k}\left\{\sum_{i=1}^{m} |\min(0, g_i(x))|^{\alpha} + \sum_{j=1}^{p} |h_j(x)|^{\beta}\right\} \tag{394}$$

using a strictly decreasing sequence of positive numbers p^k. Defining x^{k*} as the optimal solution of (EP^k), we construct a sequence of points $\{x^{k*}\}$ which, under rather mild conditions on (P), has a subsequence converging to an optimum of (P).

The foregoing derivation of exterior penalty functions can be generalized. Let r be a continuous real-valued function of the variable $p \in R$ such that $p^1 > p^2 > 0$

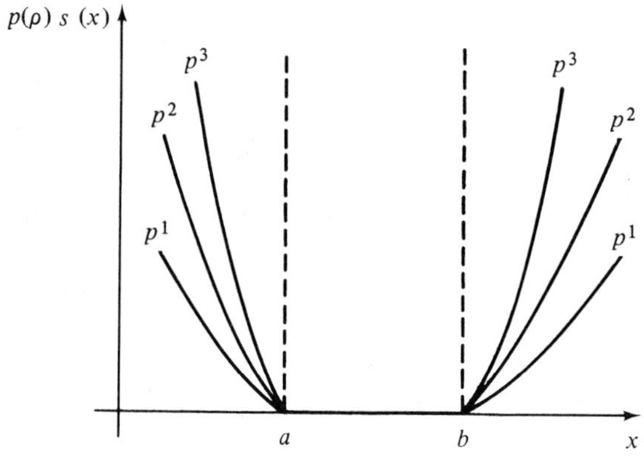

FIG. 16. Exterior penalty function.

implies $r(p^2) > r(p^1) > 0$ and every strictly decreasing sequence of positive numbers $\{p^k\}$ with the property that

$$\lim_{k\to\infty} \{p^k\} = 0 \tag{395}$$

implies

$$\lim_{k\to\infty} \{r(p^k)\} = +\infty \tag{396}$$

Also, let s be any continuous function satisfying (391) and (392). Then $r(p^k)s(x)$ is an <u>exterior penalty function</u>, and

$$(\mathrm{EP}^k) \qquad \min_x F(x, p^k) = f(x) + r(p^k)s(x) \tag{397}$$

is the corresponding unconstrained optimization problem. In Fig. 16 we illustrate this idea for a simple case in which the feasible set X is the closed interval [a, b] on the real line.

<u>Example 20</u>. We seek the minimum of $f(x) = (x)^2$, $x \in R$, subject to the constraint $x \geq 1$. The optimal solution is clearly $x^* = 1$. Let us form the augmented objective function as in (394) with $\alpha = 2$. We have the unconstrained optimization problem

$$\min_{x\in R} F(x, p^k) = (x)^2 + \frac{1}{p^k}[\min(0, x - 1)]^2 \tag{398}$$

For any given $p^k > 0$, the function F is convex and its minimum is at the point

$$x^{k*} = \frac{1}{p^k + 1} \tag{399}$$

Note that, for every $p^k > 0$, this point is infeasible for the original problem. As $\{p^k\} \to 0$, the points x^{k*} approach x^* from outside the feasible set. Of course, in any real problem the unconstrained minimization of $F(x, p^k)$ must be carried out by some numerical algorithm, such as those presented earlier.

Zangwill [113] has investigated the convergence properties of the exterior penalty method. His results indicate that this method is mainly useful for convex programs. Convergence to local minima of nonconvex programs by the exterior penalty function algorithm is discussed in the book of Fiacco and McCormick [17], where many more results on penalty function methods are also presented.

Interior Penalty Functions

Here inequality constrained nonlinear programs are solved through a sequence of unconstrained optimization problems whose minima are at points that strictly satisfy the constraints—that is, in the interior of the feasible set. Staying in the interior can be ensured, as we shall see, by formulating a "barrier" function by which an infinitely large penalty is imposed for crossing the boundary of the feasible set from the inside. Since the algorithm requires the interior of the feasible set to be nonempty, no equality constraints can be handled by the method described below.

Consider, therefore, the nonlinear program

(PI) $$\min f(x) \tag{400}$$

subject to

$$g_i(x) \geq 0, \quad i = 1, \ldots, m \tag{401}$$

As before, assume that f, $g_1, \ldots, g_m$ are continuous functions on R^n. Denote by X the feasible set for program (PI) and let X^0 be the interior of X.

In formulating the penalty function, let q be a real function of $x \in R^n$ such that q is continuous at every point of X^0. And if x^k is any sequence of points in X that converges to some $\hat{x}$ on the boundary of X—that is,

$$I(\hat{x}) = \{i: g_i(\hat{x}) = 0\} \neq \emptyset \tag{402}$$

then

$$\lim_{k\to\infty} \{q(x^k)\} = +\infty \tag{403}$$

Also let t be a real function of $p \in R$ such that

$$p^1 > p^2 > 0 \quad \text{implies} \quad t(p^1) > t(p^2) > 0 \tag{404}$$

and

$$\lim_{k\to\infty} \{p^k\} = 0 \quad \text{implies} \quad \lim_{k\to\infty} \{t(p^k)\} = 0 \tag{405}$$

The function $t(p^k)q(x)$ is called the interior penalty or barrier function.

The interior penalty method can be stated as follows: For $k = 0, 1, \ldots,$ define G to be the augmented objective function minimized in a sequence of unconstrained optimization problems (IP^k), given by

(IP^k) $$\min_x G(x, p^k) = f(x) + t(p^k)q(x) \tag{406}$$

Let $x^0 \in X^0$ be the starting point and assign a positive value to p^0. Solve problem (IP^0) by some unconstrained minimization technique, starting at x^0, and let x^{0*} be a solution of (IP^0). Presumably $x^{0*} \in X^0$. Decrease p^0 to p^1 and solve problem (IP^1), starting at x^{0*}. Denote the optimal solution of (IP^1) by x^{1*}. In this way continue solving (IP^k) for a strictly decreasing sequence p^k, starting always at x^{k-1*}.

The most common choices for the functions t and q are

$$t(p) = (p)^{\alpha}, \qquad \alpha = 1 \text{ or } 2 \tag{407}$$

$$q(x) = \sum_{i=1}^{m} \frac{1}{[g_i(x)]^{\beta}}, \qquad \beta = 1 \text{ or } 2 \tag{408}$$

$$q(x) = -\sum_{i=1}^{m} \log g_i(x) \tag{409}$$

$$q(x) = \sum_{i=1}^{m} \frac{1}{\max\,[0,\ g_i(x)]} \tag{410}$$

<u>Example 21</u>. Consider the following small problem in one variable

$$\min f(x) = \tfrac{1}{2}x \tag{411}$$

subject to

$$g(x) = x - 1 \geq 0 \tag{412}$$

The optimal solution is clearly at $x^* = 1$ and $f(x^*) = \frac{1}{2}$. Suppose that we choose $\alpha = 2$ and $\beta = 1$, as given above, for the barrier function. Then

$$G(x, p^k) = \tfrac{1}{2}x + (p^k)^2 (x - 1)^{-1} \tag{413}$$

The reader can easily verify that the unconstrained minimum of G is given by

$$x^{k*} = 1 + \sqrt{2}\,p^k \tag{414}$$

and

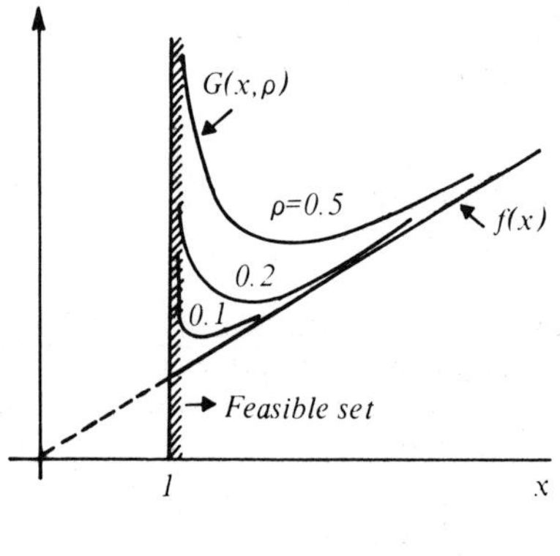

FIG. 17. Interior penalty method.

$$f(x^{k*}) = \frac{1}{2} + \frac{p^k}{\sqrt{2}} \tag{415}$$

Thus the optimal unconstrained minima are all in X^0 and converge to x^* as the values of p^k are successively reduced. The original and the unconstrained problems for a few values of p^k are illustrated in Fig. 17.

Interior penalty methods are based on an idea proposed by Carroll [115] for transforming a constrained nonlinear program into a sequence of unconstrained minimizations by using the t and q functions given above. Carroll's idea was subsequently formalized, extended, and thoroughly studied by Fiacco and McCormick [114, 116], who developed the Sequential Unconstrained Minimization Technique (SUMT), perhaps the best-known penalty function method to date.

The SUMT method has some interesting primal-dual features in cases where program (PI) is a standard convex program and the f, $g_1, \ldots, g_m$ are continuously differentiable. Consider (IP^k), the k-th unconstrained minimization problem:

$$(IP^k) \qquad \min_x G(x, p^k) = f(x) + p^k \sum_{i=1}^{m} \frac{1}{g_i(x)} \tag{416}$$

If $x^{k*} \in X^0$ minimizes (IP^k), then

$$\nabla G(x^{k*}, p^k) = \nabla f(x^{k*}) - p^k \sum_{i=1}^{m} \frac{\nabla g_i(x^{k*})}{[g_i(x^{k*})]^2} = 0 \tag{417}$$

Taking the Lagrangian corresponding to (PI)

$$L(x, \lambda) = f(x) - \sum_{i=1}^{m} \lambda_i g_i(x) \tag{418}$$

suggests defining multipliers λ_i^k as

$$\lambda_i^k = \frac{p^k}{[g_i(x^{k*})]^2}, \qquad i = 1, \ldots, m \tag{419}$$

With this choice of the λ_i^k we have

$$\nabla G(x^{k*}, p^k) = \nabla L(x^{k*}, \lambda^k) = 0 \tag{420}$$

Note that although $\lambda_i^k \geq 0$, the Kuhn-Tucker necessary conditions for convex programs as stated in Theorem 24 do not hold for (PI) at x^{k*}, since

$$\lambda_i^k g_i(x^{k*}) = \frac{p^k}{g_i(x^{k*})} \neq 0, \qquad i = 1, \ldots, m \tag{421}$$

It can be shown, however, that the vector $\lambda^k = (\lambda_1^k, \ldots, \lambda_m^k)^T$, where the λ_i^k are defined by (419), is feasible for the dual of program (PI). This dual program is given by (217) as

$$\text{(DPI)} \qquad \max_{\lambda \geq 0} \{\inf_x f(x) - \sum_{i=1}^{m} \lambda_i g_i(x)\} = \max_{\lambda \geq 0} \{\inf_x L(x, \lambda)\} \tag{422}$$

It follows by the convexity of f, the concavity of g_i, the definition of λ_i^k, and (420) that

$$L(x^{k*}, \lambda^k) = \inf_x L(x, \lambda^k) \tag{423}$$

and λ^k is feasible for (DPI). Consequently, by Theorem 27, the value $L(x^{k*}, \lambda^k)$ is a lower bound on the minimal value of f in program (PI). As the x^{k*} approach x^*, the optimal solution of (PI), the difference between $f(x^{k*})$ and $L(x^{k*}, \lambda^k)$ becomes smaller and smaller until equality is attained. Thus every successful solution of (IP^k) also provides an estimate on how far $f(x^{k*})$ is from its minimal value. This information can be useful in terminating the sequence of unconstrained minimizations. A detailed exposition of the penalty function approach to constrained optimization can be found in Fiacco and McCormick [114] and in Fiacco [117].

Augmented Lagrangian Methods

One of the difficulties in implementing the penalty function methods described above is inherent in their nature. As the parameters approach their limit, numerical computations become more difficult because of discontinuities on the boundary of the feasible set and because the Hessian matrix of the penalty function becomes ill-conditioned. Thus it could be important to derive methods in which the parameters would need to assume moderate values only. The techniques presented below are based on this fact, but as we shall see, they are also closely related to Lagrangian functions.

Consider first a nonlinear program with equality constraints

$$\text{(PE)} \qquad \min f(x) \tag{424}$$

subject to

$$h_j(x) = 0, \quad j = 1, \ldots, p \tag{425}$$

Suppose that $f, h_1, \ldots, h_p$ are twice continuously differentiable functions and that $x^* \in R^n$ is a local solution of (PE), such that the sufficient conditions for a strict local minimum as given in Theorem 5 are satisfied.

Hestenes [118] and Powell [119] have independently suggested a penalty function type method for the solution of this program. Their method, called the augmented Lagrangian method, or method of multipliers, has some nice properties which make it more attractive for computations than the original penalty function technique. Define the augmented Lagrangian $M(x, \mu)$ by

$$M(x, \mu) = f(x) - \sum_{j=1}^{p} \mu_j h_j(x) + \tfrac{1}{2}c \sum_{j=1}^{p} (h_j(x))^2 \tag{426}$$

$$= L(x, \mu) + \tfrac{1}{2}c \sum_{j=1}^{p} (h_j(x))^2 \tag{427}$$

where c is a positive number. Note that here the quadratic penalty term is added to the Lagrangian L corresponding to (PE) and not to the objective function f, as in the penalty function method. We then have Theorem 34.

Theorem 34. Suppose that x^* and μ^* satisfy the sufficiency conditions of optimality for x^* to be a strict local minimum of (PE), as given in Theorem 5. Then there exists a number $c^* > 0$ such that, for all $c \geq c^*$, the point x^* is a local unconstrained minimum of $M(x, \mu^*)$. Conversely, if $h_j(x^0) = 0$, $j = 1, \ldots, p$, and x^0 is an unconstrained minimum of $M(x, \mu^0)$ for some μ^0, then x^0 solves (PE).

We can see from the last theorem that if the values of the Lagrange multipliers μ_j^* and a sufficiently large constant c are available, then an unconstrained minimum of M yields an optimal solution of (PE). The difficulty, of course, lies in determining the correct values of the μ_j and c.

In the Hestenes-Powell augmented Lagrangian method the multipliers μ_j are updated at each iteration and sometimes the constant c as well. Suppose that we choose a sufficiently large value of c and that at iteration k we have some estimate μ^k of the vector of Lagrange multipliers. We minimize $M(x, \mu^k)$ and let the optimal solution be x^{k*}. Observe that

$$\nabla_x M(x^{k*}, \mu^k) = f(x^{k*}) - \sum_{j=1}^{p} [\mu_j^k - ch_j(x^{k*})]\nabla h_j(x^{k*}) = 0 \tag{428}$$

The multipliers are updated by the formula

$$\mu_j^{k+1} = \mu_j^k - ch_j(x^{k*}), \quad j = 1, \ldots, p \tag{429}$$

and we proceed to the next iteration in which $M(x, \mu^{k+1})$ is minimized.

The purpose of the iterative scheme in (429) is to approach μ^*, the vector of true Lagrange multipliers. The important aspect of the augmented Lagrangian method is that the sequence of points x^{k*} may converge to the optimal solution of (PE) without increasing the penalty parameter c to infinity.

Example 22 [120]. Consider the solution of the following quadratic program:

$$\min \tfrac{1}{2}(x_1)^2 + \tfrac{1}{6}(x_2)^2 \tag{430}$$

subject to

$$x_1 + x_2 = 1 \tag{431}$$

The augmented Lagrangian function is given by

$$M(x, \mu) = \tfrac{1}{2}(x_1)^2 + \tfrac{1}{6}(x_2)^2 - \mu(x_1 + x_2 - 1) + \tfrac{1}{2}c(x_1 + x_2 - 1)^2 \tag{432}$$

Minimizing $M(x, \mu^k)$ with respect to x yields

$$x_1^{k*} = \frac{c + \mu^k}{1 + 4c}, \quad x_2^{k*} = \frac{3(c + \mu^k)}{1 + 4c} \tag{433}$$

and the updating formula for the multiplier is

$$\mu^{k+1} = \frac{c + \mu^k}{1 + 4c} \tag{434}$$

The reader can verify that for any $c > 0$ we obtain convergence to $x^* = (0.25, 0.75)$ and $\mu^* = 0.25$. However, it is also easy to see that the number of iterations required to attain a given accuracy in the solution decreases with an increasing value of c.

There is again an interesting primal-dual relationship in the iterative procedure of the augmented Lagrangian method. We shall follow here the exposition of Bertsekas [121]. Consider problem (PE), given by (424) and (425), and satisfying the assumptions mentioned there. Define the perturbed problem

$$\phi(x, w) = \begin{cases} f(x), & \text{if } h_j(x) = w_j, \quad j = 1, \ldots, p \\ +\infty, & \text{otherwise} \end{cases} \tag{435}$$

It can be shown that there exist positive numbers β and δ such that for every w with $\|w\| < \beta$ the problem of minimizing $\phi(x, w)$ with respect to x has a unique solution $x(w)$ in the neighborhood $N_\delta(x^*)$ with a corresponding Lagrange multiplier $\mu(w)$ satisfying $\|\mu(w) - \mu^*\| < \delta$. The functions $x(w)$, $\mu(w)$ are continuously differentiable for $w \in N_\beta(0)$ and satisfy $x(0) = x^*$, $\mu(0) = \mu^*$. Define

$$\Phi(w) = \inf_x \phi(x, w) \tag{436}$$

and consider the augmented perturbation function

$$\Phi_c(w) = \Phi(w) + \tfrac{1}{2}cw^T w \tag{437}$$

There is a $\bar{c} > 0$ such that for all $c \geq \bar{c}$, Φ_c is a strictly convex function on $N_\beta(0)$. We can write

$$\theta(w) = \begin{cases} \Phi_c(w), & \text{if } c \geq \bar{c} \text{ and } w \in N_\beta(0) \\ +\infty, & \text{otherwise} \end{cases} \tag{438}$$

and the dual problem of this convex program is given by

$$\max_\mu -\theta^*(0, \mu) \tag{439}$$

where

$$-\theta^*(0, \mu) = \inf_{\substack{w \in N_\beta(0) \\ c \geq \bar{c}}} [\Phi_c(w) - \mu^T w] \tag{440}$$

The dual objective function, $-\theta^*$, is continuously differentiable on R^p. It is twice continuously differentiable on

$$A = \{\mu: \mu = \nabla \Phi_c(w),\ w \in N_\beta(0)\} \tag{441}$$

For every $\bar{\mu} \in A$ the infimum in (440) is attained at a unique point $w(\bar{\mu})$ and

$$\frac{\partial \theta^*(0, \bar{\mu})}{\partial \mu_j} = w_j(\bar{\mu}) = h_j(x(\bar{\mu})), \quad j = 1, \ldots, p \tag{442}$$

The function $-\theta^*$ has a unique maximizing point at μ^*, the vector of Lagrange multipliers for (PE).

Comparing (429) and (442) we finally conclude that the updating formula for the multipliers μ_j^k is equivalent to a fixed step length move in the negative gradient (steepest ascent) of $-\theta^*$, the dual objective function to be maximized. This insight into properties of the Hestenes-Powell method enables one to improve it by using the steps of more efficient optimization methods, such as quasi-Newton methods. For more information on such approaches see Tapia [122].

Augmented Lagrangian methods for solving equality constrained nonlinear programs were also extended to inequality constrained problems; see Bertsekas [123], Buys [124], Fletcher [125], Kort and Bertsekas [126, 127], and Rockafellar [128, 129].

Methods of Restricted Movement

In unconstrained optimization techniques the selection of directions for movement and the step lengths taken are generally unrestricted as long as some improvement in the function value is obtained. If constraints are present, these methods can be modified by restricting movement such that the constraints are taken into account. We present here a few representatives of different approaches for constrained optimization of this type.

Feasible Direction Methods

Methods of feasible directions have been derived by Zoutendijk [130-132]. Here we restrict our discussion to the case of the linearly constrained nonlinear program

$$\text{(LCP)} \quad \min f(x) \tag{443}$$

subject to

$$\sum_{j=1}^{n} a_{ij}x_j - b_i = 0, \quad i = 1, \ldots, m \tag{444}$$

$$\sum_{j=1}^{n} a_{ij}x_j - b_i \geq 0, \quad i = m+1, \ldots, p \tag{445}$$

Two major decisions are required for each iteration of a feasible direction method:

1. Selection of a feasible descent direction
2. Selection of a step length along the feasible direction

Beginning with the first, it can be shown that, given a point $x^k \in X$—that is, satisfying (444) and (445)—a feasible descent direction z from x^k must be in $Z(X, x^k) \cap Y(x^k)$, where

$$Z(X, x^k) = \{z: \sum_{j=1}^{n} a_{ij}z_j = 0,\ i = 1, \ldots, m;\ \sum_{j=1}^{n} a_{ij}z_j \geq 0,\ i \in I(x^k)\} \tag{446}$$

$$Y(x^k) = \{z: z^T \nabla f(x^k) < 0\} \tag{447}$$

and

$$I(x^k) = \{i: \sum_{j=1}^{n} a_{ij}x_j^k = b_i, \quad i = m+1, \ldots, p\} \tag{448}$$

Generally there are many feasible descent directions from a given point $x^k \in X$, and so the steepest feasible descent direction can be chosen from among them by solving the optimization problem

$$\text{(SFD)} \quad \min z^T \nabla f(x^k) \tag{449}$$

subject to $z \in Z(X, x^k)$, and $\|z\| \leq 1$. Since $Z(X, x^k)$ is a cone, it is necessary to bound $\|z\|$ in order to obtain a finite solution. Note that $\|z\|$, the Euclidean norm of z, is a nonlinear function, and $\|z\| \leq 1$ is a nonlinear constraint. It can also be shown that $Z(X, x^k) \cap Y(x^k)$ is empty—that is, there is no feasible discent direction from x^k—if and only if there exist vectors $\lambda \geq 0$ and $\mu \in R^m$ such that

$$\nabla f(x^k) = \sum_{i=1}^{m} \mu_i a^i + \sum_{i \in I(x^k)} \lambda_i a^i \tag{450}$$

where $a^i = (a_{i1}, \ldots, a_{in})^T$. In other words, it is possible to find a feasible descent direction from x^k if and only if $\nabla f(x^k)$ is not in the cone spanned by the vectors a^i that correspond to the active constraints. Suppose that the latter condition holds. Let t^{k+1} be the point of the cone that is closest to $\nabla f(x^k)$—that is,

$$(t^{k+1})^T [t^{k+1} - \nabla f(x^k)] = 0 \tag{451}$$

The steepest feasible descent vector that solves (SFD) is given by

$$z^{k+1} = \frac{t^{k+1} - \nabla f(x^k)}{\|t^{k+1} - \nabla f(x^k)\|} \tag{452}$$

The problem of finding the steepest feasible descent direction can be solved by a very special type of quadratic program in which the least distance from the cone spanned by the active constraints to the point $\nabla f(x^k)$ is sought. This program is given by

$$\min_{\lambda, \mu} \|\nabla f(x^k) - \sum_{i=1}^{m} \mu_i a^i - \sum_{i \in I(x^k)} \lambda_i a^i\|^2 \tag{453}$$

subject to

$$\lambda_i \geq 0, \quad i \in I(x^k) \tag{454}$$

It is possible to extend the direction-selecting step outlined above to a method that resembles constrained variable metric algorithms if, instead of using $\nabla f(x^k)$

in the formulas, we premultiply it by a matrix H_k that approximates the inverse Hessian of f.

Having found a feasible descent direction, we must select a step length α^*_{k+1}. Consequently, define

$$\alpha^*_{k+1} = \min \{\alpha'_{k+1}, \alpha''_{k+1}\} \tag{455}$$

where

$$\alpha'_{k+1} = \max \{\alpha_{k+1}: (z^{k+1})^T \nabla f(x^k + \alpha_{k+1} z^{k+1}) \leq 0\} \tag{456}$$

$$\alpha''_{k+1} = \max \{\alpha_{k+1}: (x^k + \alpha_{k+1} z^{k+1}) \in X\} \tag{457}$$

For $\alpha'_{k+1} < +\infty$, we have $\nabla f(x^k + \alpha_{k+1} z^{k+1}) = 0$. Setting $x^{k+1} = x^k + \alpha^*_{k+1} z^{k+1}$, we return to the direction-selecting step, and continuing in this way we have a complete algorithm, provided that a starting feasible point is given. Unfortunately, such a method would not only have a very slow convergence rate but the constraints can also prevent convergence to an optimum because of so-called zigzagging or jamming [130].

Feasible direction algorithms, including "antizigzagging" procedures, have been suggested by McCormick [133], Polak [134], and others.

Reduced-Gradient Algorithms

First we review the reduced-gradient method for linear constraints and then its generalizations for problems with nonlinear constraints. Consider the linearly constrained problem

$$\text{(LEP)} \quad \min f(x) \tag{458}$$

subject to

$$Ax = b \tag{459}$$

$$x \geq 0 \tag{460}$$

where f is a continuously differentiable real function, A is an m×n matrix, b is an m vector, and $m \leq n$. The vector of variables x can be partitioned into two subvectors $x = (x^B, x^N)^T$, where $x^B = (x_1^B, \ldots, x_m^B)^T$ is the vector of <u>basic</u>, or dependent, variables and $x^N = (x_1^N, \ldots, x_{n-m}^N)$ is the vector of <u>nonbasic</u>, or independent, variables. Accordingly, the matrix A is also partitioned into A = [B, C], where we assume, without loss of generality, that the first m columns of A correspond to the basic variables. Further assume that B, the m×m submatrix of A that corresponds to the components of the vector x^B, is nonsingular. Then we can write

$$Bx^B + Cx^N = b \tag{461}$$

and

$$x^B = B^{-1}b - B^{-1}Cx^N \tag{462}$$

In addition, we also assume that the basic variables are nondegenerate; that is, $x^B > 0$. The nonbasic variables are called independent, since by assigning some numerical values to them, we obtain a unique solution of (461).

The basic idea of reduced-gradient methods is to eliminate x^B (as a function of x^N) via (462) and consider the optimization problem only in terms of x^N. This idea was used in the differential algorithms derived by Wilde [135] and Wilde and Beightler [136] through the notion of "constrained derivatives," in the reduced-gradient method of Wolfe [137, 138], and in the convex-simplex method of Zangwill [139].

From (462) we obtain the reduced gradient $r \in R^{n-m}$ by the formula

$$r(x^N) = \nabla_{x^N} f(x^B(x^N), x^N) - (B^{-1}C)^T \nabla_{x^B} f(x^B(x^N), x^N) \tag{463}$$

Now if we could make a small move from a current value of x^N in the direction of the negative reduced gradient without violating the nonnegativity constraints on the vector x, a decrease in the function value of f would occur. This step is accomplished as follows: Given a feasible x^k, compute for $i = 1, \ldots, n - m$

$$z_i^{N,k+1} = \begin{cases} 0, & \text{if } x_i^{N,k} = 0 \text{ and } r_i(x^{N,k}) > 0 \\ -r_i(x^{N,k}), & \text{otherwise} \end{cases} \tag{464}$$

and let

$$z^{B,k+1} = -B^{-1}Cz^{N,k+1} \tag{465}$$

Then $z^{k+1} = (z^{B,k+1}, z^{N,k+1})^T$. The next point, $x^{k+1} = (x^{B,k+1}, x^{N,k+1})^T$, is given by

$$x^{k+1} = x^k + \alpha^*_{k+1} z^{k+1} \tag{466}$$

where α^*_{k+1} is computed from the relations

$$\alpha^1_{k+1} = \max \{\alpha_{k+1} : x^{B,k} + \alpha_{k+1} z^{B,k+1} \geq 0\} \tag{467}$$

$$\alpha^2_{k+1} = \max \{\alpha_{k+1} : x^{N,k} + \alpha_{k+1} z^{N,k+1} \geq 0\} \tag{468}$$

and

$$f(x^k + \alpha^*_{k+1} z^{k+1}) = \min_{\alpha_{k+1}} \{f(x^k + \alpha_{k+1} z^{k+1}) : 0 \leq \alpha_{k+1} \leq \min(\alpha^1_{k+1}, \alpha^2_{k+1})\} \tag{469}$$

If $\alpha^*_{k+1} < \alpha^1_{k+1}$, let x^{k+1} be defined by (466). Otherwise

$$x_\ell^{B,k} + \alpha^1_{k+1} z_\ell^{B,k+1} = 0 \tag{470}$$

for some ℓ, and x_ℓ^B is dropped from the vector of basic variables in exchange for the largest positive nonbasic variable. The algorithm terminates if $\|z^{k+1}\| \leq \epsilon$, where $\epsilon > 0$ is some small predetermined number.

Example 23. Let us illustrate the reduced-gradient method on the problem

$$\min f(x) = (x_1)^2 + 4(x_2)^2 \tag{471}$$

subject to

$$x_1 + 2x_2 - x_3 = 1 \tag{472}$$

$$-x_1 + x_2 + x_4 = 0 \tag{473}$$

$$x \geq 0 \tag{474}$$

by carrying out a few iterations. Suppose that we start at $x^0 = (2, 1, 3, 1)$ and let $x^{B,0} = (x_1, x_4)^T$, $x^{N,0} = (x_2, x_3)^T$. Then

$$B_0 = \begin{bmatrix} 1 & 0 \\ -1 & 1 \end{bmatrix}, \quad (B_0)^{-1} = \begin{bmatrix} 1 & 0 \\ 1 & 1 \end{bmatrix}, \quad C_0 = \begin{bmatrix} 2 & -1 \\ 1 & 0 \end{bmatrix} \tag{475}$$

and $\nabla f(x^0) = (4, 8, 0, 0)^T$. Let us compute the reduced gradient

$$r(x^{N,0}) = \begin{pmatrix} 8 \\ 0 \end{pmatrix} - \begin{bmatrix} 2 & 1 \\ -1 & 0 \end{bmatrix} \begin{bmatrix} 1 & 1 \\ 0 & 1 \end{bmatrix} \begin{pmatrix} 4 \\ 0 \end{pmatrix} = \begin{pmatrix} 0 \\ 4 \end{pmatrix} \tag{476}$$

Hence $z^{N,1} = (0, -4)^T$ and

$$z^{B,1} = -\begin{bmatrix} 1 & 0 \\ 1 & 1 \end{bmatrix} \begin{bmatrix} 2 & -1 \\ 1 & 0 \end{bmatrix} \begin{pmatrix} 0 \\ -4 \end{pmatrix} = \begin{pmatrix} -4 \\ -4 \end{pmatrix} \tag{477}$$

Now we compute the step length along z^1. First we find α_1^1 from (467). It is the largest α_1 satisfying

$$\begin{pmatrix} 2 - 4\alpha_1 \\ 1 - 4\alpha_1 \end{pmatrix} \geq 0 \tag{478}$$

and we obtain $\alpha_1^1 = \frac{1}{4}$. Similarly, we find α_1^2 from

$$\begin{pmatrix} 0 + 0\alpha_1 \\ 3 - 4\alpha_1 \end{pmatrix} \geq 0 \tag{479}$$

and $\alpha_1^2 = \frac{3}{4}$. Hence $\alpha_1^1 < \alpha_1^2$. Minimizing $f(x^0 + \alpha_1 z^1)$ with respect to α_1, we obtain from (469) $\alpha_1^* = \frac{1}{4}$. The new point is then given by $x^1 = x^0 + \alpha_1^* z^1$ and $x^1 = (1, 1, 2, 0)$. At this point, $\nabla f(x^1) = (2, 8, 0, 0)^T$. Since $\alpha_1^* = \alpha_1^1$, we change basis and x_3 enters the basis, replacing x_4. Hence

$$B_1 = \begin{bmatrix} 1 & -1 \\ -1 & 0 \end{bmatrix}, \quad (B_1)^{-1} = \begin{bmatrix} 0 & -1 \\ -1 & -1 \end{bmatrix}, \quad C_1 = \begin{bmatrix} 2 & 0 \\ 1 & 1 \end{bmatrix} \tag{480}$$

The new reduced gradient is given by

$$r(x^{N,1}) = \begin{pmatrix} 8 \\ 0 \end{pmatrix} - \begin{bmatrix} 2 & 1 \\ 0 & 1 \end{bmatrix} \begin{bmatrix} 0 & -1 \\ -1 & -1 \end{bmatrix} \begin{pmatrix} 2 \\ 0 \end{pmatrix} = \begin{pmatrix} 10 \\ 2 \end{pmatrix} \tag{481}$$

and $z^{N,2} = (-10,\ 0)^T$ by (464). Consequently,

$$z^{B,2} = -\begin{bmatrix} 0 & -1 \\ -1 & -1 \end{bmatrix}\begin{bmatrix} 2 & 0 \\ 1 & 1 \end{bmatrix}\begin{pmatrix} -10 \\ 0 \end{pmatrix} = \begin{pmatrix} -10 \\ -30 \end{pmatrix} \tag{482}$$

Next we compute the step length along z^2. The inequalities

$$\begin{pmatrix} 1 - 10\alpha_2 \\ 2 - 30\alpha_2 \end{pmatrix} \geq 0 \tag{483}$$

and $1 - 10\alpha_2 \geq 0$ must hold, yielding $\alpha_2^1 = \frac{1}{15}$, $\alpha_2^2 = \frac{1}{10}$, and $\alpha_2^1 < \alpha_2^2$. The unrestricted step length minimizing $f(x^1 + \alpha_2 z^2)$ with respect to α_2 is $\alpha_2 = \frac{1}{10} > \alpha_2^1$. Thus we set $\alpha_2^* = \frac{1}{15}$.

The next point is given by $x^2 = (\frac{1}{3},\ \frac{1}{3},\ 0,\ 0)$ and $\nabla f(x^2) = (\frac{2}{3},\ \frac{8}{3},\ 0,\ 0)^T$. We change the basis and x_2 becomes a basic variable instead of x_3. Now $x^{B,2} = (x_1,\ x_2)^T$, $x^{N,2} = (x_3,\ x_4)^T$. The new partitions of the A matrix are

$$B_2 = \begin{bmatrix} 1 & 2 \\ -1 & 1 \end{bmatrix}, \quad (B_2)^{-1} = \begin{bmatrix} \frac{1}{3} & -\frac{2}{3} \\ \frac{1}{3} & \frac{1}{3} \end{bmatrix}, \quad C_2 = \begin{bmatrix} -1 & 0 \\ 0 & 1 \end{bmatrix} \tag{484}$$

The reduced gradient at this point is given by

$$r(x^{N,2}) = \begin{pmatrix} 0 \\ 0 \end{pmatrix} - \begin{bmatrix} -1 & 0 \\ 0 & 1 \end{bmatrix}\begin{bmatrix} \frac{1}{3} & \frac{1}{3} \\ -\frac{2}{3} & \frac{1}{3} \end{bmatrix}\begin{pmatrix} \frac{2}{3} \\ \frac{8}{3} \end{pmatrix} \begin{pmatrix} \frac{10}{9} \\ -\frac{4}{9} \end{pmatrix} \tag{485}$$

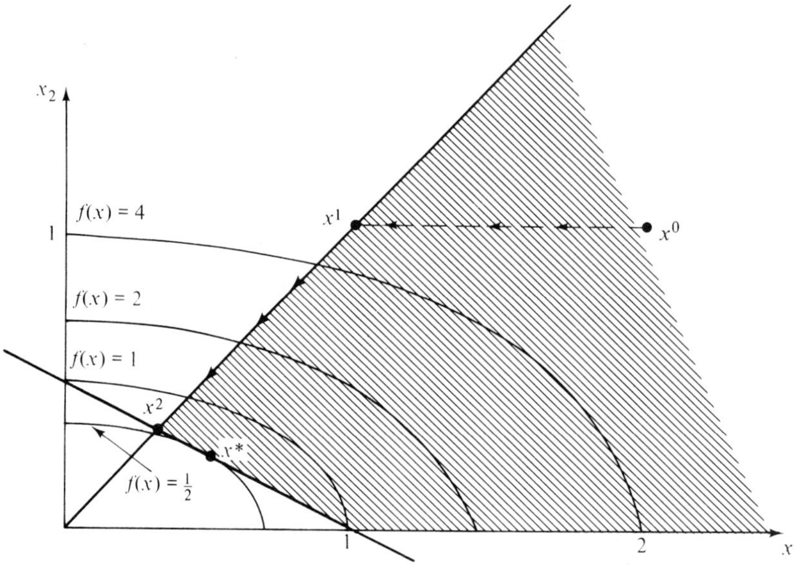

FIG. 18. The reduced-gradient method.

and $z^{N,3} = (0, \frac{4}{9})^T$. Thus

$$z^{B,3} = -\begin{bmatrix} \frac{1}{3} & -\frac{2}{3} \\ \frac{1}{3} & \frac{1}{3} \end{bmatrix}\begin{bmatrix} -1 & 0 \\ 0 & 1 \end{bmatrix}\begin{pmatrix} 0 \\ \frac{4}{9} \end{pmatrix} = \begin{pmatrix} \frac{8}{27} \\ -\frac{4}{27} \end{pmatrix} \tag{486}$$

Continuing the computations we can observe convergence to the optimum at $x^* = (\frac{1}{2}, \frac{1}{4}, 0, \frac{1}{4})$. The progress of the algorithm (in x_1, x_2 space) is illustrated in Fig. 18.

The reduced-gradient method, as described above, can fail to converge to a point satisfying the Kuhn-Tucker conditions because of zigzagging. Several "anti-zigzagging" techniques were suggested to overcome this difficulty. One suggestion is to modify the selection of $z_i^{N,k+1}$ by setting it to zero if $x_i^{N,k} \leq \epsilon$ and $r_i(x^{N,k}) > 0$, where $\epsilon > 0$ is some predetermined value, instead of the formula given by (464).

Abadie and Carpentier [140] generalized the reduced-gradient method to problems with nonlinear equality constraints. Consider the bounded-variable problem

$$\text{(NEP)} \quad \min f(x) \tag{487}$$

subject to

$$h_i(x) = 0, \quad i = 1, \ldots, p \tag{488}$$

$$\beta \geq x \geq \alpha \tag{489}$$

where the f, $h_1, h_2, \ldots, h_p$ are assumed to be continuously differentiable functions of $x \in R^n$ and $p \leq n$.

Every feasible vector x^0 is assumed to be nondegenerate, which means that x^0 can be partitioned, as before, into two subvectors, $x^0 = (x^{B,0}, x^{N,0})^T$, $x^{B,0} \in R^p$, $x^{N,0} \in R^{n-p}$ and

$$\beta^{B,0} > x^{B,0} > \alpha^{B,0} \tag{490}$$

where $\alpha = (\alpha^{B,0}, \alpha^{N,0})^T$, $\beta = (\beta^{B,0}, \beta^{N,0})^T$. The vectors

$$\nabla_{x^B} h_i(x^0) = \left[\frac{\partial h_i(x^0)}{\partial x_1^{B,0}}, \ldots, \frac{\partial h_i(x^0)}{\partial x_p^{B,0}}\right]^T, \quad i = 1, \ldots, p \tag{491}$$

are linearly independent. That is, the p×p matrix $\Delta^B(x^0)$ whose columns are the vectors $\nabla_{x^B} h_i(x^0)$ is nonsingular. It can easily be verified that the Kuhn-Tucker necessary conditions of optimality at x^0 for problem (NEP) are the existence of vectors $\mu^0 \in R^p$, $\lambda^0 \in R^{n-p}$ such that

$$\nabla_{x^N} f(x^0) - \sum_{i=1}^{p} \mu_i^0 \nabla_{x^N} h_i(x^0) - \lambda^0 = 0 \tag{492}$$

$$\nabla_{x^B} f(x^0) - \sum_{i=1}^{p} \mu_i^0 \nabla_{x^B} h_i(x^0) = 0 \tag{493}$$

where

$$\lambda_j^0 \geq 0, \quad \text{if } x_j^{N,0} = \alpha_j^{N,0} \tag{494}$$

$$\lambda_j^0 = 0, \quad \text{if } \beta_j^{N,0} > x_j^{N,0} > \alpha_j^{N,0} \tag{495}$$

$$\lambda_j^0 \leq 0, \quad \text{if } x_j^{N,0} = \beta_j^{N,0} \tag{496}$$

Hence

$$\mu^0 = [\Delta^B(x^0)]^{-1}\nabla_{x^B} f(x^0) \tag{497}$$

and

$$\lambda^0 = \nabla_{x^N} f(x^0) - \Delta^N(x^0)[\Delta^B(x^0)]^{-1}\nabla_{x^B} f(x^0) \tag{498}$$

where $\Delta^N(x^0)$ is the $(n - p)\times p$ matrix whose columns are the vectors $\nabla_{x^N} h_i(x^0)$. Note that λ^0 is the reduced gradient, analogous to (463), if the constraints are linear. It follows that, for every feasible x^0, we can compute a vector λ^0 by the last equation; and if λ^0 also satisfies (494) to (496), then the necessary optimality conditions for (NEP) hold at x^0. The generalized reduced-gradient method is based on iterative moving from a feasible x^0 to feasible x^1, x^2, and so on, until a point x^ℓ is reached where a λ^ℓ computed from (498) satisfies (494) to (496). Suppose that we have a feasible x^k and the corresponding λ^k does not satisfy these three relations. Then we change the current value of the nonbasic vector by

$$x^{N,k+1} = x^{N,k} + \theta_{k+1} z^{k+1}, \qquad \theta_{k+1} \geq 0 \tag{499}$$

where $z_j^{k+1} = 0$ if

$$x_j^{N,k} = \alpha_j^{N,k} \quad \text{and} \quad \lambda_j^k > 0, \quad \text{or} \quad x_j^{N,k} = \beta_j^{N,k} \quad \text{and} \quad \lambda_j^k < 0 \tag{500}$$

Otherwise we set either $z_j^{k+1} = -\lambda_j^k$ or $z_j^{k+1} = 0$, according to rules that depend on the particular version of the method used:

1. In the GRG version, we set $z_j^{k+1} = -\lambda_j^k$ whenever (500) does not hold. Note that this choice of z_j^{k+1} is identical to the one used above for linear constraints.
2. In the GRGS version, we first find an index s by

$$|\lambda_s^k| = \max_j |\lambda_j^k| \tag{501}$$

where the indices j range over those for which (500) does not hold. Then we let

$$z_j^{k+1} = \begin{cases} 0, & j \neq s \\ -\lambda_s^k, & j = s \end{cases} \tag{502}$$

This version coincides with the simplex method if the optimization problem is actually a linear program.

3. In the GRGC version, we cyclically set $z_j^{k+1} = -\lambda_j^k$. A cycle consists of n iterations. For $k = 1, \ldots, n$, we set

$$z_j^{k+1} = \begin{cases} 0, & j \neq k \\ -\lambda_j^k, & j = k \end{cases} \tag{503}$$

unless k is the index of a basic variable or (500) holds, in which case iteration k is omitted. After $k = n$ we return to $k = 1$ and so on.

Having determined a $z^{k+1} \in R^{n-p}$ by one of these versions, we take a trial value of the step length θ_{k+1}, say $\theta_{k+1}^1 > 0$, and define

$$\tilde{x}_j^{N,k} = \begin{cases} \beta_j^{N,k}, & \text{if } x_j^{N,k} + \theta_{k+1}^1 z_j^{k+1} > \beta_j^{N,k} \\ \alpha_j^{N,k}, & \text{if } x_j^{N,k} + \theta_{k+1}^1 z_j^{k+1} < \alpha_j^{N,k} \\ x_j^{N,k} + \theta_{k+1}^1 z_j^{k+1}, & \text{otherwise} \end{cases} \tag{504}$$

To maintain feasibility, we solve the set of p nonlinear equations

$$h_i(x^B, \tilde{x}^{N,k}) = 0, \qquad i = 1, \ldots, p \tag{505}$$

in p variables $x_1^B, \ldots, x_p^B$ by an iterative procedure, such as Newton's method. Then, at iteration ℓ of Newton's method we have the formula

$$x^{B,\ell+1} = x^{B,\ell} - [\Delta^N(x^{B,\ell}, \tilde{x}^{N,k})]^{-1} h(x^{B,\ell}, \tilde{x}^{N,k}) \tag{506}$$

where $h = (h_1, \ldots, h_p)^T$ and $x^{B,\ell} = x^{B,k}$ for $\ell = 0$. Formula (506) is used until one of the following cases occurs:

1. The norm $\|h(x^{B,\ell}, \tilde{x}^{N,k})\|$ increases in a few successive iterations. In this case we reduce θ_{k+1}^1 by some factor, recompute (504), and return to solve (505).
2. We obtain

$$f(x^{B,\ell}, \tilde{x}^{N,k}) > f(x^{B,\ell-1}, \tilde{x}^{N,k}) \tag{507}$$

The correction is the same as in Case 1.

3. For some ℓ the point $(x^{B,\ell}, \tilde{x}^{N,k})$ is outside the bounds defined by (489). We find a point $\tilde{x}^{B,\ell}$ on the line segment joining $x^{B,\ell-1}$ and $x^{B,\ell}$ such that $\tilde{x}_r^{B,\ell} = \alpha_r^{B,\ell}$ or $\tilde{x}_r^{B,\ell} = \beta_r^{B,\ell}$ for some r and make a change in the current basis by replacing the variable $x_r^{B,\ell}$ by, say $x_s^{N,\ell}$, where the index s is determined by some rule similar to the one used in the simplex method. For example, we can use the relation (omitting the iteration index)

$$|\Delta^N(\tilde{x})_s (\Delta^B(\tilde{x}))_r^{-1}| v_s = \max_q \{|\Delta^N(\tilde{x})_q (\Delta^B(\tilde{x}))_r^{-1}| v_q\} \tag{508}$$

where $\Delta^N(\tilde{x})_q$ is the row vector $(\partial h_1(\tilde{x})/\partial x_q^N, \ldots, \partial h_p(\tilde{x})/\partial x_q^N)$, evaluated at $\tilde{x} = (x^{B,\ell}, \tilde{x}^{N,k})$, $(\Delta^B(\tilde{x}))_r^{-1}$ is the r-th column of the matrix $(\Delta^B(\tilde{x}))^{-1}$, and

$$v_q = \min \{(\tilde{x}_q^{N,k} - \alpha_q^{N,k}), (\beta_q^{N,k} - \tilde{x}_q^{N,k})\} \tag{509}$$

The solution of (505) is now attempted with the new basis.

4. The iterations converge. That is,

$$\| h(x^{B,\ell^*}, x^{N,k}) \| \leq \epsilon \tag{510}$$

for some ℓ^*, where ϵ is a small positive number. If

$$\beta^{B,\ell^*} > x^{B,\ell^*} > \alpha^{B,\ell^*} \tag{511}$$

we set $x^{k+1} = (x^{B,\ell^*}, \tilde{x}^{N,k})$ and find a new search direction. If some component r of x^{B,ℓ^*} is at a lower or upper bound, we change basis as outlined in Case 3. The computations terminate when a point that satisfies the necessary optimality conditions is found.

The actual way in which the computations are performed involves many important considerations, and as the reader has undoubtedly recognized, the implementation of this reduced-gradient method for nonlinear constraints is not trivial. For computational details see Lasdon, Fox, and Ratner [141].

Cutting Plane Methods

So far we have been studying computational methods in which convexity does not play a major role. Nevertheless, many algorithms can be easier implemented for convex programs in which the limit points of the iterations usually satisfy necessary and sufficient conditions of optimality, an important feature not shared by every nonlinear program. The methods discussed here apply to certain nonlinear convex programs. The underlying principle of cutting plane methods is to approximate the feasible set of a nonlinear program by a finite set of closed half spaces and to solve a sequence of approximating linear programs.

Consider the problem

$$\text{(CPP)} \quad \min f(x) = c^T x \tag{512}$$

subject to

$$g_i(x) \geq 0, \quad i = 1, \ldots, m \tag{513}$$

where $x \in R^n$ and $g_1, \ldots, g_m$ are closed proper concave functions on R^n. Let

$$X = \{x: g_i(x) \geq 0, \ i = 1, \ldots, m\} \tag{514}$$

and assume that X is contained in a compact set $T \subset R^n$, defined by a finite set of linear inequalities

$$T = \{x: x \in R^n, \ Ax \geq b\} \tag{515}$$

For $x \in T$, let $\partial g_i(x) \subset R^n$ be the set of subgradients or subdifferential of g_i; that is, $\xi \in \partial g_i(x)$ implies

$$g_i(y) \leq g_i(x) + \xi^T(y - x) \tag{516}$$

for every $y \in R^n$. If g_i is differentiable at x, then $\xi = \nabla g_i(x)$. We shall assume that

$$\partial g_i(x) \neq \emptyset, \quad i = 1, \ldots, m \tag{517}$$

for $x \in T$ and that the sets

$$\bigcup_{x \in T} \partial g_i(x), \quad i = 1, \ldots, m \tag{518}$$

are bounded. This assumption is satisfied if, for example, the g_i are continuously differentiable. We can now state a <u>cutting plane algorithm</u> for solving (CPP), derived by Kelley [142] and Cheney and Goldstein [143], which we shall call the KCG algorithm:

1. Solve the linear program of minimizing $f(x) = c^T x$, subject to $x \in T$, and let x^0 be the optimal solution. If x^0 is contained in the set

   $$X(\epsilon) = \{x: x \in T,\ g_i(x) \geq -\epsilon,\ i = 1, \ldots, m\} \tag{519}$$

   where ϵ is a small positive number, stop; an optimum of (CPP) has been reached. Otherwise let k = 0 and go to Step 2.
2. Given an $x^k \in T$ such that $x^k \notin X(\epsilon)$, find the index s_k by

   $$g_{s_k}(x^k) = \min \{g_i(x^k),\ i = 1, \ldots, m\} < 0 \tag{520}$$

   and select a $\xi^k \in \partial g_{s_k}(x^k)$. Solve the linear program

   $$\min c^T x \tag{521}$$

   $$\tilde{g}_{s_h}(x, x^h) = g_{s_h}(x^h) + (\xi^h)^T(x - x^h) \geq 0, \quad h = 0, 1, \ldots, k \tag{522}$$

   $$x \in T \tag{523}$$

3. Let x^{k+1} be the optimal solution of the preceding linear program. If $x^{k+1} \in X(\epsilon)$, stop. Otherwise set k = k + 1 and return to Step 2.

Note that (516), which follows from the concavity of g_i, and constraints (522) explain the name cutting plane. Suppose that $x^k \notin X(\epsilon)$. Then there is at least one constraint such that $g_s(x^k) < -\epsilon$ for some index s. The half space defined by

$$g_s(x^k) + \xi^T(x - x^k) \geq 0 \tag{524}$$

where $\xi \in \partial g_s(x^k)$, cuts off the point x^k, which does not satisfy (524), from the feasible set X that is contained in the half space defined by (524). Denote by S_k the feasible set of the linear program solved in Step 2 of iteration k. These sets are nested—that is,

$$S_k \subset S_{k-1} \subset \cdots \subset S_0 \tag{525}$$

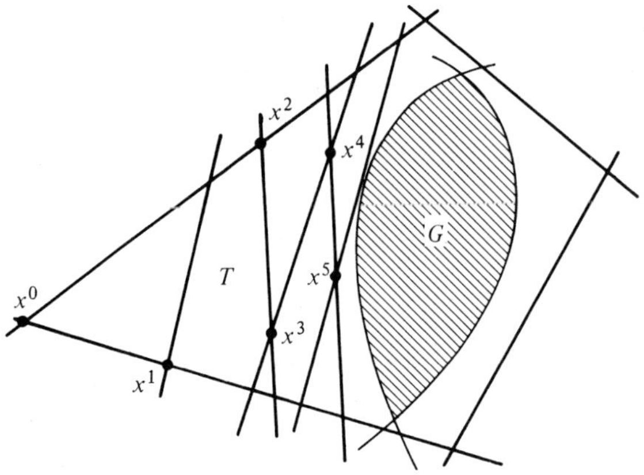

FIG. 19. The KCG cutting plane algorithm.

At each iteration a new linear inequality constraint is added to the set of constraints already present from the previous iteration. This addition of a new constraint at each iteration suggests using a dual simplex method for solving the linear programming problems. Let us state the convergence of the KCG method.

Theorem 35. Let $g_1, \ldots, g_m$ be closed concave functions on the compact convex set $T \subset R^n$ such that at every point $x \in T$ the sets of subgradients $\partial g_i(x)$ are nonempty for $i = 1, \ldots, m$ and there exists a K such that

$$\sup \{ \|\xi^i\| : \xi^i \in \partial g_i(x), \ i = 1, \ldots, m; \ x \in T\} \leq K \tag{526}$$

Further assume that X, the feasible set of (CPP), is nonempty and contained in T. Let

$$S_k = S_{k-1} \cap \{x: \tilde{g}_{s_k}(x, x^k) \geq 0\} \tag{527}$$

where $S_0 = T$. If $x^{k+1} \in S_k$ is such that

$$f(x^{k+1}) = c^T x^{k+1} = \min \{c^T x: x \in S_k\} \tag{528}$$

then the sequence $\{x^k\}$ contains a subsequence that converges to an optimal solution of (CPP).

A typical trajectory of points of the sequence $\{x^k\}$ found by successive construction of cutting planes and solution of approximating linear programs is shown in Fig. 19.

Perhaps the most important step that influences the efficiency of the method just described is the selection of a cutting plane. Several modifications of the basic KCG method have been proposed. Veinott [144] extended the cutting plane method to problems having generalized concave constraint functions that define a convex feasible set. The construction of cutting planes is carried out by actually finding supporting hyperplanes to the feasible set.

One clear disadvantage of the cutting plane methods is that the size of the linear programming subproblems increases from iteration to iteration, since cutting plane constraints are always added to the existing set of constraints but are never deleted. Some later works on cutting plane methods center on the question of dropping inactive constraints or, more generally, approximating the feasible set of the nonlinear program by linear constraints without nesting. Topkis [145] has derived such a method and proved convergence under assumptions similar to those stated above. Eaves and Zangwill [146] have studied cutting plane methods in a more general framework which contains the KCG, Veinott, Topkis, and several other versions of cutting plane algorithms as special cases.

ACKNOWLEDGMENT

The illustrations are reproduced from M. Avriel, Nonlinear Programming: Analysis and Methods, by permission of Prentice-Hall, Inc.

REFERENCES

1. M. Avriel, Nonlinear Programming: Analysis and Methods, Prentice-Hall, Englewood Cliffs, New Jersey, 1976.
2. H. Hancock, Theory of Maxima and Minima, Dover Publications, New York, 1960.
3. R. G. Bartle, The Elements of Real Analysis, 2nd ed., Wiley, New York, 1976.
4. A. V. Fiacco, Second order sufficient conditions for weak and strict constrained minima, SIAM J. Appl. Math. 16, 105-108 (1968).
5. H. W. Kuhn and A. W. Tucker, Nonlinear programming, in Proceedings of the Second Berkeley Symposium on Mathematical Statistics and Probability (J. Neyman, ed.), University of California Press, Berkeley, California, 1951.
6. W. Karush, Minima of Functions of Several Variables with Inequalities as Side Conditions, Master's Thesis, University of Chicago, 1939.
7. H. W. Kuhn, Nonlinear programming: A historical view, in Nonlinear Programming (SIAM-AMS Proceedings Vol. 9), (R. W. Cottle and C. E. Lemke, eds.), American Mathematical Society, Providence, Rhode Island, 1976.
8. J. Abadie, On the Kuhn-Tucker theorem, in Nonlinear Programming (J. Abadie, ed.), North Holland, Amsterdam, 1967.
9. A. Y. Dubovitskii and A. A. Milyutin, Extremum problems in the presence of restrictions, USSR Comp. Math. Math. Phys. 5(3), 1-80 (1965).
10. F. J. Gould and J. W. Tolle, A necessary and sufficient qualification for constrained optimization, SIAM J. Appl. Math. 20, 164-172 (1971).
11. F. J. Gould and J. W. Tolle, Geometry of optimality conditions and constraint qualifications, Math. Prog. 2, 1-18 (1972).
12. L. W. Neustadt, Optimization: A Theory of Necessary Conditions, Princeton University Press, Princeton, New Jersey, 1976.

13. K. J. Arrow, L. Hurwicz, and H. Uzawa, Constraint qualifications in maximization problems, Naval Res. Log. Q. 8, 175-191 (1961).
14. M. S. Bazaraa, J. J. Goode, and C. M. Shetty, Constraint qualifications revisited, Manage. Sci. 18, 567-573 (1972).
15. F. John, Extremum problems with inequalities as subsidiary conditions, in Studies and Essays, Courant Anniversary Volume (K. O. Friedrichs et al., eds.), Interscience, New York, 1948.
16. O. L. Mangasarian and S. Fromovitz, The Fritz John necessary optimality conditions in the presence of equality and inequality constraints, J. Math. Anal. Appl. 17, 37-47 (1967).
17. A. V. Fiacco and G. P. McCormick, Nonlinear Programming: Sequential Unconstrained Minimization Techniques, Wiley, New York, 1968.
18. G. P. McCormick, Second order conditions for constrained minima, SIAM J. Appl. Math. 15, 641-652 (1967).
19. E. J. Messerli and E. Polak, On second order necessary conditions of optimality, SIAM J. Control 7, 272-291 (1969).
20. A. V. Fiacco, Second order sufficient conditions for weak and strict constrained minima, SIAM J. Appl. Math. 16, 105-108 (1968).
21. M. R. Hestenes, Optimization Theory: The Finite Dimensional Case, Wiley, New York, 1975.
22. W. Fenchel, Convex Cones, Sets and Functions, mimeographed lecture notes, Princeton University, Princeton, New Jersey, 1951.
23. A. W. Roberts and D. E. Varberg, Convex Functions, Academic, New York, 1973.
24. R. T. Rockafellar, Convex Analysis, Princeton University Press, Princeton, New Jersey, 1970.
25. J. Stoer and C. Witzgall, Convexity and Optimization in Finite Dimensions I, Springer, Berlin, 1970.
26. H. G. Eggleston, Convexity, Cambridge University Press, Cambridge, England, 1958.
27. O. L. Mangasarian, Nonlinear Programming, McGraw-Hill, New York, 1969.
28. F. A. Valentine, Convex Sets, McGraw-Hill, New York, 1964.
29. O. Hölder, Über einen Mittelwertsatz, Göttinger Nachr., pp. 38-47 (1889).
30. J. L. W. V. Jensen, Sur les fonctions convexes et les inégalités entre les valeurs moyennes, Acta Math. 30, 175-193 (1906).
31. B. Bernstein and R. A. Toupin, Some properties of the Hessian matrix of a strictly convex function, J. Reine Angew. Math. 210, 67-72 (1962).
32. M. Slater, Lagrange Multipliers Revisited: A Contribution to Nonlinear Programming, Cowles Commission Discussion Paper, Math. 403, November 1950.
33. G. B. Dantzig, Linear Programming and Extensions, Princeton University Press, Princeton, New Jersey, 1963.
34. D. Gale, H. W. Kuhn, and A. W. Tucker, Linear programming and the theory of games, in Activity Analysis of Production and Allocation (T. C. Koopmans, ed.), Wiley, New York, 1951.
35. W. S. Dorn, A duality theorem for convex programs, IBM J. Res. Dev. 4, 407-413 (1960).
36. P. Huard, Dual programs, IBM J. Res. Dev. 6, 137-139 (1962).
37. O. L. Mangasarian, Duality in nonlinear programming, Q. Appl. Math. 20, 300-302 (1962).

38. P. Wolfe, A duality theorem for non-linear programming, Q. Appl. Math. 19, 239-244 (1961).
39. W. Fenchel, On conjugate convex functions, Can. J. Math. 1, 73-77 (1949).
40. R. Courant and D. Hilbert, Methods of Mathematical Physics, Interscience, New York, Vol. I (1953), Vol. II (1962).
41. I. M. Gelfand and S. V. Fomin, Calculus of Variations, Prentice-Hall, Englewood Cliffs, New Jersey, 1963.
42. A. M. Geoffrion, Duality in nonlinear programming: A simplified applications-oriented development, SIAM Rev. 13, 1-37 (1971).
43. M. Hamala, Geometric Programming in Terms of Conjugate Functions, Center for Operations Research and Econometrics, Université Catholique de Louvain Discussion Paper No. 6811, Louvain, Belgium, June 1968.
44. R. T. Rockafellar, Duality in nonlinear programming, in Mathematics of the Decision Sciences, Part 1 (G. B. Dantzig and A. F. Veinott, eds.), American Mathematical Society, Providence, Rhode Island, 1968.
45. R. W. Cottle, Symmetric dual programs, Q. Appl. Math. 21, 237-243 (1963).
46. J. B. Dennis, Mathematical Programming and Electrical Networks, Technology Press, Cambridge, Massachusetts, 1959.
47. W. S. Dorn, Duality in quadratic programming, Q. Appl. Math. 18, 155-162 (1960).
48. W. S. Dorn, Self-dual quadratic programs, J. SIAM 9, 51-54 (1961).
49. K. J. Arrow and A. C. Enthoven, Quasi-concave programming, Econometrica 29, 779-800 (1961).
50. M. Avriel, r-Convex functions, Math. Prog. 2, 309-323 (1972).
51. M. Avriel and I. Zang, Generalized convex functions with applications to nonlinear programming, in Mathematical Programs for Activity Analysis (P. Van Moeseke, ed.), North Holland, Amsterdam, 1974.
52. A. Ben-Tal, On generalized means and generalized convexity, J. Optimization Theory Appl. 21, 1-14 (1977).
53. H. J. Greenberg and W. P. Pierskalla, A review of quasi-convex functions, Oper. Res. 19, 1553-1570 (1971).
54. D. G. Luenberger, Quasi-convex programming, SIAM J. Appl. Math. 16, 1090-1095 (1968).
55. O. L. Mangasarian, Pseudo-convex functions, J. SIAM Control, Ser. A 3, 281-290 (1965).
56. J. Ponstein, Seven kinds of convexity, SIAM Rev. 9, 115-119 (1967).
57. B. Martos, Nonlinear Programming, Theory and Applications, American Elsevier, New York, 1975.
58. W. C. Davidon, Variable Metric Method for Minimization, AEC Research and Development Report ANL-5990 (Rev.), November 1959.
59. J. Kiefer, Sequential minimax search for a maximum, Proc. Am. Math. Soc. 4, 502-506 (1953).
60. M. Avriel and D. J. Wilde, Optimality proof for the symmetric Fibonacci search technique, Fibonacci Q. 4, 265-269 (1966).
61. D. J. Wilde, Optimum Seeking Methods, Prentice-Hall, Englewood Cliffs, New Jersey, 1964.
62. R. P. Brent, Algorithms for Minimization without Derivatives, Prentice-Hall, Englewood Cliffs, New Jersey, 1973.
63. W. Spendley, G. R. Hext, and F. R. Himsworth, Sequential application of simplex designs in optimisation and evolutionary operation, Technometrics 4, 441-461 (1962).

64. J. A. Nelder and R. Mead, A simplex method for function minimization, Comput. J. 7, 308-313 (1965).
65. R. W. H. Sargent, Minimization without constraints, in Optimization and Design (M. Avriel, M. J. Rijckaert, and D. J. Wilde, eds.), Prentice-Hall, Englewood Cliffs, New Jersey, 1973.
66. R. Hooke and T. A. Jeeves, Direct search solution of numerical and statistical problems, J. Assoc. Comput. Mach. 8, 212-221 (1961).
67. B. Noble and J. W. Daniel, Applied Linear Algebra, 2nd ed., Prentice-Hall, Englewood Cliffs, New Jersey, 1977.
68. M. J. D. Powell, An efficient method for finding the minimum of a function of several variables without calculating derivatives, Comput. J. 7, 155-162 (1964).
69. W. I. Zangwill, Minimizing a function without calculating derivatives, Comput. J. 10, 293-296 (1967).
70. P. E. Gill and W. Murray, Quasi-Newton methods for unconstrained optimization, J. Inst. Math. Appl. 9, 91-108 (1972).
71. P. E. Gill, W. Murray, and R. A. Pitfield, The Implementation of Two Revised Quasi-Newton Algorithms for Unconstrained Optimization, National Physical Laboratory DNAC Report No. 11, April 1972.
72. J. Greenstadt, A quasi-Newton method with no derivatives, Math. Comput. 26, 145-166 (1972).
73. J. Greenstadt, Improvements in a QNWD Method, IBM Palo Alto Scientific Center Tech. Report No. 320-3306, October 1972.
74. R. Mifflin, A superlinearly convergent algorithm for minimization without evaluating derivatives, Math. Prog. 9, 100-117 (1975).
75. G. W. Stewart, A modification of Davidon's minimization method to accept difference approximations of derivatives, J. Assoc. Comput. Mach. 14, 72-83 (1967).
76. D. P. Bertsekas and S. K. Mitter, A descent numerical method for optimization problems with nondifferentiable cost functions, SIAM J. Control 11, 637-652 (1973).
77. A. A. Goldstein, Optimization with corners, in Nonlinear Programming 2 (O. L. Mangasarian, R. R. Meyer, and S. M. Robinson, eds.), Academic, New York, 1975.
78. C. Lemarechal, An extension of "Davidon" methods to nondifferentiable functions, Math. Prog. Study 3, 95-109 (1975).
79. P. Wolfe, A method of conjugate subgradients for minimizing nondifferentiable functions, Math. Prog. Study 3, 145-173 (1975).
80. P. E. Gill and W. Murray, Newton-type methods for unconstrained and linearly constrained optimization, Math. Prog. 7, 311-350 (1974).
81. A. Cauchy, Méthode générale pour la résolution des systéms d'equations simultanées, C. R. Acad. Sci. pp. 536-538 (1847).
82. H. B. Curry, The method of steepest descent for non-linear minimization problems, Q. Appl. Math. 2, 258-261 (1944).
83. L. Armijo, Minimization of functions having Lipschitz continuous first partial derivatives, Pac. J. Math. 16, 1-3 (1966).
84. R. Fletcher and C. M. Reeves, Function minimization by conjugate gradients, Comput. J. 7, 149-154 (1964).
85. J. W. Daniel, The conjugate gradient method for linear and nonlinear operator equations, SIAM J. Numer. Anal. 4, 10-26 (1967).
86. H. W. Sorenson, Comparison of some conjugate direction procedures for function minimization, J. Franklin Inst. 288, 421-441 (1969).

87. P. Wolfe, The Method of Conjugate Gradients, lectures delivered at the NATO Advanced Study Institute on Mathematical Programming in Theory and Practice, Figueira da Foz, Portugal, June 1972.
88. E. Polak and G. Ribière, Note sur la convergence de méthodes de directions conjugées, Rev. Fr. Inform. Rech. Oper. 16-R1, 35-43 (1969).
89. B. T. Polyak, The conjugate gradient method in extremal problems, USSR Comp. Math. Math. Phys. 9(4), 94-112 (1969).
90. R. Fletcher and M. J. D. Powell, A rapidly convergent descent method for minimization, Comput. J. 6, 163-168 (1963).
91. M. J. D. Powell, On the convergence of variable metric algorithms, J. Inst. Math. Appl. 7, 21-36 (1970).
92. H. Y. Huang, Unified approach to quadratically convergent algorithms for function minimization, J. Optimization Theory Appl. 5, 405-423 (1970).
93. L. C. W. Dixon, Quasi-Newton algorithms generate identical points, Math. Prog. 2, 383-387 (1972).
94. L. C. W. Dixon, Quasi-Newton techniques generate identical points II. The proofs of four new theorems, Math. Prog. 3, 345-358 (1972).
95. C. G. Broyden, Quasi-Newton methods and their application to function minimisation, Math. Comput. 21, 368-381 (1967).
96. C. G. Broyden, The convergence of a class of double-rank minimization algorithms 1. General considerations, J. Inst. Math. Appl. 6, 76-90 (1970).
97. C. G. Broyden, The convergence of a class of double-rank minimization algorithms 2. The new algorithm, J. Inst. Math. Appl. 6, 222-231 (1970).
98. R. Fletcher, A new approach to variable metric algorithms, Comput. J. 13, 317-322 (1970).
99. D. Goldfarb, A family of variable-metric methods derived by variational means, Math. Comput. 24, 23-26 (1970).
100. D. F. Shanno, Conditioning of quasi-Newton methods for function minimization, Math. Comput. 24, 647-656 (1970).
101. W. C. Davidon, Variance algorithm for minimization, Comput. J. 10, 406-410 (1968).
102. B. A. Murtagh and R. W. H. Sargent, A constrained minimization method with quadratic convergence, in Optimization (R. Fletcher, ed.), Academic, London, 1969.
103. M. J. D. Powell, Rank one methods for unconstrained optimization, in Integer and Nonlinear Programming (J. Abadie, ed.), North Holland, Amsterdam, 1970.
104. C. G. Broyden, The convergence of single-rank quasi-Newton methods, Math. Comput. 24, 365-382 (1970).
105. C. G. Broyden, A class of methods for solving nonlinear simultaneous equations, Math. Comput. 19, 577-593 (1965).
106. W. C. Davidon, Optimally conditioned optimization algorithms without line searches, Math. Prog. 9, 1-30 (1975).
107. I. Fried, N-step conjugate gradient minimization scheme for nonquadratic functions, AIAA J. 9, 2286-2287 (1971).
108. D. H. Jacobson and W. Oksman, An algorithm that minimizes homogenous functions of N variables in N + 2 iterations and rapidly minimizes general functions, J. Math. Anal. Appl. 38, 535-552 (1972).
109. J. S. Kowalik and K. G. Ramakrishnan, A numerically stable optimization method based on a homogeneous function, Math. Prog. 11, 50-66 (1976).

110. C. A. Botsaris and D. H. Jacobson, A Newton-type curvilinear search method for optimization, J. Math. Anal. Appl. 54, 217-229 (1976).
111. J. P. Vial and I. Zang, Unconstrained optimization by approximation of the gradient path, Math. Oper. Res. 2, 253-265 (1977).
112. M. Avriel and J. P. Dauer, A homotopy based approach to unconstrained optimization, Applicable Anal., to appear.
113. W. I. Zangwill, Non-linear programming via penalty functions, Manage. Sci. 13, 344-358 (1967).
114. A. V. Fiacco and G. P. McCormick, The sequential unconstrained minimization technique without parameters, Oper. Res. 15, 820-827 (1967).
115. C. W. Carroll, The created response surface technique for optimizing nonlinear restrained systems, Oper. Res. 9, 169-184 (1961).
116. A. V. Fiacco and G. P. McCormick, The sequential unconstrained minimization technique for nonlinear programming: A primal-dual method, Manage. Sci. 10, 360-366 (1964).
117. A. V. Fiacco, Barrier methods for nonlinear programming, in Encyclopedia of Computer Science and Technology, Vol. 3 (J. Belzer, A. G. Holzman, and A. Kent, eds.), Dekker, New York, 1976.
118. M. R. Hestenes, Multiplier and gradient methods, J. Optimization Theory Appl. 4, 303-320 (1969).
119. M. J. D. Powell, A method for nonlinear constraints in minimization problems, in Optimization (R. Fletcher, ed.), Academic, London, 1969.
120. D. P. Bertsekas, Multiplier methods: A survey, Automatica 12, 133-145 (1976).
121. D. P. Bertsekas, On penalty and multiplier methods for constrained minimization, in Nonlinear Programming 2 (O. L. Mangasarian, R. R. Meyer, and S. M. Robinson, eds.), Academic, New York, 1975.
122. R. A. Tapia, Diagonalized multiplier methods and quasi-Newton methods for constrained optimization, J. Optimization Theory Appl. 22, 135-194 (1977).
123. D. P. Bertsekas, On penalty and multiplier methods for constrained minimization, SIAM J. Control Optimization 14, 216-235 (1976).
124. J. D. Buys, Dual Algorithms for Constrained Optimization Problems, Doctoral Dissertation, University of Leiden, The Netherlands, 1972.
125. R. Fletcher, An ideal penalty function for constrained optimization, in Nonlinear Programming 2 (O. L. Mangasarian, R. P. Meyer, and S. M. Robinson, eds.), Academic, New York, 1975.
126. B. W. Kort and D. P. Bertsekas, A new penalty function method for constrained minimization, in Proceedings of 1972 IEEE Conference on Decision and Control, New Orleans, December 1972.
127. B. W. Kort and D. P. Bertsekas, Combined primal-dual and penalty methods for convex programming, SIAM J. Control Optimization 14, 268-294 (1976).
128. R. T. Rockafellar, The multiplier method of Hestenes and Powell applied to convex programming, J. Optimization Theory Appl. 12, 555-562 (1973).
129. R. T. Rockafellar, A dual approach to solving nonlinear programming problems by unconstrained optimization, Math. Prog. 5, 354-373 (1973).
130. G. Zoutendijk, Methods of Feasible Directions, Elsevier, Amsterdam, 1960.
131. G. Zoutendijk, Some algorithms based on the principle of feasible directions, in Nonlinear Programming (J. B. Rosen, O. L. Mangasarian, and K. Ritter, eds.), Academic, New York, 1970.
132. G. Zoutendijk, Mathematical Programming Methods, North Holland, Amsterdam, 1976.

133. G. P. McCormick, Anti-zigzagging by bending, Manage. Sci. 15, 315-320 (1969).
134. E. Polak, Computational Methods in Optimization, Academic, New York, 1971.
135. D. J. Wilde, Jacobians in constrained nonlinear optimization, Oper. Res. 13, 848-856 (1965).
136. D. J. Wilde and C. S. Beightler, Foundations of Optimization, Prentice-Hall, Englewood Cliffs, New Jersey, 1967.
137. P. Wolfe, Methods of nonlinear programming, in Recent Advances in Mathematical Programming (R. L. Graves and P. Wolfe, eds.), McGraw-Hill, New York, 1963.
138. P. Wolfe, Methods of nonlinear programming, in Nonlinear Programming (J. Abadie, ed.), North Holland, Amsterdam, 1967.
139. W. I. Zangwill, Nonlinear Programming: A Unified Approach, Prentice-Hall, Englewood Cliffs, New Jersey, 1969.
140. J. Abadie and J. Carpentier, Generalization of the Wolfe reduced gradient method to the case of nonlinear constraints, in Optimization (R. Fletcher, ed.), Academic, London, 1969.
141. L. S. Lasdon, R. L. Fox, and M. W. Ratner, Nonlinear optimization using the generalized reduced gradient method, Rev. Fr. Automa. Inf. Rech. Oper. V-3, 73-104 (1974)
142. J. E. Kelley, The cutting-plane method for solving convex programs, J. SIAM 8, 703-712 (1960).
143. E. W. Cheney and A. A. Goldstein, Newton's method for convex programming and Tchebycheff approximation, Numer. Math. 1, 253-268 (1959).
144. A. F. Veinott, The supporting hyperplane method for unimodal programming, Oper. Res. 15, 147-152 (1967).
145. D. M. Topkis, Cutting-plane methods without nested constraint sets, Oper. Res. 18, 404-413 (1970).
146. B. C. Eaves and W. I. Zangwill, Generalized cutting plane algorithms, SIAM J. Control 9, 529-542 (1971).

INDEX